Introduction to Cultural Ecology

Introduction to Cultural Ecology

MARK Q. SUTTON AND E. N. ANDERSON

A Division of
ROWMAN & LITTLEFIELD PUBLISHERS, INC.
Walnut Creek • Lanham • New York • Toronto • Oxford

To our wives, Melinda and Barbara

AltaMira Press
A division of Rowman & Littlefield Publishers, Inc.
1630 North Main Street, #367
Walnut Creek, CA 94596
www.altamirapress.com

Rowman & Littlefield Publishers, Inc.
A wholly owned subsidiary of The Rowman & Littlefield Publishing Group, Inc.
4501 Forbes Boulevard, Suite 200
Lanham, MD 20706

PO Box 317
Oxford
OX2 9RU, UK

British Library Cataloguing in Publication Information Available

Library of Congress Cataloging-in-Publication Data

Sutton, Mark Q.
Introduction to cultural ecology / Mark Q. Sutton and E. N. Anderson.
p. cm.
Including bibliographical references and index.
ISBN 0-7591-0530-8 (alk. paper)—ISBN 0-7591-0531-6 (pbk. : alk. paper)
1. Human ecology. 2. Social ecology. I. Anderson, Eugene N. (Eugene Newton), 1941– II. Title.

GF50.S88 2004
304.2—dc22

2003022110

Printed in the United States of America

∞™ The paper used in this publication meets the minimum requirements of American National Standard for Information Sciences—Permanence of Paper for Printed Library Materials, ANSI/NISO Z39.48-1992.

Contents

List of Figures and Tables

FIGURES

TABLES

Preface

Cultural ecology is one of the two major subdivisions of human ecology, the other being human biological ecology. We felt that the books available for a class in cultural ecology focused either too heavily on general ecology or too much on human biological ecology or not enough on cultural ecology. Thus, we faced a "Goldilocks" dilemma: None of the usual textbooks were "just right" for our introductory classes. Therefore, we decided to create a new book. We begin with the assumption that the student has no prior knowledge of anthropology or ecology and try to build an understanding from the ground up.

All peoples and cultures are faced with a number of major environmental issues, problems that can be addressed by anthropology and cultural ecology. How have other people faced and dealt with the same basic problems that face us all today? How can we improve our situation? What can anthropology and cultural ecology contribute to the future?

The key is understanding what the options are, what works, and what does not. This understanding requires a great deal of knowledge, which must be obtained through the study of other groups, including the documentation of their environments and adaptations. Next we must analyze what we have learned in order to develop alternative responses to environmental situations. We must also understand the consequences of the choices that have been made; we can learn from the successes and mistakes of others rather than having to repeat those same mistakes.

We do not attempt to cover all aspects of the incredibly complex and diverse field of the relationships between humans and the environment. We thought it important to provide a reasonably comprehensive introduction to ecological theory in a simple format, combined with discussions of various human cultures.

We spend more time defining the concepts and classifying things typologically than most treatments do. We have concentrated on things we thought would be appropriate for an introductory student, including traditional food production, but have not included sophisticated materials beyond the introductory level. Many such concepts are important, but it is impossible to do everything in one book. We plead for charity. However, we do spend some time discussing and critiquing evolutionary ecology, primarily because we feel it is so widely used and misunderstood.

We have also had to be highly selective in using and citing the incredibly large body of literature that now treats even narrow and specialized questions within the field. We can only abjectly apologize to those experts (the vast majority, alas) who find themselves uncited. We have tried, insofar as possible, to confine citations to easily located, basic works or other literature readily accessible to students.

Our main goal is to try to communicate the anthropological side of ecological matters. It is not our intent to cover all the ecological issues or problems of the world or to deal in detail with modern matters of pollution, climate change, environmental degradation, and the like; these issues are now at the center of the world stage (though perhaps not of the U.S. government). Our aim is to explore how traditional cultures operate and adapt to their environments, how they function, and what the Western world can learn from them.

Assuming that the student has no prior knowledge of the subject, we begin with a very basic introduction to anthropology and to scientific inquiry and end chapter 1 with a brief history of the development of cultural ecological theory. Chapter 2 provides an introduction to the concepts and terms used in general ecology, many of which are heard on the news and seen in newspapers and magazines daily yet rarely defined in any detail. Human biological ecology (chapter 3) is then discussed as a background to understanding and distinguishing cultural adaptations, which are the subject of chapter 4. We thought it important to clearly distinguish between human biological ecology and cultural ecology, as the two tend to get mixed up in much of the literature, creating a source of confusion for everyone.

The next five chapters (chapters 5–9) deal with discussions of the cultural ecology of the two broad and generalized economic strategies that are the subject of much anthropological study: hunting and gathering and food production, the latter having three basic adaptations: horticulture, pastoralism, and intensive agriculture. This is a rather traditional approach, and we recognize the problems with the pigeonholes in which we place societies. However, we feel that it is a sound approach at the introductory level. Finally, we close (chapter 10) with some discussion of contemporary environmental problems and the role traditional cultures play in them.

The philosophical position of this book is not value free. One of the primary goals of most cultural ecological work is to use the knowledge in an effort to stem global catastrophe. However, we believe—deeply—that to do this we must study all cultures, past and present, and learn from each of them. We try to impress on the reader that no one culture has a monopoly on environmental care or on environmental carelessness. Only by combining the best of many plans will the human species save itself.

ACKNOWLEDGMENTS

We acknowledge the intellectual contributions of Eugene P. Odum (1975, 1993) in ecology and of John Bennett (1976), Roy Ellen (1982), Robert M. Netting (1986), Carole Crumley (1994), and Bernard Campbell (1995) in understanding the interrelationships of culture and ecology.

Among those who have helped with information, suggestions, and materials, it is necessary to single out Paul Buell, Scott Fedick, Alan Fix, Jean Hudson, Michael Kearney, Bonnie McCay, Endre Nyerges, Richard Osborne, Evelyn Pinkerton, Philip Silverman, and Maria Cruz Torres. Many others have helped with information and encouragement, and we thank them all.

We appreciate the advice and comments of Jill K. Gardner, Brian E. Hemphill, Becky Orfila, Kristin D. Sobolik, Melinda B. Sutton, Linda Wells, and a number of anonymous reviewers, several of whom provided extensive and useful comments. We greatly appreciate the work of Gina Bahr and Janet Gonzales at the California State University, Bakersfield, Library, who contributed greatly to this project by their hard work in obtaining the many books and articles we ordered through interlibrary loan.

We also thank Francesca Bray, Ping-ti Ho, Philip Huang, Richard Pearson, Pierre-Etienne Will, and Bin Wong for providing unpublished materials and discussion of their work. The many persons who helped ENA in field research, especially Marja L. Anderson, Chow Hung-fai, and Choi Kwok-tai, are also gratefully acknowledged.

1

Introduction

In the four or five million years since their development, humans have colonized practically every terrestrial environment on the planet. Humans everywhere are virtually the same biologically but have been able to adapt to the enormous environmental diversity of Earth through **culture**, an immensely flexible and adaptive mechanism that other animals lack. Thus, humans have been a very successful species. Human activity has a wide range of impacts on the environment, however, from exceedingly minor to catastrophic. Now, human activities are having huge impacts on the very environment on which we humans depend, ultimately threatening our own existence. Understanding and dealing with these challenges is a daunting, but essential, task.

People in Western societies tend to hold the view that humans are separate from the environment, above it in some way. This view can be traced back to the Bible, which tells us the world (the environment) was created first, and then "man"—an entity separate from and superior to nature—was created and given the task of subduing nature (Gen. 1:28; also see Wilson 1967:1205). Other cultures have similar stories. Western philosophy continues to include the view that it is the goal and mission of people to "conquer" nature. Thus, many people today still believe that humans are not participants in the environment but that we must overcome it and bend it to our will. This conviction continues to permeate Western thought and action; consider, for example, the way we strive to separate ourselves from our natural surroundings by creating closed environments, including our homes, offices, and cars.

One could argue that many traditional societies (non-Western, non-industrialized cultures) do not hold this Western view and that they are somehow "ecologists" living in harmony with their environment (e.g., White 1997; but also see Krech 1999). It is true that the activities of many societies have less impact on the environment than those of others, and it is also true that some of these groups hold a more ecologically friendly philosophy of life than Westerners as a whole do. However, it has also been argued that traditional cultures generally have less impact on the environment only because their technology is less complex and their populations smaller. Given the right conditions and incentives, the argument goes, they would do as Westerners do. In support of this argument, one can point to the destruction of the habitat on Easter Island (e.g., Diamond 1995), the deforestation of most of Europe during the Neolithic era, and the Norse degradation of the Northern Islands (McGovern et al. 1988), among other examples.

Fortunately, traditional people can in general be trusted to take care of their resources, not out of fuzzy, New-Age love for Mother Earth, but out of solid, hardheaded good sense, often shored up by traditional religion and morality (Anderson 1996; Berkes 1999; Lentz 2000). Biologists are beginning to realize this truth. Kent Redford, who coined the sarcastic (and racist) term "the ecologically noble savage" (Redford 1990; see the superb refutation of that article by Lopez 1992), has since repented and now takes a more balanced and reasonable position (Redford and Mansour 1996; Sponsel 2001).We must first of all recognize that humans and their cultures are an integral part of the matrix of the environment and are not separate from it in either cause or effect. Human activity affects the environment, which is then altered, in turn affecting human activities, and so forth. The shape and form of the environment are dependent on its history, a history that includes humans. Yet it is also important to realize that humans are not just another animal running around the landscape. Humans are self-aware, cooperative, technological, and highly social. This unique combination does separate humans from other organisms, making their interactions with the environment very complex and fascinating.

WHAT IS CULTURAL ECOLOGY?

Ecology is the study of the interaction between living things and their environment. **Human ecology** is more specific, being the study of the relationships and interactions between humans, their biology, their cultures, and their physical environments. In the 1950s the overall field was known as cultural ecology, but it is now more commonly referred to as human ecology (following *Human Ecology*, the leading journal in the field). Human ecology is sometimes referred to by other names, including ecological anthropology (which may include aspects of biological anthropology), culture and environment, and even (still) cultural ecology.

Several comprehensive treatments of the field are available. Most impressive is *Human Adaptive Strategies: Ecology, Culture, and Politics* (Bates 1998). The title suggests a grounding in the new knowledge and also the way Bates integrates the field around the concepts of adaptation and strategizing. In addition, Bates and Susan Lees, longtime editors of *Human Ecology*, have produced a collection of articles from that journal, *Case Studies in Human Ecology* (1996). Patricia Townsend has brought out a brief but extremely well-targeted overview, *Environmental Anthropology* (2000), which covers basically the same ground from a very similar point of view but at an entry level.

Human ecologists study many aspects of culture and environment, including how and why cultures do what they do to solve their subsistence problems, how groups of people understand their environment, and how they share their knowledge of the environment. The broad field of human ecology includes two major subdivisions (see fig. 1.1). **Human biological ecology** is the study of the biological aspect of the human/environment relationship, and **cultural ecology** is the study of the ways in which culture is used by people to adapt to their environment.

The primary focus of this book is the subdivision of cultural ecology, which encompasses everything from pet dogs to the Fall of Rome, an ecological catastrophe caused in part by misuse of resources (see, e.g., Ponting 1991). This book examines, among other things, salmon ceremonies among Northwest Coast Indians, Maya agriculture today and in the past, sacred groves in southern China, and the use of various foods. Insects, for example, are an abundant and nutritious source of food, yet many cultures consider them pests. Some of the very cultures that loathe insects see shrimps and lobsters as delicacies, although all three are very similar biologically. Why is there such a *cultural* difference in the way they are regarded?

Consider a food question that is far more serious. Significant deforestation has resulted from the creation of cattle pasture. As much as Americans may feel they depend on their hamburgers, beef is really a luxury item, producing relatively

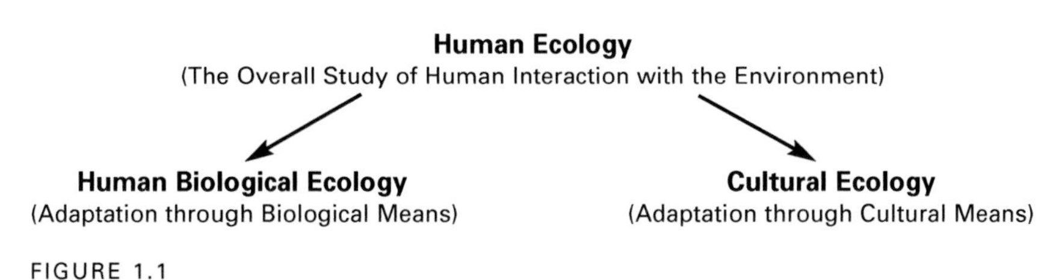

FIGURE 1.1
The general relationship of the subdivisions within human ecology.

little protein at huge expense. To produce it, millions of acres of land that were once covered in forest or in farmland growing food for local people have been converted into pasture. The devastation to wildlife and biological diversity is bad enough; the impoverishment, and frequently the starvation, of local people may be considered even more serious. Thus, cultural beliefs about food can dramatically affect the world environment.

The study of the relationships between culture and environment is not just academic; it is vital, not simply because it is interesting, but because it offers understanding of and possible solutions to important contemporary problems. Issues of deforestation (e.g., Anderson 1990), loss of species (e.g., Blaustein and Wake 1995), food scarcity (Brown 1994, 1996), and soil loss (Pimentel et al. 1995) are on the minds of many and are beginning to be addressed by human ecologists. (For general discussions of human impact on the environment, see Goudie 1994; Ehrlich and Ehrlich 1996; Meyer 1996; Redman 1999; and Molnar and Molnar 2000, among others.) Some of these issues reflect overexploitation of resources and require conservation measures. Such measures may threaten certain short-term economic activities, such as unrestricted logging, and many leaders launch verbal, and sometimes even physical, attacks on conservationists (see Helvarg 1994) to protect their short-term interests.

Cultural ecologists record other traditional and local knowledge that is of value to the wider world. Thousands of useful drugs used by Westerners have been derived from traditional medicines, and more are being tested and developed almost daily. Ancient crops of the Andes and Tibet, such as potatoes and barley, respectively, are becoming important worldwide. Long-established land management techniques used in Indonesia and Guatemala—multilayered and multicropped orchard gardens, for example—are inspiring changes in the greater arena. The accumulated cultural knowledge of billions of people over tens of thousands of years is available and is a tremendous resource for our resource-short world.

ANTHROPOLOGY

Human ecology is generally included within the discipline of **anthropology**, the study of human beings. Anthropology includes the study of human biology, language, prehistory, religion, social structure, economics, evolution, and anything else that applies to people. Thus, anthropology is a very broad discipline, holistic in its approach and comparative, or cross-cultural, in its analyses. Anthropologists generally concentrate their work on small-scale cultures and tend to have considerable personal contact with the people of those cultures.

Culture, *learned* and *shared* behavior in humans, is the fundamental element that sets humans apart from the other animals. The vast complexities of human behavior are largely related to culture and, to a lesser extent, biology. Culture is largely transmitted through language, which, as far as we know, is unique to humans. In addition, every person belongs to **a culture**, a group of people who share the same basic pattern of learned behavior—the same values, views, language, and identity. Each culture holds an identity unto itself, such as the Cheyenne, the Germans, or the Yanomamo, and its members recognize that they are different from other cultures.

Anthropologists traditionally follow a set of basic beliefs in their study of other cultures. First, they recognize that all cultures are at least a bit **ethnocentric**: in other words, people believe their culture is superior to others (although many envy the more rich or powerful). Americans tend to view non-Americans as being inferior, less cultured, or backward. Germans have the same view of non-Germans, as do the Chinese of non-Chinese. In fact, every culture holds this view; it is a normal part of the self-identification process. However, ethnocentrism has often been used to rationalize mistreatment of peoples. Virtually all colonial powers exploited native populations in the belief that they were inferior, and the colonizers used this belief to justify enslaving and murdering the natives. In North America, the natives were considered "savages" who were "in the way" of "civilization." The Native Americans were thus moved, incarcerated, or killed with government approval. A similar situation currently exists in a number of Third World countries attempting to "develop."

Anthropologists are usually from a culture other than the one being studied. Thus, researchers often view the culture through the lens of their own culture, in essence an outsider's view. One's perspective, whether insider or outsider, influences what is observed and ultimately what can be learned. Anthropologists deal with this problem as best they can.

A basic conviction in anthropology is **cultural relativism**, the belief that cultures and cultural practices should not be judged. This term has been misunderstood to imply that anthropologists approve of anything practiced in any culture. More correctly, it means that anthropologists study cultures without trying to show that one is "better" than another and without trying to impose their culture on other people. This relativity is methodological, not moral. Indeed, anthropologists have traditionally taken a very strong stand against genocide and "culturocide," or forcing people to give up their culture against their will. Anthropologists attempt to avoid being ethnocentric and believe that all people and cultures are valid; that people have the right to exist, to have their own culture and practices, and to speak their own language; and that individuals have fundamental human rights (Nagengast and Turner 1997; Merry 2003).

Anthropology can be divided into many subdisciplines, perhaps dozens, depending on how it is defined and who is defining it. Here we follow a traditional division of the field into four subdisciplines: cultural anthropology, biological (or physical) anthropology, anthropological linguistics, and archaeology.

Cultural Anthropology

Cultural anthropology, sometimes called social or sociocultural anthropology, is the study of existing peoples and cultures. Cultural anthropologists conduct two major types of studies: **ethnography**, the study of a particular group at a particular time, and **ethnology**, the comparative study of culture. Cultural anthropologists strive to learn everything they can about a culture, such as kinship systems, marriage rules, economics, language, and politics. Cultural anthropologists generally live with a group under study, observe and record its members' activities and behavior, and ask people questions. Cultural anthropologists can sometimes even participate in, and so can better record, community activities. As such, cultural anthropologists can get a rich, though still incomplete, record of a group. The weakness of cultural anthropology is that, despite the detail of the information, little time depth is reflected, making change difficult to detect.

Biological Anthropology

Biological anthropology is the study of the biology and evolution of people, as well as the study of the biology, evolution, and behavior of nonhuman primates and other animals for clues to understanding humans. While humans are all very similar biologically, some differences between groups do exist. These differences may take a variety of forms, including stature, blood type, and adaptations to cold or high altitude. An understanding of past human biology can help us understand evolution and suggest relationships with other populations, such as intermarriage and/or migration, changes in past environment, and changes in subsistence.

Anthropological Linguistics

Anthropological linguistics is the study of human language. Its scope includes the historical relationships between languages, common "ancestors" of languages and language groups, syntax, and meaning. Cultural anthropologists are interested in linguistics because a great deal about a particular culture can be learned by looking at its language. Archaeologists are interested in linguistics, especially historical linguistics, because languages (and cultures) can be traced back in time.

Archaeology

Archaeology is the study of the human past. There is some overlap between archaeology and cultural anthropology, as archaeologists want to learn the same things about past cultures that cultural anthropologists do about living ones. Archaeologists often study the present as well, either to find clues for interpreting the past or to investigate present-day problems by using archaeological methods.

The major differences between archaeology and cultural anthropology are in the available data and the methods used to obtain those data. The material remains with which archaeologists work are limited, partly due to excavation techniques; as a result, archaeologists do not obtain the entire picture of a past culture. However, archaeologists are able to detect change over long periods of time, can identify broad trends, and can examine transitions, such as the change of some cultures from hunting and gathering to agriculture. In addition, an archaeologist can detect the traces of behavior that a cultural anthropologist might not see. This access to "hidden behavior" is another advantage of archaeology. In addition, understanding the ecology of past peoples is a major goal in archaeology (see Butzer 1982; Dincauze 2000).

THE STUDY OF HUMAN ECOLOGY

Human ecology, and all of anthropology, is a Western empirical science and adheres to the procedures and rules of Western science, including the use of the scientific method. All cultures have some form of science; other approaches are discussed in chapter 4.

Science

The goal of any **science** is to generate new knowledge, to learn new things, however that is done. Many sciences are empirical and employ data that are objective, observable, measurable, and reproducible. Information not meeting the standards of objectivity, measurability, and reproducibility is nonempirical.

Western science is an **empirical science** and employs the scientific method, a specific, systematic set of rules of scientific inquiry. In this method, empirical data are first observed and then recorded. Next, testable hypotheses are formed to account for the relationships among the data. An experiment is then formulated to test the hypothesis using additional empirical data; an untestable hypothesis is dismissed immediately. Based on the results of the test, the hypothesis is either supported or rejected. If the hypothesis is rejected, it may be abandoned or revised and retested. If supported, the hypothesis is tested again with even more data. A set of interrelated hypotheses is called a theory, which is then subjected to yet more testing. Even generally accepted hypotheses and theories get tested over and over, making science self-correcting. If a theory survives considerable and repeated testing, it may then be called a law. In practice, scientists do not always act so systematically (Kuhn 1962). They have their own personal agendas and even biases. Even the best of them make mistakes—sometimes honest ones, sometimes mistakes based on their biases. This is why constant testing, especially by scientists of other theoretical persuasions, is valuable. Thus, science still manages to function (Kitcher 1993).

Evolution and Adaptation

Fundamental to any inquiry in human ecology are the concepts of change and adaptation to change. All environments are dynamic, and changes will vary in both time and space. As environments change, organisms must adapt to those changes, a process that can entail a variety of mechanisms. Humans use both biological and cultural mechanisms.

The concept of **evolution** is widely misunderstood. Quite simply, evolution is change. All things change, and so all things evolve. Biological anthropologists define evolution more specifically as the change in gene frequency in populations from generation to generation. Other disciplines might define evolution in different ways, but in essence it is simply change.

Many also believe that evolution has direction. It is commonly thought that as something evolves, it advances up some kind of evolutionary ladder, that it somehow "advances" toward a goal or an ideal of some sort, that it embodies some sort of "progress." These notions are false. While it is true that some things become more complex over time, not all things do; complexity itself is not necessarily an advantage. A simple amoeba living today is as "evolved" as any human being—not as complex, to be sure, but certainly as evolved—as it has a long evolutionary history, and its continued existence reflects biological success. In the same vein, all living human cultures are equally evolved, just to different environments. As there is no direction in evolution, there is no such thing as devolution, no such thing as more or less "advanced," and no external scale of progress.

In biological evolution, species adapt to their environment through natural selection, the process by which some traits are selected "for"—by allowing their bearers to leave more descendants—and retained in the gene pool of the next generation while deleterious traits are selected "against" (with some neutral traits going along to the next generation for the ride). For selection to occur, variation (i.e., differing traits to be selected for or against) must exist within the population. With the exception of clones or identical twins, all individuals vary. Variation is ultimately attributable to mutation, or accidental changes in a gene. Most mutations are deleterious, and most are quickly selected against and deleted from the population. However, some mutations are advantageous and are selected for.

An example of human biological evolution is provided by lactose intolerance. By about the age of six, most humans cease producing the enzyme lactase, which allows digestion of lactose, the sugar found in milk. Thereafter, milk upsets their stomachs. By sheer genetic accident, some humans continue producing lactase throughout life. The gene for continued lactase production has been selected for in milk-drinking areas, notably western Europe and eastern Africa. Most people whose ancestors come from these regions can happily drink milk all their lives (Huang 2002). In most of the rest of the world, people must ferment the milk into yogurt, using *Lactobacillus* bacteria to break the lactose down for them.

In addition to evolving biologically, humans evolve culturally, and cultural evolution has been an important concept in anthropology. Cultural anthropologists could view cultural evolution as differential persistence of behaviors through time. We know that this sort of cultural evolution does occur, although we do not know exactly why or how. Some analytical approaches, such as evolutionary ecology, equate cultures to organisms and apply the concepts of biological evolution (see chapter 3).

As environmental conditions change, some sort of response is necessary. That response, or **adaptation**, is an ongoing process because environmental conditions are always dynamic. The variability within an organism allows for an appropriate response to be selected, and the greater the variation, the more likely an adequate adaptation can be made.

For most organisms, adaptation is purely biological and ultimately regulated by natural selection. However, for humans, adaptation will also be cultural, a mechanism that can act in a much shorter time. Culture is a way in which *groups* of people can adapt to the environment through collective behavior and/or technology. This concept is discussed in greater detail in chapter 4.

Solutions

If problems presented by the environment are solved, the organism adapts. Any problem may have multiple possible solutions involving the adoption of traits and/or behaviors. In some cases, the organism may find the best available solution, and we would say it adapts well. In other cases, a solution may be bad, perhaps due to poor decision making or other factors, and the organism becomes extinct. In many cases, a solution may be "adequate," good enough to get by. If the truth were known, probably most solutions would turn out to be just good enough. To win, a football team does not necessarily have to score every time it has the ball; it just has to score more points than the other team.

In some environments, a limited set of solutions is possible. For example, in the climate of the Arctic, the Eskimo (the Yuit and Inuit) developed a cultural solution to the cold environment, along with some physical adaptations, including the use of animal skins for clothing, building houses with snow, and obtaining food from hunting the animals of the region. Given the technology, the Eskimo solution was as good as it could be. When Euro-Americans entered the region, they adapted to the environment using *their* culture and technology (as well as by borrowing a few things, such as dogsleds). Their adaptations included clothing made from artificial materials, housing made from wood and metal, heat and light from imported and processed fuels (gasoline), and foods imported from other areas. Although the two adaptations were quite different, they were both successful.

In other cases, there seems little connection between solutions. When we learn that the ancient Irish and ancient Haida of the Northwest Coast of North America, who lived in similar environments, both ate salmon, we are not too surprised. That they both viewed wolves as symbols of power and ferocity seems a little less obvious but hardly surprising. But when we find they both regarded the tiny winter wren as a powerful supernatural being, we are genuinely surprised. There was little reason to predict that. Only intimate knowledge of both cultures and of the winter wren solves the mystery: both the Irish and Haida value song greatly, and the tiny, delicate winter wren fills the whole forest with loud and triumphant song during the most violent winter storms. The shared belief in the supernatural power of this bird now makes sense because the ancient Irish and Haida both believed that power over song is part of a wider power over all things.

Some General Theoretical Approaches to Human Ecology

Human ecology, like much of anthropology, is an eclectic science. As scientists, we want to learn, understand, and apply the knowledge about how people interact with their environment. We will utilize any theory or idea that might help us learn about how people adapt and why they do things in a particular way. Many approaches can be employed in the study of human ecology.

For a significant number of human ecologists, including many cultural ecologists, people are seen as animals much like any other animal (Park 1936), concerned solely or mainly with obtaining food and mates by the most efficient means possible. This general approach, embodied in evolutionary ecology (see chapter 3), directs our attention toward serious studies of food getting, among other things, and has produced much useful research. It also directs our attention toward serious consideration of the environment: what resources it offers, how difficult it is to obtain those resources, and any other problems it may present. However, most human ecologists find this model inadequate because it predicts neither the wide variety of cultures observed in the world nor the existence of art, music, poetry, and all the other things people have and do that other animals do not.

A second approach regards humans as rational choosers. In this model, humans set a variety of goals, not merely the necessities. They then seek, methodically and rationally, to reach those goals. This model directs our attention to individual choice. It assumes that people choose carefully and seriously on the basis of good information. This model has been shown to be very useful in many situations. However, people do not always have good information about their environment. More important, human choice is greatly affected by emotion, by social pressures, by cultural traditions, and by plain, ordinary mistakes. Thus, this model alone is also inadequate.

A third approach looks at political processes, from individual negotiation to worldwide political forces. This model directs our attention most especially to power differentials, from the power of village authorities to the far greater power of multinational agencies and corporations. This model has a number of major empirical successes to its credit, but it does not adequately deal with human long-term goals.

Other approaches and models will be discussed throughout this book. We take a somewhat flexible position toward all these theories. We suggest at the outset that understanding will come only from combining models, both existing and new. People have biological needs, and they have to fulfill them. People choose, and they make the best choices they can—and mistakes cannot be ignored or denied. They have to negotiate with others; they cannot do what they please in a social vacuum. Cooperation and competition are the common lot of social life.

To comprehend ecological practices, we must first understand the history of those practices. We must look at the whole chain of specific events, including pure chance, that actually causes behavior to become established (Vayda 1996). For example, could any rational choice theorist, in the absence of previous knowledge, predict that most Americans would celebrate December 25 by piling gifts around an evergreen tree? December 25 was not the actual birthday of Jesus Christ; the date and tree were originally part of a pagan northern European Yule festival that Christianity took over as it expanded northward. If we want to explain why Americans cut down millions of trees every year for a holiday, we must look at history. The individual choices that brought us to this ecological adjustment may have been rational, but no rational choice theorist could ever have predicted the present situation on the basis of theory alone.

A HISTORY OF THOUGHT ON CULTURE AND ENVIRONMENT

Scientists and laypersons alike have long been interested in how people lived in and utilized the natural environment. Throughout history, many theories on the interaction of culture and environment have been proposed, accepted as fact, disproved, resurrected, and codified in mythology. We still live with, and must respond to, many of these old, prejudiced, ethnocentric, downright wrong ideas.

Ecological anthropology of today has proposed, or drawn on, several useful and innovative theories. All of these theories have been valuable contributions; their limitations are those expected of theories in a developing field. Because science grows through translating new facts into more comprehensive theories, the value of a theory often lies in the stimulus it provides for further research and thought. Sometimes, the best theory is thus the first to be superseded.

It has been proposed (Kormondy 1996:383–385; also see Wilson 1967) that the development of the field of human ecology be divided into three historical traditions: imperialist, arcadian, and scientific. The imperialist tradition holds that humans are superior to, and hold dominion over, nature (e.g., Wilson 1967:1205). This tradition is of long standing but has gained great power since the industrial revolution and the expansion of Western culture across the globe at the expense of the environment and traditional cultures. Many still adhere to this tradition, and ecological imperialism by governments and corporations is still very widespread.

The arcadian tradition advocates that people should live in satisfaction, harmony, and idyllic contentment with nature. The ancient Greeks idealized in this way the Vale of Arcady, actually a poor and hardscrabble region. The tradition has taken on new life in the last couple of centuries with the rise of industrial society (Kormondy 1996:384).

The scientific tradition is a long-standing approach, one that dominates the field of human ecology today. The first scientific theories regarding culture and environment date from thousands of years ago. By about the fourth century B.C., the Greeks developed an explanatory view of culture and environment in which people and their potential were classified based on climate. The view was that cold climates made for "stupid" people, warm climates made for "perfect" people; and hot climates made people listless and lazy. It was no coincidence that Greece was located in a warm climate.

At about this time, the Chinese philosopher Mencius (see Mencius, tr. Lau 1979:164–165) pleaded for conservation, recounting how a certain mountain was deforested by woodcutters and the new brush eaten away by livestock. The mountain then eroded and seemed as if it had always been bare. Mencius drew a parallel with people who were bad; they, too, were once good, but poor management had corrupted them. Mencius, as well as many other early Chinese writers, provided a great deal of information on environmental management, showing that it was already a highly evolved science in China at that time.

By the seventeenth century, western Europe took a commanding lead in the study of natural science through its universities, learned societies, open publishing, open debate, and rewards and grants for science. These innovations led to an explosive increase in scientific activity. Many ideas about the relationship between culture and environment were proposed, including a notion of geographic determinism quite similar to the early Greek view (Montesquieu 1949, orig. 1748), and a concept of cultural evolution with stages progressing from savagery (hunting and gathering) to herding to agriculture to states (Smith 1920, orig. 1776)— an idea that was widely accepted throughout the nineteenth century. Adam Smith, along with Thomas Malthus, developed the ideas of competition in nature and in human affairs that later fed into contemporary evolutionary theory.

These and other thinkers of the Enlightenment era were the real founders of all social science, including anthropology. The main contribution of this period was the basic concept of systematic, comparative studies of human society. Studies became more objective, more systematic and classificatory, and a bit more tolerant of non-European practices.

Environmental Determinism

The first major theory regarding the interaction between culture and environment, one that has been in circulation since the time of classical Greece, is **environmental determinism**, or environmentalism. This idea basically states that environment mechanically "dictates" how a culture adapts. In the twentieth century, the idea was championed by Huntington (1945), who added detail about the importance of rainfall and drought. Many still believe that the environment does dictate cultures, at least those that are viewed as somehow being more closely tied to "nature."

Environmental determinism is attractive due to its simplicity, but there are obvious problems with the approach. The first is the belief that the environment and the life within it are fixed and unchanging, a view held for thousands of years. This premise is now known to be false; environments are constantly changing. The second major problem is the depreciation of the role of culture and the assignment of a compulsory role to environment. While this second premise seems to have merit in some environments with very few options, most environments allow a variety of alternatives, resulting in a large set of possible choices. If the environment dictates responses, then responses should be the same for different cultures in the same environment, and the same response should not be present in different environments.

For example, following environmental determinism, the Eskimo must hunt seals and live in snow houses because they live in the Arctic. The Polynesians must fish and live in grass huts because they live on tropical islands. Anyone who has been to the Arctic or Polynesia knows that some people do live as described above, but others live very differently. The difference is culture (including technology), not just environment, and this is why environmental determinism fails.

The Culture Area Concept

Somewhat related to environmental determinism is the idea of **culture areas**: large-scale geographic regions where environment and culture are similar to each other, particularly in economics. Culture areas were recognized in the 1890s, first in North and South America (Mason 1894), then in other regions of the world. For example, a number of culture areas have been defined for North America, and various schemes have been proposed and argued; the current consensus recognizes ten culture areas (eleven if you separate the Prairies from the Plains) in North America (fig. 1.2).

The use of the culture area concept gives anthropologists the opportunity to broadly compare cultures within generally similar environments and to determine the influence from cultures outside the culture area, such as by means of diffusion or migration. Nevertheless, the concept has many weaknesses, including the definition of a single area that contains considerable environmental and cultural diversity, the use of somewhat arbitrary defining criteria, the assumption of a static cultural situation, and the tendency to equate environment with cause (see Forde 1934:467; Kroeber 1939). However, the concept continues to be useful as a unit of comparison or reference, and most anthropologists use it, even if informally, to refer to geographic regions and general culture traits.

A good example of the concept is the Plains of North America. Geographically, the Plains is a relatively flat grassland extending north from the Gulf of Mexico to southern Canada and west from the Mississippi River to the Rocky Mountains. Relatively few trees and little water are present on the Plains, except in the rivers

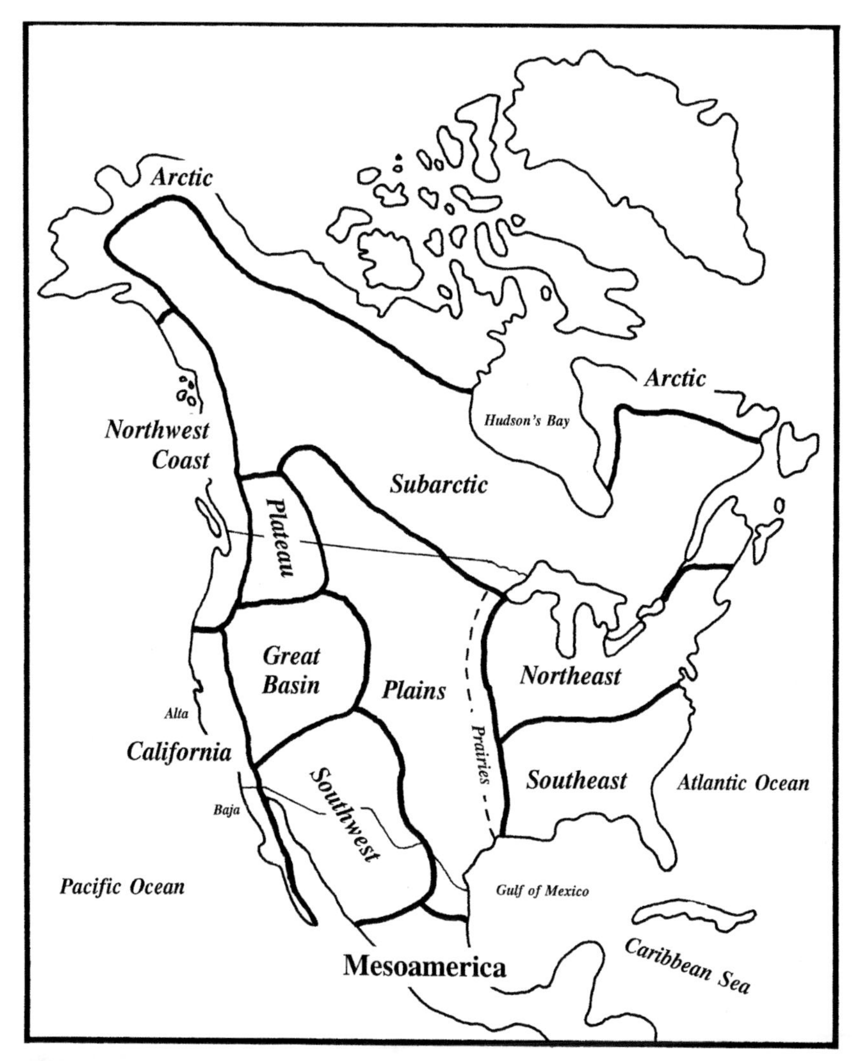

FIGURE 1.2
Culture areas of North America (presented as a guide rather than a representation of actual territories). (Adapted from *Handbook of North American Indians*, Vol. 4, *History of Indian-White Relations*, W. E. Washburn, ed., p. ix; copyright © 1988 by the Smithsonian Institution. Used by permission of the publisher)

that cross the region from west to east. The dominant animal on the Plains was the bison (*Bison bison bison*), often called the buffalo. Prior to the acquisition of the horse in historical times, the human population on the Plains was relatively small, and bison were hunted on foot. With the arrival of horses, new groups entered the Plains, some giving up farming to do so, and quickly developed a general culture based on bison hunting on horseback. The pre- and posthorse cultural patterns were similar to each other, particularly the subsistence (bison hunting) system. Even though technology had dramatically changed, the basic Plains economic pattern remained the same.

Cultural Evolution

Scientists had come to understand that cultures changed over time but not how or why they changed. While the notion of a cultural evolutionary framework had been proposed by Smith (1920, orig. 1776), it was not until Charles Darwin proposed his new theory of biological evolution in 1859 that a comprehensive theory of cultural evolution developed.

Unilinear Cultural Evolution

The first major theory in anthropology was the concept that cultures evolved upward along a single line. It was developed by Lewis H. Morgan and amplified by Edward B. Tylor and others and later known as **unilinear cultural evolution** (UCE). It proposed that cultures evolved progressively through three basic stages: from "savagery" (hunting and gathering) to "barbarism" (pastoralism to agriculture) and then up to "civilization" (see esp. Morgan 1851, 1871, 1877, 1882), a view that encompassed the nineteenth-century notion of "Progress." Morgan saw human life as a search for "livelihood," i.e., subsistence: food, clothing, and shelter. He was aware that humans need social life and a sense of control over their world but thought that ways of obtaining food, and therefore technology, varied according to local creativity and local environment. Morgan held that certain inventions, such as the bow and arrow, pottery, and agriculture, were keystones in cultural evolution.

At about the same time, Karl Marx and Friedrich Engels (see Engels 1942) proposed a six-stage theory of unilinear cultural evolution, but with politics and economics, rather than technology, as the most important factors. They proposed that societies would ultimately evolve to advanced communism, the pinnacle of development.

From the point of view of contemporary human ecology, perhaps the most important contribution of the evolutionary ideas of Marx and Engels was the assertion of the creativity and resourcefulness of human beings. Earlier thinkers (e.g., Montesquieu) gave nature pride of place and claimed that nature determined culture. Marx and Engels gave human creativity pride of place over nature, and this is a point that has come to underlie much contemporary human ecological work.

The theory of UCE was accepted throughout the social sciences in the nineteenth century, only to be disproved in the early twentieth century (see Boas

1927, 1940), partly through the understanding that technology alone does not dominate cultures and partly through the realization that historical process was an important factor. However, it was recognized that technological innovations have played a major role in cultural change and that certain innovations are more important than others. The use of UCE's broad categories of hunting-gathering, pastoralism, and agriculture has also survived, and they are still widely employed by anthropologists, but now only as a classificatory scheme (as in this book), not as a description of evolutionary progress.

Whatever their details, cultural evolutionary theory proposed relationships between culture and environment, including natural, political, social, and technological environments. Even today, the details of how cultures evolve remain unresolved, but it is clear that there is a relationship between environment and cultural change.

Multilinear Cultural Evolution

By the early twentieth century, the UCE model was in trouble. Key postulates, such as the idea that herding preceded agriculture, were not standing up under investigation. Worse, the simple scheme of "progress" as a procession of neat, regular stages was utterly inadequate to deal with the accumulating ethnographic data. Investigators also realized that food was not the only thing people got from the environment. Early theories (just as contemporary ones such as optimal foraging theory) dealt almost exclusively with food. But in fact other activities, such as art and religion, also draw on resources.

It soon became obvious that evolution was not always unidirectional. It was discovered that some groups had abandoned agriculture to become hunter-gatherers (e.g., on the Plains of North America). Some broad ethnic categories included elements of each of the supposedly distinct evolutionary stages: hunter-gatherers, agriculturalists, and "advanced" townsfolk, all trading with each other and giving every appearance of being economically specialized subgroups of one broad social formation rather than mixtures of ancestors and evolved descendants. Some hunter-gatherers, such as those of the Pacific Coast of North America, had exceedingly complex social and technological systems—much more complex than those of many farmers. It was realized that if cultural evolution were to survive as a theory, it would have to accommodate multiple lines of evolution (see Steward 1955).

Neoevolution

Despite the rejection of cultural evolutionary models in the early 1900s, by the middle of the twentieth century many anthropologists began to accept the reality that cultural evolution had occurred, even if in a multilinear way. Leslie

White, one of the founders of the ecological tradition in anthropology, made an attempt to revive unilinear cultural evolution by framing it in a new way (neo-evolution). White (1949) argued that cultures evolved as they increased their control of energy sources from fire to animal power, then to coal, then to oil, then electricity, and finally to thermonuclear power. At every stage, we become more adept at using greater and greater amounts of energy. Contemporary theorists would add that we increase in ability to use energy more efficiently and to control it better. White expressed this in summary form, C = E x T, where C is culture, E is energy, and T is technology. It was not intended to be taken as literal mathematics; White did not argue that the United States is twice as advanced as Sweden because it uses twice as much energy per capita. But White was arguing—rightly or wrongly—that energy and the means of harnessing it are basic to a culture in a way that art styles or dynastic genealogies are not. White also held that symbols were the basis of culture, and humans were symboling animals. However, he saw this characteristic as equally typical of all humans and thus not a cause of change or "evolution" per se.

Steward (1955; also see Harding et al. 1960; Service 1962; Johnson and Earle 2000) introduced an evolutionary scheme based on increasing sociocultural complexity. The least complex was the **band**, consisting of small, relatively mobile hunting and gathering groups with informal leaders. Next was the **tribe**, consisting of larger, more or less settled groups of hunter-gatherers or incipient agriculturalists (horticulturalists or pastoralists) with several settlements and relatively formal decision-making authority but still no centralized authority. Third was the **chiefdom**, which had large, sedentary populations (usually of agriculturalists), elites, some social stratification, and leaders with the authority to impose their will. Last was the **state** (sometimes called "civilization," a highly loaded term), a large and complex system based on grain agriculture and with larger and more dense populations, complex social and political structures, elaborate record keeping, urban centers or cities, central authority, monumental architecture, and specialization. While there has been much criticism of this scheme, it is still widely utilized by anthropologists to describe political entities and, in some sense, ecological adaptations.

There now seems to be a growing acceptance of some sort of unilinear developmental scheme of increasing complexity, at least sociopolitical complexity. This view would have band-level hunter-gatherers as the original societies, with some evolving to tribe-level groups. With the incorporation of agriculture as the economic base, some tribes evolved into chiefdoms, and some of those, under certain circumstances, evolved into states. While there is ample archaeological and other evidence of this general trend, there is no reason to believe that such evolution has been directional, either good or bad, or that any "progress" has occurred.

Possibilism

As anthropologists began to accumulate more general knowledge of culture and detailed knowledge of specific cultures, it became apparent that culture was highly adaptive, that most environments had been modified by humans, that various responses to most environmental situations were possible, and that cultures were considerably influenced by other cultures. Clearly environment influenced culture, but it had become clear that environment did not dictate culture. In **possibilism**, the environment is seen as a limiting factor rather than a determining factor. To be sure, the environment may deny certain possibilities, such as the use of snow houses in Arabia, but it will open a variety of other possibilities, such as houses of wood, grass, mud, cloth, stone, or skins, all of which occur in Arabia. The culture makes the choice of which of the possibilities to employ.

The culture also has limiting factors, including technology, belief systems, and extracultural relations. In our housing example above, metal houses are a possibility offered by the environment (i.e., iron ore exists), but if a culture does not possess the technology to mine, process, and fabricate metal, metal houses are not really a choice. However, if that culture has access to metal through trade, perhaps it could be a choice. Possibilism is really an interactive process between culture and the environment. The choices available in the environment may be limited by the capabilities of the culture, or vice versa, and as culture and the environment evolve (change), the interplay also changes.

While the culture area concept related similarities between cultures and environment, it was recognized (Wissler 1926; Kroeber 1953; also see Meggers 1954) that the same environment (or culture area) might include cultures with quite different ecological adaptations. For example, the Southwest culture area contained Pueblo agriculturalists, hunting-and-gathering Apaches, and sheepherding Navajo. These groups coexisted by filling complementary niches. Thus, culture shaped environmental response. The environment did not control behavior; it merely made some behaviors more reasonable than others. Another classic study along these lines was the work of Birdsell (1953) on the relationship between rainfall and population density in aboriginal Australia.

Some would view possibilism as the opposite of determinism. In fact, however, possibilism is deterministic from the standpoint that some options are excluded, and the solutions are limited to a subset of all possibilities, which amounts to a determination of which are possible and which are not. Possibilism seems much more realistic than strict determinism because the role of culture is considered to some extent. Human culture has a penchant for changing the conditions and rules; as human cultural institutions and technology become more complex, the environment seems to play less and less of a role in limiting or determining human responses and adaptation. Thus, possibilism has been frequently criticized (e.g., Smith 1991) as not being a theory at all, as it predicts nothing specific. It is difficult to deny this charge.

THE RISE OF HUMAN ECOLOGY

While always embedded in general anthropological theory (e.g., Adams 1935; Park 1936), human ecology came into its own after the late 1930s, primarily through the work of Julian Steward. He began his career working with the Paiute and Shoshone people of the Great Basin in western North America but later worked in South America and eventually in Puerto Rico, a colonial-type society in the contemporary world. He was one of the first anthropologists to look at complex societies and their place in the even more complex world of today. Steward also drew on the concept of possibilism. Rather than being subject to environmental determinism, societies could adapt in any of a number of possible directions.

Steward's Ecology

Steward was the first to combine four approaches in studying the interaction between culture and environment: (1) an explanation of culture in terms of the environment where it existed rather than as just an association of geography with economy; (2) an understanding of the relationship between culture and environment as a process (not just a correlation); (3) a consideration of small-scale environment rather than culture-area-sized regions; and (4) an examination of the connection of ecology and multilinear cultural evolution.

Steward's landmark ecological work, *Basin-Plateau Aboriginal Sociopolitical Groups* (1938; also see Steward 1936), dealt with native peoples of the Great Basin. In that work, Steward first described the general environment, listed important resources, and then discussed how those resources were utilized. He then discussed the sociopolitical patterns of the people and how those patterns related to technology, the environment, and the distribution of resources. His approach was groundbreaking. Steward's (1955) primary arguments were that (1) cultures in similar environments may have similar adaptations; (2) all adaptations are short-lived and are constantly adjusting to changing environments; and (3) changes in culture can elaborate existing culture or result in entirely new ones. Steward (1955:5, 30) coined the term "cultural ecology" to describe his approach and is frequently referred to as the father of ecological studies in anthropology.

Steward (1955:31) recognized that the ecology of humans had distinct biological and cultural aspects, although they were intertwined. He argued that the cultural aspect was associated with technology, which set humans and their cultures above and separate from the rest of the environment. While Steward was correct in recognizing the difference between the biological and cultural aspects of human ecology, he was wrong to view humans as separate from the rest of the environment.

The "New Ecology"

While Steward tied culture into the environment, a new approach tied culture into the emerging science of systems ecology (e.g., Vayda and Rappaport 1968). It was argued that human cultures were not unique but formed only one of the population units interacting "to form food webs, biotic communities, and ecosystems" (Vayda and Rappaport 1968:494). This approach placed humans within a "unified science of ecology" such that "generalizations concerning human behavior [would now] have a broader scope and applicability" (Brush 1975:803).

This approach has had the effect of moving the analysis of human behavior from strictly qualitative ethnography to quantitative science, leading to a whole new way to look at humans. The weakness of this approach is that analysis is based on data that describe situations at a single point in time. While variables can be measured and compared to each other, and relationships between variables can be described and modeled, it is difficult to model culture change and evolution using such data. This problem persists in much of the more recent work, such as the use of optimization models (see chapter 3).

Cultural Materialism

Cultural materialism is a practical, rather straightforward, "functionalist" approach to anthropology, with a focus on the specific hows and whys of culture. It is based on the idea that "human social life is a response to the practical problems of earthly existence" (Harris 1979:ix) and that these issues can be studied in a practical way. Cultural materialism emphasizes very empirical phenomena, such as technology, economy (e.g., food), environment, and population, takes an evolutionary perspective, and has an unwavering commitment to the rules of Western science.

Marvin Harris (1966, 1968) espoused a concept of "techno-environmental materialism," which initially held that all cultural institutions could be explained by direct material payoff. Harris did not claim that this concept always provided a total explanation; he saw it as a "research strategy." One starts by looking for a direct material payoff—in calories of food—for a cultural institution. If that explanation is inadequate, look for a payoff in protein or in shelter. Only when all material payoffs have been eliminated should one investigate psychological and sociological factors. This approach has proved exceedingly useful in research (e.g., see Smith 1991), but it probably never produces an altogether adequate account.

Materialists tend to look at specifics rather than trends, at distinctive traits rather than general ones. The task, then, is to explain a trait and why it is done in its particular way. Detractors would consider this approach to be biased away from nonmaterial aspects of culture, often overlooking important, if not critical, information. Proponents would argue that the overlooked "information" was not empirical and so not science. Nonetheless, cultural materialism has formed the basis for much anthropological research since the 1960s.

An excellent example of the functionalist/materialist approach is the analysis of the role of sacred cows in India (Harris 1966, 1974). To Hindus, cows are sacred and cannot be eaten. Hindu religion includes a belief in reincarnation, and so it is possible that a relative may have been reincarnated as a cow and that eating the cow would be the equivalent of cannibalism. To many Westerners who eat beef on an almost daily basis, it seems silly for starving people to refuse to eat their cows.

However, an analysis of the function of cows in Indian society revealed that they were simply too important to eat. First, cows provided labor for plowing; few farmers could afford a tractor, and there was no infrastructure for the support of such machines. Next, cow dung was used as fertilizer and fuel; no substitutes were available, and the fields had to have some fertilizer to maintain productivity. Cows did not have to be fed; they survived by foraging trash and weeds, helping to keep the area clean. In addition, they provided milk, a renewable resource.

Thus, slaughtering the cows for food would provide a wonderful few weeks of steak and ribs but would also result in no labor to pull plows, no fertilizer, no fuel, no milk, no weed or trash disposal, the rapid collapse of the entire agricultural system, and the death of millions by famine. As it turns out, the cows are eventually eaten. When cows die naturally, members of the "untouchables," the lowest caste in Indian society, butcher them, eat the meat, and manufacture useful products from their skins and other parts. From a purely practical standpoint, it is a system that functions well under the circumstances. However, it does not necessarily explain the origin of the practice.

A Note on Function and Origin

To many cultural materialists, explanation is fundamentally about the question of function and origin. If something serves a specific function, the rationale goes, it must have originated or been designed to fulfill that function. In human ecology, this line of thinking appears to be even more prevalent, perhaps because human ecologists tend to focus their research on functional matters, such as food procurement or technology. However, it is a mistake to assume that function must be equated with origin. Some, if not most, things have multiple functions. Choosing one and then "determining" the origin will likely result in error. Also, some things may have multiple origins and may have been recombined to serve different functions. Technology and culture are constantly being modified, changed, and adjusted to fit new conditions. It may be that the origin of a particular practice is lost in all the changes.

The reverse investigative approach also is true; knowing an origin does not necessarily mean the function is known. The function of things may change over time, so their origin may have little to do with their current use. Sleeve buttons once used to attach gloves are now mere decoration, and old engine parts are now paperweights.

Rational Choice Theory

Currently, one paradigm in environmental social science is some form of **rational choice theory** (Elster 1987; also see Frank 1988; Green and Shapiro 1994). This theory, popular in economics and political science as well as some fields of anthropology, asserts that people decide how to achieve their goals on the basis of deliberate, individual consideration of all available information, that they seek out better information as required, and that they are good calculators of their chances; that they know where to hunt deer, which crops will grow, and how to trade off the potential yields of hunting deer versus cultivating crops. Some would consider rational choice theory to be related to evolutionary ecology, in that people take what information they have and make the best (optimal) decision in a rational manner. Poor choices would be subject to negative selective pressure.

It is true that much conventional behavior is rationally chosen. However, some behavior is not rational, even if a seemingly reasonable argument can be made for such conduct. In practice, people rationalize irrational behavior. It seems obvious that cultures vary in their approaches to adaptation and that if rational choice were always correct, much less variation would be expected. However, remember that each culture has different goals, different technologies, and different concepts of what is rational, so the rational choice of Group A will likely be different from that of Group B, even in the same environment.

In addition, people have many of their choices made for them before they are old enough to choose for themselves. People take on many traits, such as language and diet, long before they are old enough to make rational choices for themselves. People do not have time to decide everything in detail. They have to take shortcuts, which usually means going with habit or imitating others. When we are forced to change, we are forced to make more-or-less-rational decisions. The rest of the time, we tend to find that the most "rational" course is to minimize the effort of making decisions; we go with our habits.

It has been argued that the whole Western attitude toward "nature" is cultural and irrational. Westerners tend to regard nature as something separate from humans, ours to exploit and ruin at will. This belief is neither scientifically sensible nor conducive to maximizing any of the many resources we get from the nonhuman world. A rational chooser would choose to believe something quite different—so goes the argument.

Rational choice is an indispensable tool of human ecological analysis, but it leaves a good deal for us to explain. In fact, it leaves almost all the content of culture for us to explain. We might freely grant that *all* cultural practices were adopted because they seemed, at one time, the most rational things to do under the existing circumstances. However, such practices often persist not because they are always rational or optimal, but because children learn them before they are old enough to have a choice.

Political Ecology

A recent development in human ecology is the rapid spread of **political ecology**, a field developed in about 1970 (Wolf 1972, 1982) and popularized after the mid-1980s (see Greenberg and Park 1994; Kottak 1999). Political ecology is concerned with the day-to-day conflicts, alliances, and negotiations that ultimately result in some sort of definitive behavior. It directs our attention to immediate processes and conflicts. It has therefore meaningfully supplemented the other branches of human ecology that tend to look at the long term, in which continued actions have an impact on the environment.

The field of political ecology rose rapidly in the late 1980s and the 1990s, heavily influenced by contemporary economic theory and stimulated by a number of other influences (Bennett 1976, 1992). Perhaps the most important of these influences was environmental politics. Worldwide battles between exploiters and conservationists have always had a serious impact on indigenous communities (see Bodley 1999). For example, by the 1990s, even remote native groups in rain forests found themselves used as pawns in power struggles between national governments, multinational companies, and international conservation organizations. Such struggles are not limited to native groups; African American communities in the southern United States have suddenly found themselves targeted as sites for toxic waste disposal (Bullard 1990). As a result, gender, ethnicity, and identity—all concepts that are notorious political battlegrounds—emerged as important topics of anthropological research. Anthropologists had never ignored these ideas, but interest in them intensified in the 1980s and 1990s.

Most political ecology falls into two broad categories. First is the work on resource management in complex contemporary societies (e.g., McCay and Acheson 1987; Pinkerton and Weinstein 1995). Much of this work involves management of resources owned by the community or not owned at all, and studies of common-property water resources have been important (Ostrom 1990). Second is research on the fate of small-scale, indigenous societies caught in the midst of "modernization" (e.g., Netting 1974, 1986; Gladwin and Truman 1988; Wilk 1991; Stonich 1993; Anderson 1994).

Recently, political ecology has retreated somewhat from its early ideals. Many studies naively demonize "globalization" and "multinationalism," regarding the international economy more or less as a single and simple thing. At its worst, political ecology has been no more than a reiteration of the ancient complaint that the rich get richer and the poor get poorer. One of the major drawbacks of this approach is that it discredits local people, and all their incredible accomplishments, by labeling them as "mere victims."

This sort of abuse of the term led to a sharp critique of the whole field by Andrew Vayda and Bradley Walters (1999). They argued that the term itself should be dropped and that we should return to a holistic, event-centered ecology. Their critique, however, touches only the highly oversimplified literature, leaving unscathed the works cited above. Thus, in spite of problems with the naive literature, the term "political ecology" seems here to stay (Anderson 2000).

Meanwhile, political ecology has broadened its view. The *Journal of Political Ecology* and new collections such as *Political Ecology* (Stott and Sullivan 2001) go far beyond politics. Biology and culture have been brought back into the fold. Political and cultural ecology continue to blend into each other.

In recent years, cultural/political ecology has been increasingly influenced by world systems theory. This theory was developed largely by Immanuel Wallerstein (1976). He began to look seriously at the interconnections of societies around the world, going beyond the simple "rich-poor" and "developed–less developed" contrasts to see how the rise of one society might lead to, or be linked with, the fall of others. He separated the world into "cores" (the rich nations: Europe, North America, and Japan today; China and the Near East a thousand years ago), "peripheries" (poor and isolated societies), and "semiperipheries." These last are the countries in between, fairly well off but with much poverty and displaying a contrast between highly developed and much less developed sectors. Mexico, Turkey, and China provide current examples; in the world of a thousand years ago, southernmost Europe qualified (the rest was "periphery"), and so did much of southeast Asia. Long cycles of empire and dominance occur, reflected in economic swings and geographic shifts of power.

Wallerstein's theory has been used to evaluate the rise and fall of cultures and the problems of smaller, more remote societies in today's world (Bodley 1999, 2001) and in their own earlier worlds, where they could create small local "world" systems (Chase-Dunn and Mann 1998). In general, world systems are unfair, and today's "globalization"—which is no new phenomenon—is perhaps the least fair of all. No one seems to have figured out a way to deal with an economy of goods and information that is inevitably global—in this age of container shipping, jet planes, and the Internet—but is dominated by a few warlike or predatory nation-states.

Historical Ecology

Recently, the number of terms in the human-ecological field has been increased by the addition of **historical ecology** (Balée 1998a; Crumley 1994; Winthrop 2001). The term has been around at least since the 1970s, when Edward Deevey directed a Historical Ecology Project at the University of Florida (Crumley 1998:xii). This field is close to environmental history (Cronon 1983), landscape history, and similar historical subfields, as well as to cultural geography. In practice, it has been something of a blend of these fields—anthropology with more historical detail than usual, or history with more holistic cultural and environmental data than usual (e.g., Ehrlich 2000).

A specific theory for this subfield was proposed by William Balée (1998b), who emphasized interactions ("dialectical" in his usage) between people and environments, with the two being active players rather than humans merely adapting to the environment. Balée (1998b:14) argued that (1) human activity has affected virtually all environments, (2) human activity does not necessarily degrade or improve environments, (3) different cultural systems have different impacts on their environments, and (4) human interaction with the environment can be understood as a total phenomenon. This view directs attention to *individual action* as opposed to such things as evolutionary dynamics, cultural ideologies, or social systems, which are abstractions that do not really interact.

Historical ecologists, like other human ecologists in recent years, have paid much attention to the influence of small-scale societies on their environments. Such people were once dismissed as "primitives" and "savages" who had minimal effect on their surroundings—who were, indeed, part of "nature" rather than "culture," according to earlier formulations. In North America, we still find Native American exhibits in museums of "natural history" rather than in museums of history or art. From such unthinking prejudice, contemporary ecological anthropology—including historical ecology—can deliver us.

In short, historical ecology focuses much more on *change, contingency*, and *human agency* than did some of the other traditions within cultural ecology. In this, it continues a long-standing agenda in the field (e.g., Anderson 1972, 1988; Bennett 1976; Netting 1986). The counteragendas include evolutionary ones, whether Darwinian or cultural, and highly determinative "adaptation" theories that view human action as more or less a reflex to the environment.

Historical ecology, along with recent archaeological theory (Ashmore in press) and cultural ecology in general (Feld and Basso 1996), has revivified the old cultural-geography concept of "landscape." The concept stems from the work of Carl Sauer (beginning with Sauer 1925). Since Sauer, geographers have used "landscape" to refer to the face of the earth as modified, created, or perceived by people. By contrast, *environment* is far wider, including everything from cosmic rays to bacteria. People may know very little about their environment, but, by definition, the landscape is what they see, know, and interact with. The effect of humans on landscape may be seen today as "good" or "bad"; it is, perhaps, hard to avoid judging.

A particularly stunning example is the recent encyclopedic trilogy by Sauer students on the cultivated landscapes of Native America (Denevan 2001; Doolittle 2000; Whitmore and Turner 2001). Denevan emphasized the success of Native cultivators; Whitmore and Turner, their failures; and Doolittle maintained a careful balance. Balée (1998b) emphasized the need to look at actual influences, without prejudging, in order to understand. This attitude is doubly important if we wish to maintain or revive traditional land management techniques (such as controlled burning). Impartial analysis and understanding must preempt the sort of naive, condemnatory judgment that we have critiqued elsewhere in this book.

Sauer was far ahead of his time, and only recently has his concept of landscape broadened its scope and exploded into widespread use. This has followed on the realization noted above: even the "simplest" peoples not only know an enormous amount about their environments but also profoundly modify them.

The "landscape" concept also has the benefit (again, emphasized by Sauer) of uniting science and humanities (cf. Balée 1998b, 1998c). Not only archaeologists, ecologists, and historians, but also students of traditional art and myth, to say nothing of phenomenological philosophers and poets, all talk about landscape (Ashmore in press) and can find here a common "ground" in a thoroughly literal sense. Human ecology profits considerably from such meetings of the minds. (It does not, however, profit from the endless multiplication of terms; one truly wonders if we need "historical ecology," "political ecology," "event ecology," "social ecology," and the rest for what are, after all, merely aspects of one discipline.)

Postmodernism

A fairly new paradigm in contemporary social thought is **postmodernism**. The postmodernists are critical of all modern things and argue that science itself is flawed. Postmodernism takes a very subjective and antiscientific stance, in opposition to the objectivism of the modern world. Postmodernists hold that science is subjective, so our interpretations of cultures are also subjective (see Clifford and Marcus 1986). Postmodernists argue that there is no objective reality and that fact lists have no place in anthropology.

This approach is not science but has evolved within studies of literature, religion, and expressive behavior. Insofar as it directs our attention to interpreting and understanding cultural practice, it is a valuable contribution. However, it has not spread far into human ecological studies, because ecologists find that humans do have to consider some blunt truths: the need for food, the need to avoid extreme cold and heat, the brutal facts of disease, broken bones, grizzly bear attacks, and the like. Humans confront these details of life in a multitude of ways, and cultural interpretation plays a vital role in our understanding of those ways. However, we cannot ignore the life-and-death matters that force humans to adjust.

Postmodernism is a philosophical position that runs counter to science. In a succession of articles, Vayda (1993, 1995, 1996) argued against the antiscience view of postmodernism and maintained that ecological anthropology should be carefully establishing cause-and-effect links (when possible). He also argued against certain views that have been broadly labeled "functionalist" and that often lead to the construction of "models" rather than cause-and-effect explanations.

THUS . . .

As noted earlier, anthropology and human ecology are eclectic sciences. Human ecology does not have its own theory; it uses ideas taken from other disciplines, modified as needed, and then applied to problems. Any approach that can be used to learn about how people and cultures interact with their environments is employed. If something does not inform us, it is abandoned and another procedure used. This flexibility in research approaches is a strength and helps us to better understand humans, their cultures, and their relationship with the environment.

CHAPTER SUMMARY

People occupy most of the diverse environments of the planet, and cultural ecology is the story of how they do that. Some do it well, others less so. In any case, humans are the key species in most environments, and whatever their practices, we can learn from them.

The study of humans' interaction with their environment is called human ecology, with human biological ecology emphasizing the biological aspect of the adaptation (including evolution) and cultural ecology emphasizing the cultural aspect. Human ecology generally falls within anthropology, the study of humans (through time and space).

Human ecology has a long history containing many ideas and concepts. The ideas of environment dictating culture (environmental determinism), of cultures evolving through stages, and of cultures operating within environmental parameters (possibilism) have been proposed, rejected, refined, argued, and reconsidered. However, modern human ecology, including cultural ecology, was founded by Julian Steward, who argued that human adaptation was an interplay between environment, biology, and culture.

Today, most human ecologists utilize the principles of empirical science in their investigations in a number of ways. These include a "humans as animals" view, which relies heavily on biological principles. Some view humans as rational choosers and use a functional and materialist approach. Others emphasize political processes and the interaction between power, modernization, and globalization.

KEY TERMS

a culture
adaptation
anthropological linguistics
anthropology
archaeology
band
biological anthropology
chiefdom
cultural anthropology
cultural ecology
cultural materialism
cultural relativism
culture
culture area
ecology
environmental determinism
ethnocentrism
ethnography
ethnology
evolution
human biological ecology
human ecology
multilinear cultural evolution
neoevolution
political ecology
possibilism
postmodernism
rational choice theory
science
state
tribe
unilinear cultural evolution

2

Fundamentals of Ecology

Ecology is the study of the relationships between organisms and their environment; the "economics" (or livelihood) of the Earth and its totality of life forms. The term "ecology" comes from the Greek word *oikos* (house or habitat) and ology (the study of). It was given its contemporary usage by Ernst Haeckel in 1866 (see Goodland 1975). Most of the concepts used in human ecology have been borrowed from biology, so an understanding of the basic notions of biological ecology is essential (Richerson 1977). For more detailed treatment of ecological concepts, see Odum (1975, 1993), Kormondy (1996), or Molles (1999).

THE ENVIRONMENT

The **environment** consists of the surroundings within which an organism interacts—a pretty broad definition. One of the problems in defining the environment is this breadth; the environment can be viewed as different things in different places and at different geographic or spatial levels, such as a pond, a valley, a continent, the Earth, the solar system, or even the universe. Perhaps the concept of the "operating environment," that area in which the population under consideration operates, is more useful.

The environment can be divided into two primary components, **abiotic** and **biotic** (see Kormondy 1996:6). The abiotic component consists of the inorganic materials present in the environment, including elements such as oxygen, nitrogen, sodium, and carbon and compounds such as water and carbon dioxide. The abiotic component also includes physical factors, such as weather, climate, geological materials, geography, time, solar radiation (the source of most energy), and even the cosmos.

The biotic component consists of all of the materials that are biological in origin: plants, animals, and microbes, either living or dead. Thus, a living tree is part of the biotic environment, as is a dead, fallen, and decomposing tree. Eventually, the tree will be broken down into its inorganic constituents, and those materials will again enter the abiotic environment.

Classifying Environments

As we noted, environments can be defined based on any number of criteria, and operational definitions differ depending on the situation. Here we define two basic divisions of the biotic environment, biomes and ecozones, differentiated on the basis of scale.

Biomes

A **biome** is a large-scale, broad region of similar temperature, rainfall, and biology. The biome concept is frequently used by ecologists, biologists, and anthropologists as a general descriptive category and as a starting point for classification and analysis. Culture areas, discussed in chapter 1, are essentially biomes with an added layer of culture, and each has a human carrying capacity and trends of economy.

It is possible to define any number of biomes (fig. 2.1). Some of the basic terrestrial biomes (see Campbell [1995:fig. 1.1], Molnar and Molnar [2000:figs. 2-1a and 2-1b], Kormondy [1996:338–379]) illustrating the diversity of environments across the globe are briefly described in table 2.1. These biomes contain considerable diversity within them. They begin at the north and south poles and generally extend toward the equator and from higher elevations to lower.

In addition to the terrestrial environments, all of which are inhabited by humans, several aquatic environments have been, and continue to be, heavily utilized although not generally inhabited by humans. Aquatic environments are divided into marine (ocean) and freshwater components (see fig. 2.2 and table 2.2). Until recently, human utilization of marine environments was largely confined to shores and shallow waters.

Environmental Zones

An environmental zone, or **ecozone**, is a geographic area defined by fairly specific biotic communities (a plant association dominated by a certain species) within biomes. Thus, a river running through a savanna would contain a riverine ecozone within the savanna biome. The definition of any particular ecozone is based on judgmental criteria that depend on the goals of the researcher. Ecozones are most commonly defined on the basis of dominant plant communities because

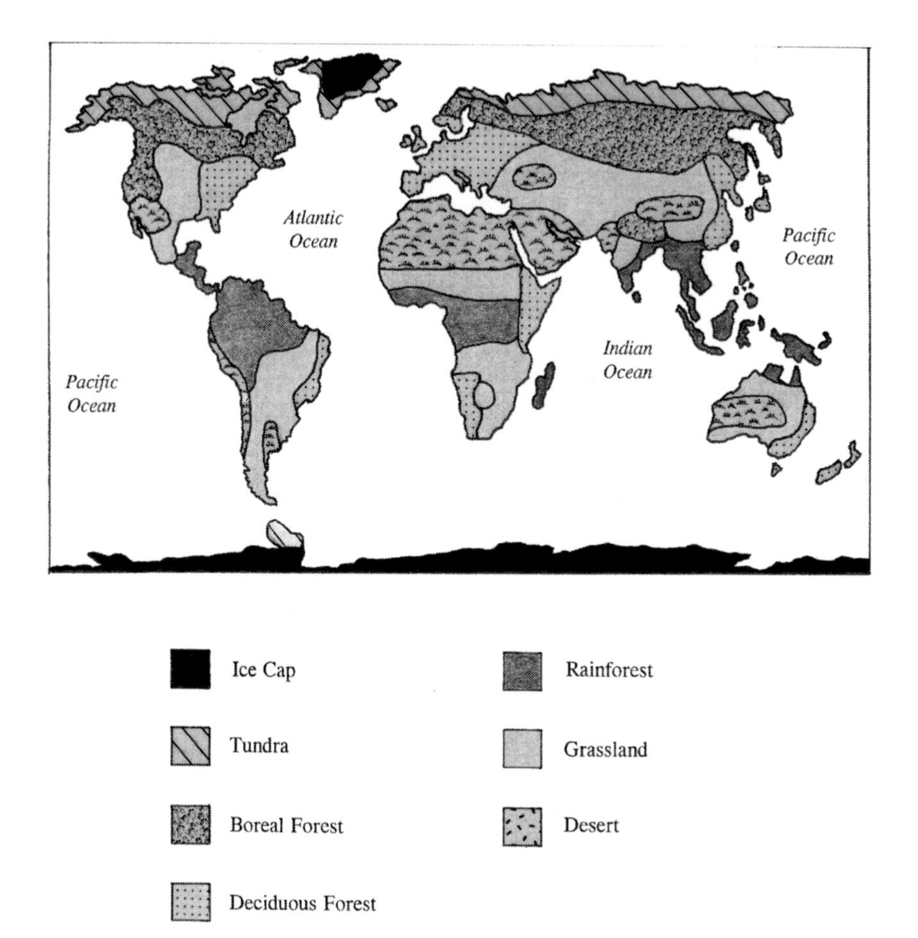

FIGURE 2.1
The general terrestrial biomes around the world. (Adapted from Campbell 1995:Fig. 1.1. Reprinted with permission by Bernard Campbell. HUMAN ECOLOGY, 2/E. New York: Aldine de Gruyter. Copyright © 1983, 1995 by Bernard Campbell)

they are easy to recognize and map (animals are harder). The "pinyon-juniper" forest, the "pine belt," and the "aspen zone" are typical examples of ecozones.

Ecotones

An **ecotone** is the geographic intersection of—as well as the transition between—ecozones (see fig. 2.3). Since ecozone boundaries may also be biome boundaries, ecotones exist between biomes as well. Examples of ecotones include estuaries (places where freshwater meets saltwater, such as where a river empties into the ocean), shorelines, and the areas where forests and grasslands meet. An ecotone is usually a more productive place than either of the individual ecozones

Table 2.1. The Major Terrestrial Biomes

Biome	*General Location*	*Primary Vegetation*	*Temperature and Precipitation*	*Comments*
Tundra	near the poles	lichens, mosses, some grass	very cold, often frozen, generally low precipitation, but little evaporation	very low biodiversity
Boreal forest	just south of the tundra in the northern hemisphere	evergreen coniferous trees (e.g., pines)	long cold winters and variable precipitation	generally low biodiversity, poor soils and short growing season (not well suited to agriculture)
Deciduous forest	midlatitudes	deciduous trees (which drop their leaves in the winter, e.g.)	moderate temperatures but cold winters, considerable rainfall	considerable biodiversity, fertile soils and long growing season (well suited to agriculture)
Tropical rain forest	equatorial zones	evergreen (not conifers) trees and shrubs	warm with abundant (often daily) rainfall, general lack of "seasons"	enormous biodiversity, generally poor soils (suited to some types of agriculture)
Grassland (aka plains, savanna, pampas, steppe)	large flat regions	grasses, a few trees	warm summers and cold winters, wet and dry seasons	moderate biodiversity, large herd animals common, soils suited to some forms of agriculture
Desert	regions of low precipitation	generally sparse	often very warm, but some deserts are cold; generally low precipitation	low biodiversity; not well suited to agriculture except in regions with rivers

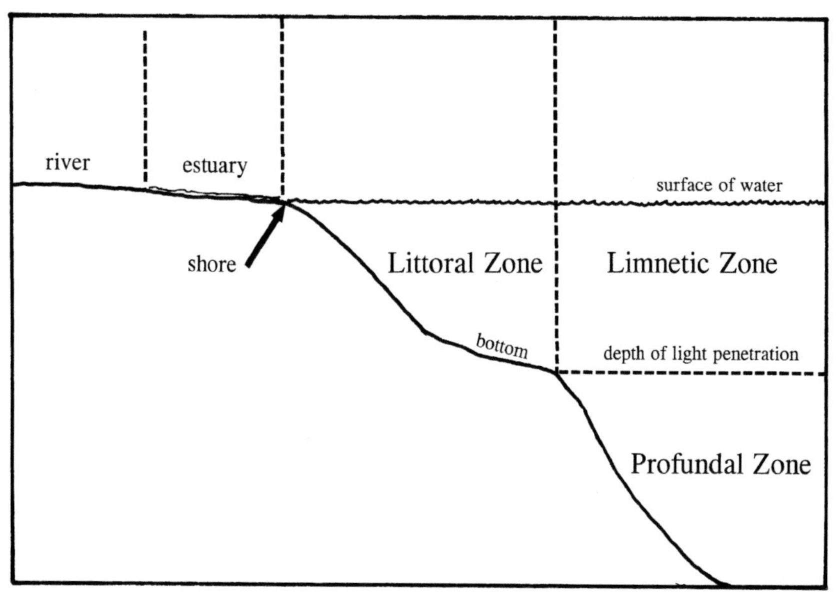

FIGURE 2.2
A general diagram of the aquatic biomes.

because species of both zones intermingle within it. Even in cases where there is less diversity, an ecotone is a good place for an organism to be located because access to both ecozones is easier.

This same concept could be applied directly to cultural systems; the border between two cultures would form a cultural ecotone. It might be an especially "culturally productive" place where ideas and goods could intermingle. Examples of such places would be trading centers, ports, and centers of learning.

Refugia

Most biomes and ecozones are defined on the basis of the current distribution of plants and animals. Researchers studying past environments also use the ecozone concept but define ecozones based on past biotic distributions. Occasionally, a remnant of a past biome or ecozone will survive into the present as sort of a "living fossil." These areas, and the life within them, are called **refugia** and can be quite valuable in the study of past environments.

For example, a number of regions that are now desert once contained different vegetation. If a small pine forest were found on top of a mountain now within a desert, the forest may be a surviving remnant of a larger forest that once covered the area. This refugium could provide clues to the past plant and animal life within the region and constitute a starting point for the reconstruction of the ecozone at that time.

Table 2.2. The Major Aquatic Environments

	Location	General Characteristics	Comments
		Marine Environments	
Estuary	where rivers and streams enter the ocean	sea water mixes with freshwater, forming an ecotone with tidal marshes and mudflats	diverse and productive areas, heavily utilized by people
Littoral zone	from the beach into coastal waters to the depth of light penetration	rocky and sandy beaches and bottoms	diverse and productive areas, heavily utilized by people
Limnetic zone	open water to the depth of light penetration (above about 300 feet)	open water away from coastlines	moderate diversity, some utilization by humans with large watercraft
Profundal zone	open water below the depth of light penetration (below about 300 feet)	dark, not inhabited by plants	unknown diversity but not often utilized by humans
		Freshwater Environments	
Still water	lakes, ponds, and marshes	shallow to deep waters, some with limnetic and profundal zones	considerable diversity, heavily used by humans
Moving water	rivers and streams	generally shallow, some very muddy, some treacherous	less diversity than still water, but some resources (e.g., salmon) heavily used by people
Riparian	along shorelines	ecotone between land and water	high diversity, heavily used by people

Ecosystems

An **ecosystem** is a geographically bounded system within which a defined group of organisms interacts with both the abiotic and the biotic components of the environment. Ecosystems are active and dynamic places and may be contained within or overlap with biomes and ecozones (see fig. 2.4). The size and scale of ecosystems can vary, depending on how and why they are defined. The largest ecosystem currently defined is called the **biosphere** and is the global environment and all of its interacting ecosystems.

The concept of ecosystem was formalized by Odum (1953), and a history of the idea has been provided by Golley (1993). Ecosystems are conceptual entities, often being defined by researchers on the basis of the goals of their research, but

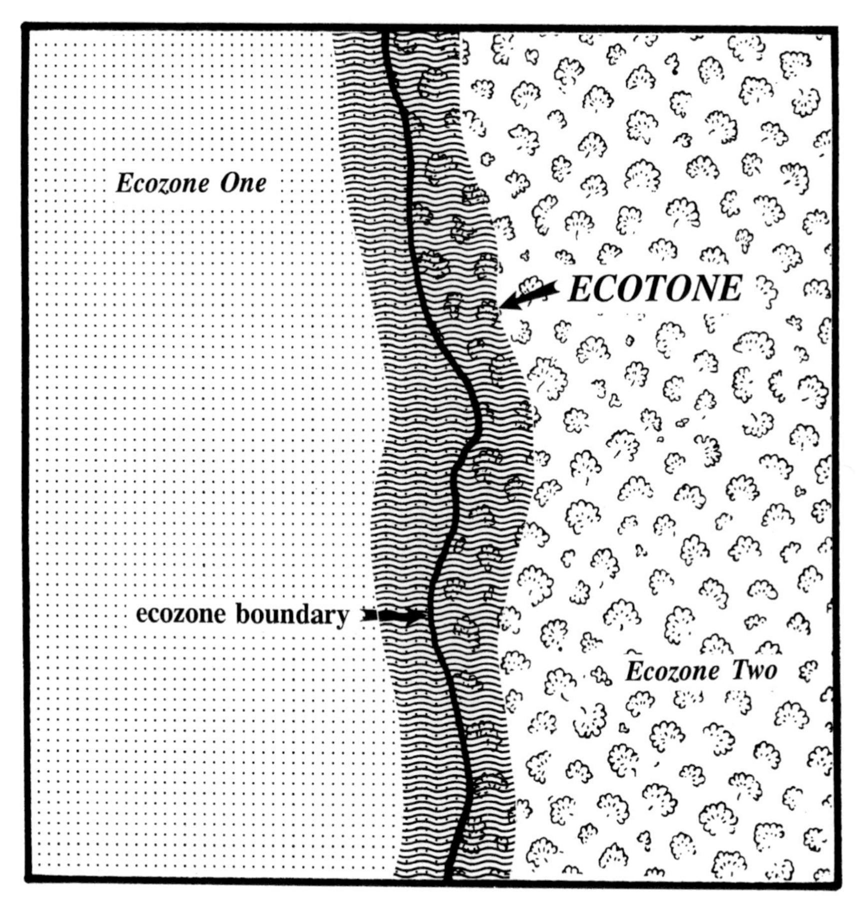

FIGURE 2.3
An example of an ecotone between two ecozones.

the phenomena are real. For example, a person studying a pond could define it as an ecosystem and would find it inhabited by a variety of species. These species would form a community and would be linked to each other in an overall symbiotic relationship. One could study the various species, how they related to each other, their population cycles, and so on. However, to fully understand the pond, one would have to realize that it was linked to other things, that it was part of a larger system. One would have to account for the source of the water and the inorganic nutrients that are critical to the system, so the watershed would have to be considered. As sunlight powers the plants, which in turn support all other life, the sun would have to be included, and so on. If a predator, such as an eagle, visited the pond and ate fish from it, the eagle would then become part of the

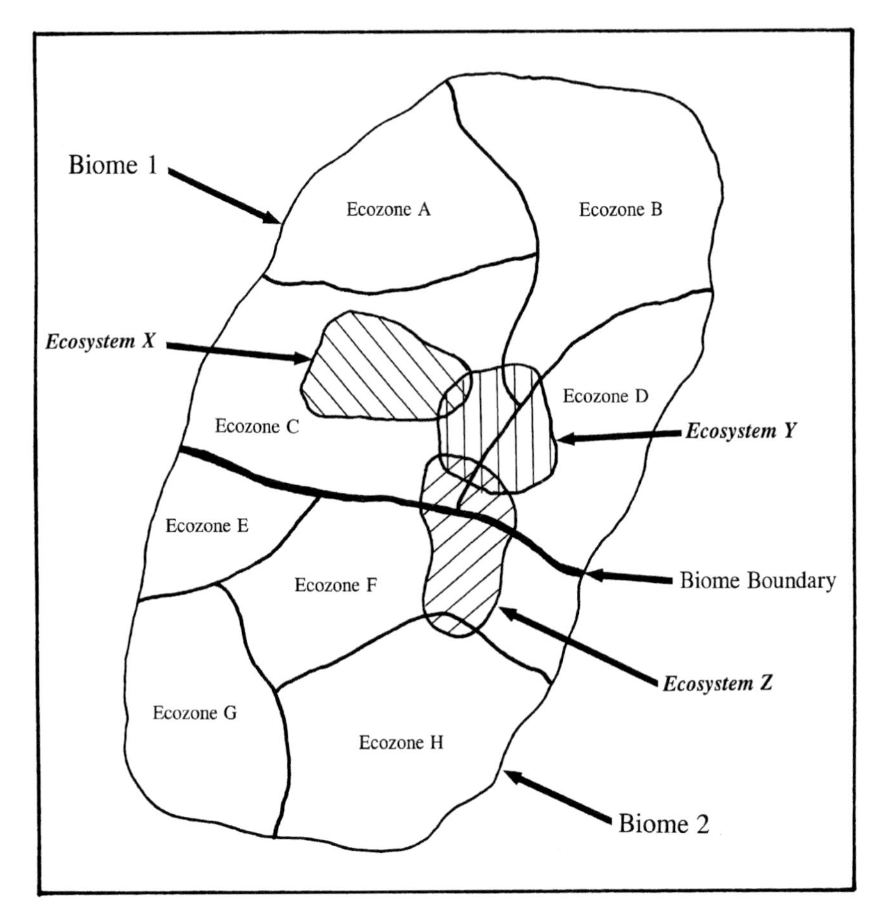

FIGURE 2.4
The general relationship between biomes, ecozones, and ecosystems (note that an ecosystem can cross biome and ecozone boundaries); no scale.

ecosystem of the pond. That same eagle might interact with other ponds that share nothing else in common and thereby link the two systems.

Thus, while ecosystems can be defined at a variety of scales for different purposes, all are ultimately linked to each other, making the separation of ecosystems somewhat arbitrary and a matter of convenience. Some ecosystems, such as islands, may be "more separate" than others and therefore are better laboratories for the analysis of ecological interaction.

Measuring and Analyzing Ecosystems and Ecozones

At least two measures of an ecosystem can be made, those of productivity and complexity. The productivity of an ecosystem is commonly assessed by measuring its **biomass**, the quantity (mass) of living matter within a specified area at a specific time (e.g., a standing crop). The amount of time it takes to replace the biomass through natural processes is called turnover and is a measure of biological activity. Another such measure is the quantity of chlorophyll per square meter. Biomass "productivity" is an important aspect of the human use of ecosystems.

The complexity of an ecosystem is measured by its **biodiversity**, an index of variety (the gross number of species; more species means greater variety) and dominance (the greater the dominance of one species, the less evenness between species) present in an area. The greater the variety and the lower the dominance, the greater the biodiversity, although it is often difficult to measure, because many species remain undocumented. High diversity provides resilience in any ecosystem, as genetic diversity does in biological evolution. If an ecosystem contained only ten species, one could say it was not very diverse. If that same system contained a hundred species (greater variety), it would have much greater diversity. If five species dominated the other ninety-five (e.g., in number or biomass), the system would be less diverse than if no species were dominant (a situation of greater evenness). Recently, biocomplexity has also entered the conceptual field. This concept implies not only biodiversity but also complex interactions among the biota. Analysis of biocomplexity involves highly sophisticated models that are only beginning to be used in human ecology.

Highly diverse ecosystems have sometimes been called generalized, while ecosystems with low diversity are called specialized. The Arctic and some desert ecosystems could be characterized as specialized. An agricultural ecosystem is almost always more specialized than the ecosystem it replaced because diversity is lowered artificially with the clearing of land, which removes many species to plant a single crop. Specialized ecosystems are always more unstable and unpredictable than generalized, more complex and therefore more resilient ecosystems.

It is necessary to consider how species are distributed within an ecosystem, a very important factor in optimization models (see chapter 3). Species can be distributed more or less evenly or clustered in patches. These two conditions form ends of a continuum, with more ecosystems closer to the patchy end than to the even end.

Humans almost always lessen the diversity of the ecosystems they inhabit. Farmers specialize in a few species, modifying ecosystems to focus on those species and so decreasing diversity through a reduction in variety and an increase in dominance. For example, a parcel of rain forest may contain a large number of species, none of which is dominant. Once that forest is cleared and planted with corn (or any other crop), only a few species will be present, and all dominated by corn.

Succession Stages

In the early twentieth century, Clements and Shelford (1939; also see Tobey 1981) defined the concept of climax vegetation: a stable, self-reproducing, interrelated plant community that will arise naturally on a site. After a forest burns, the vegetation goes through stages of **succession** (fig. 2.5): grass and herbs, then shrubs and vines, then weedy trees that require a large amount of sun, and finally the "climax forest," which grows up under the weedy trees. The climax vegetation then reproduces itself until the next fire. However, it was quickly realized that this was an overly simplistic, static model. Environmental change is constant (and sometimes sudden and drastic), and a place that has been stable long enough to develop a full, mature, self-reproducing forest is rare. Even redwoods and Sierra sequoias have proved to be "weeds," sun-loving, disturbance-following plants that cannot reproduce in old-growth forests. Yet their life spans are long enough to stretch across major climatic fluctuations. Studies of contemporary forests have demonstrated that many (if not most) of

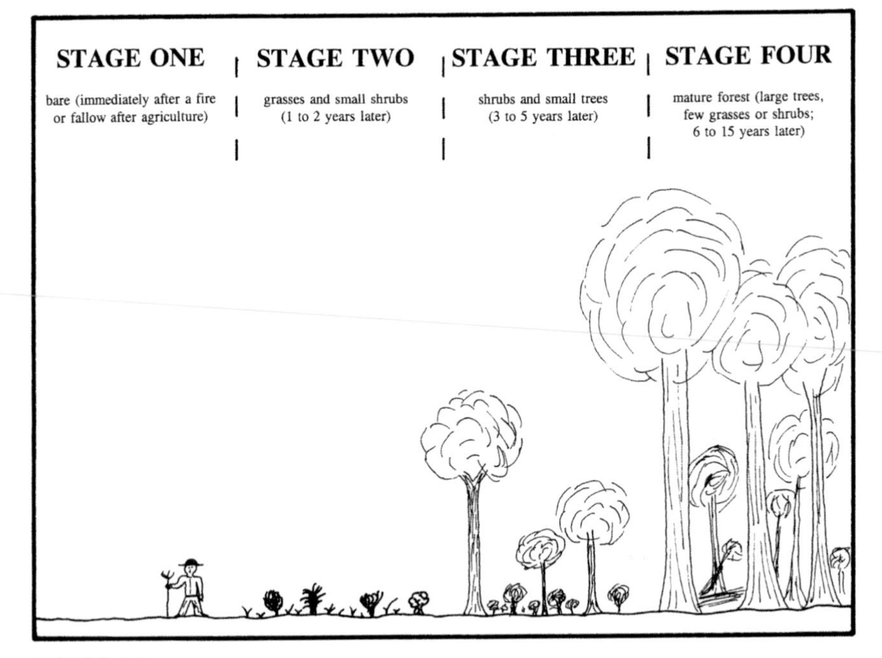

FIGURE 2.5
A hypothetical series of forest succession stages.

them, because of the massive climatic changes at the end of the Ice Age, have not stabilized. Moreover, it turns out that virtually all land areas have been significantly affected by humans. Climax communities are almost nonexistent because systems are in constant flux.

However, the concept of succession is useful. As a community evolves, it often follows an orderly and predictable pattern through developmental stages to maturity, with each stage having its own degree of productivity and diversity. Succession stages relate to the classification system of a particular group and its use of the environment. For example, some groups of contemporary Maya recognize six stages of forest regrowth after the land is cleared for agriculture. Understanding this succession and its associated agricultural implications is critical to the timing of reuse of the land for agriculture. In some cases, succession can be very rapid; some tropical and Mediterranean trees often grow six feet a year, and a fifteen-year-old forest in southern Mexico is a most impressive forest indeed.

Studying Ecosystems

Ecosystems are very complex things. To study them, investigators make measurements and construct models of how they are thought to work. Many of these models are highly sophisticated and deal with the flow of nutrients and energy within lakes, watersheds, islands, and other fairly well-bounded systems. The goal for such models is to understand the effects of each component of an ecosystem on the other components. Models provide ways to describe and measure this flow within systems and to understand the relationship between organisms and the physical environment.

However, change is so universal and rapid that it is very difficult to construct a model that is accurate for very long. If a model were constructed for a particular lake, the model would probably not be applicable to the next lake. If a watershed were modeled, the model would have to change as soon as weeds from the next watershed invaded or a flood swept through the area. Fires occur frequently, and a fire during a wet spring will have a totally different effect than a fire in a dry summer does. (The first will merely char things, while the second could be catastrophic.)

Trying to utilize ecosystem models to deal with humans has proved even more problematic. Culture is constantly changing, and people's reactions, desires, and wants also change. Even societies that are viewed as "changeless" and "ageless," such as China, have gone through profound ecological changes at frequent intervals (see Anderson 1988). Stable, predictable systems are rare.

NICHE AND HABITAT

Each species occupies a **niche**, the role it plays within its environment, community, or ecosystem. A niche is defined by what the species eats, how it reproduces, and what it does. For example, some species eat just a few things while others, such as humans, will eat almost anything. Some organisms (e.g., fish) have large numbers of offspring but provide no care, but the number of offspring is so large that at least some survive. Other species, such as humans, have far fewer offspring but give them a great deal of care. Niche can also be defined by what a species does, such as humans being either hunter-gatherers or farmers.

The geographic location where a species lives and operates is called its **habitat**. Niche and habitat are interrelated and somewhat dependent on each other, because the type of habitat influences the possible niches present. For example, some species of primates eat only fruit, others only leaves, and others just about anything. All three species of primate could coexist in the same habitat because they occupy three different niches. In a forest, different species may live in the canopy while others live on the ground. In this case, one could have different habitats (high and low) on the same tree.

Ecosystems will contain innumerable niches. If a beetle lives in the bark of a tree and eats wood, and is then eaten by a bird, the bird and the beetle have quite different niches but share a portion of the same tree habitat. While a niche is specific and can be occupied by only one organism in the same geographic place—although several may be competing to occupy a niche—the same geographic space may serve as habitat to a large number of species in different niches. In different geographically isolated habitats, the "same" niche may be occupied by different organisms. For example, the medium-size terrestrial carnivores in North America are placental mammals: wolves, coyotes, and dogs. In Australia, that same niche was filled by marsupial mammals until they were replaced by dogs after the arrival of humans.

As systems change, niches come into and go out of existence. The niche of a high-canopy leaf-eater does not exist in a young forest, so that forest would not serve as habitat to such an animal. As the forest matures, that habitat comes into being, and the niche then exists. A high-canopy leaf-eater may migrate and settle into the niche in the new habitat, or the niche may remain unoccupied. Alternatively, an existing species may evolve an adaptation to the niche as behavioral changes or random mutations that result in the ability to utilize new resources become more advantageous and are selected for.

Humans have come to occupy and dominate most terrestrial habitats and, through the use of technology, will modify a habitat to suit their needs and will even create artificial habitats, such as cities. Adding to human adaptability is the capability of rapidly altering practices and of eating many different fungi, plants, and animals.

Humans occupy a very broad niche, eating just about everything except cellulose. However, under a definition of niche slightly different from that used by biologists, humans may be seen as occupying different niches depending on their general subsistence systems, such as hunter-gatherer, herder, or farmer (Barth 1956:1088; also see Hardesty 1975, 1977), as some of the resources they use are different, and they live in different habitats. Indeed, it could be argued that contemporary human culture has created a new niche, an "urban-industrial niche" (Molnar and Molnar 2000:xiii).

RESOURCES

A **resource** is something that is actually used by an organism. If some material exists but is not used, it is not a resource, although it may be a potential resource if conditions were to change. Something may be a resource to one entity (organism or culture) but not to another, even at the same place and time. Some resources, such as water, are universal, while others change over time as technology and/or customs change. For example, many traditional cultures consider insects to be food resources (see Bodenheimer 1951; Sutton 1988), while most industrialized societies consider them pests. The food value of these animals does not change, only the opinion and use of the food.

Natural resources are those that are not manufactured by humans, including land, water, air, and time. Renewable natural resources are those of "unlimited" quantity; that is, they can be used and replaced within a relatively short period of time. These include solar radiation, water, firewood, and food. Nonrenewable natural resources, such as fossil fuels, minerals, and metals, are of limited quantity and can be replaced only over a considerable period of time.

Resources generally have two dimensions: time and space. Most resources are available only at certain times, for example, when the seeds or nuts of specific plants ripen or when certain animals migrate through an area. Other resources, such as stone, may not have time limitations. The other dimension is space, that is, where resources are located in the landscape. No resource is randomly distributed (air might come pretty close), and so the organism has to know where resources are.

All species are limited in population size by the availability and distribution of resources, most notably water. The distribution of many species is limited by the availability of water; in other words, the species are **tethered** to water. While frogs can leave the water, they cannot be out too long or they will die. Their tether is the distance they can travel (round trip) from water without drying out and dying. Humans are also tethered to water, although they have overcome many of the resulting limitations through technology such as aqueducts, water containers, and wells. Other resources, such as certain kinds of food, may also create tethers.

Carrying Capacity

Carrying capacity can be defined in a number of ways (see Brush 1975; Glassow 1978; Dewar 1984) but is commonly viewed as a measure of the maximum number of individuals of a species that can be supported within a specific ecosystem for a specific period of time. Some define time as indefinitely large (e.g., Ellen 1982:41; also see Baumhoff 1981), but as all systems are dynamic, carrying capacity will fluctuate.

Carrying capacity applies to all organisms. It is possible, for example, to determine the carrying capacity of trout in a lake using an analysis of oxygen content of the water, number and density of prey species, water temperature, and other factors. If these conditions change, however, the carrying capacity also will change. Thus, the measure of carrying capacity is dynamic and dependent on conditions as well as the definition of the geographic boundaries of the ecosystem.

In human carrying capacity, the variables of culture must be factored into the determination. The human carrying capacity of a specific area may be a hundred people with a hunting and gathering economy but may be a thousand people if their subsistence system is agriculture (a cultural change, not a change in the natural environment). However, if rainfall were low, the carrying capacity might be greater for hunter-gatherers than for farmers; it all depends on environmental conditions, technology, and cultural practice. In addition, a single human group usually occupies more than one ecosystem, and so carrying capacity would be variable depending on conditions in each ecosystem; an average would be calculated to describe the total carrying capacity.

Liebig's Law of the Minimum

The carrying capacity of a region can be only as high as its most limiting resource will permit. This principle of a limiting resource is known as **Liebig's Law of the Minimum** (or the "Convoy principle," as the speed of the slowest ship determines the speed of the convoy). The limiting factor can be anything, such as food, water, temperature, or space. For example, if there is enough food to support fifty individuals, enough land for seventy-five individuals, but enough water for only thirty individuals, the carrying capacity is thirty despite the abundance of food and land. In this case, water is the limiting or minimum factor; if the amount of water were increased, the carrying capacity would be correspondingly increased to the level of the next minimum resource; that is, if water were increased to support eighty, the carrying capacity would rise to fifty, the limit imposed by the amount of food.

All aspects of ecosystems, including their resources and environmental and cultural conditions, are dynamic, so the "minimum" is often changing. In addition, human societies can be very innovative in seeking solutions to minimums.

Boom and Bust Cycles

As noted above, carrying capacity is the measure of the maximum number of individuals that can be supported in a particular system for a specific amount of time. Carrying capacity will vary seasonally, annually, and over longer periods. Some species, such as many plants and some rodents, will enter boom cycles when resources are abundant, substantially increasing their populations. If the abundance is short-lived and the carrying capacity falls, the population will be too large, and a bust cycle will result, with individuals starving until the population falls below the new carrying capacity (see fig. 2.6). As the carrying capacity fluctuates, these boom and bust cycles can be very common.

Human populations do not often go through such cycles, although they may for a variety of reasons, including wars and embargoes. Humans tend to stay below the carrying capacity of an area for several reasons. People can manipulate the environment and thus "control" their carrying capacity to some extent. Human culture provides a variety of solutions for resource shortages, including

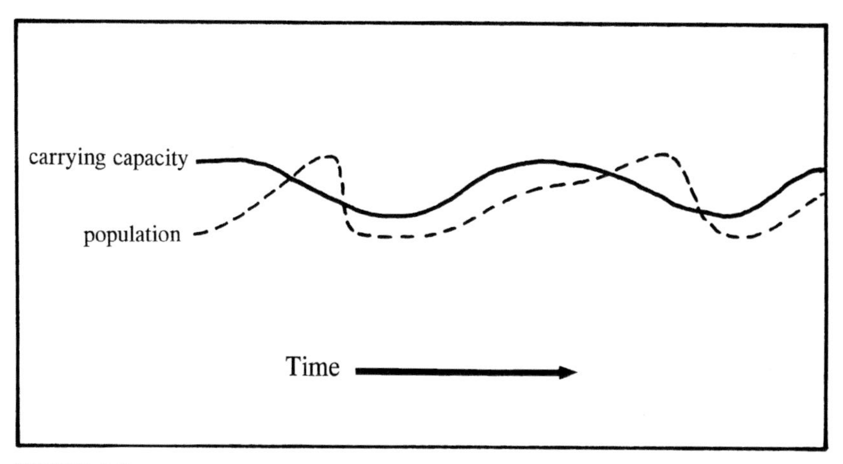

FIGURE 2.6
A model of boom and bust cycles.

storage, trade, kinship assistance, and warfare. Humans almost never eat everything that can be consumed in any particular environment, so they have the capability of "expanding" their diet if the need arises.

ENERGY

All systems depend on energy. The first law of thermodynamics states that energy can be transformed but not created or destroyed, although it can move around within or between systems, such as when plants convert light energy to chemical bond energy. The second law of thermodynamics holds that conversion is never 100 percent efficient; some energy always escapes in the conversion process. This energy is usually lost as heat, such as the heat generated within the body by muscles that do not completely convert their fuels. Thus, energy flow in a system is one-way.

The vast majority of the energy used in natural systems is derived from the sun, although there are a few natural systems that are not solar powered, such as those fueled by heat from thermal vents in the ocean. Plants combine solar radiation with nutrients and water to manufacture matter, some of which is then consumed by animals, and both the plants and the animals are eventually consumed by decomposers. The nutrients are recycled in the system. Thus, energy and matter are converted in a series of trophic levels (explained below).

Humans consume plants and other animals for food, converting that material into the energy required by the individual. This metabolic energy is used both for body maintenance and as fuel for any work performed, or it is converted into stored energy, such as fat (see Giampietro et al. 1993:242).

Energy in human systems is usually measured as "food energy," or the number of calories consumed. This number would be far less than the total available energy in a system, because humans cannot consume all possible foods (for example, people cannot eat cellulose, but termites can). While calories are the items generally measured, others (e.g., protein) are also important factors in human systems. One must distinguish between subsistence energy expenditure and total energy expenditure.

Humans also utilize supplemental energy, which is derived from biological materials such as firewood, oil, and gas (see Giampietro et al. 1993:242). For most of human history, such energy was derived from plants and animals used by people to accomplish various tasks. For example, in 1850, about 91 percent of the energy used in the United States came from burning wood and other biological materials (Pimentel et al. 1994:207). In about 2002, most of the supplemental energy used by humans originated from fossil fuels (such as gas, coal, and oil) used to power machinery and comprised some 81 percent of the energy used in the United States. Thus, Americans now operate using a fuel-subsidized system in which manufactured fuels are used for various energy needs, such as gasoline used to power plows. Efforts are being made to develop alternative energy sources to replace fossil fuels, which pollute the environment and are nonrenewable. However, some (e.g., Price 1995) are very pessimistic about our ability to find an appropriate alternative and predict the demise of industrialized culture as a result.

The Trophic Pyramid

The **trophic pyramid** (fig. 2.7) describes the levels of relationship between producers, heterotrophs (explained below), and decomposers: what is eaten and how many conversions from solar energy have taken place. Humans utilize food resources from several trophic levels, generally recognize the differences in energy, and usually use those differences to their advantage.

Producers are species that can synthesize their own food, green plants being the most common examples. Green plants take energy directly from the sun and combine it with water, gases, and minerals to manufacture food in a process called photosynthesis. Ultimately, we all depend on green plants to capture solar energy and convert it into compounds that humans can consume or utilize. Producers are the initial level of the trophic pyramid and have the best energy conversion rate. Interestingly, a nonsolar energy source, hydrothermal vents in the ocean floor, has recently been discovered and may support nonphotosynthesizing producers, such as vent worms. Studies currently are under way to determine the role of these vents as energy sources separate from the sun.

Heterotrophs are species that consume other species as food. Some examples are cows, which eat plants, and people, who eat cows. Heterotrophs cannot directly utilize energy from the sun and are able to obtain energy only by consuming producers or other heterotrophs.

The decomposers (or microconsumers; technically also heterotrophs) are small species, such as bacteria, that feed on dead organisms. They break down organic materials and allow the nutrients to be recycled to producers.

As one moves up the trophic pyramid, the number of species and individuals decreases; for example, there are fewer animals than plants and fewer carnivores than herbivores. The ratio of energy consumed to energy converted to body mass is called the conversion ratio and is based on the second law of thermodynamics. For example, plants convert about 1 percent of the energy they receive from the sun to actual matter (a 100:1 ratio). Cows generally have a 10:1 conversion ratio

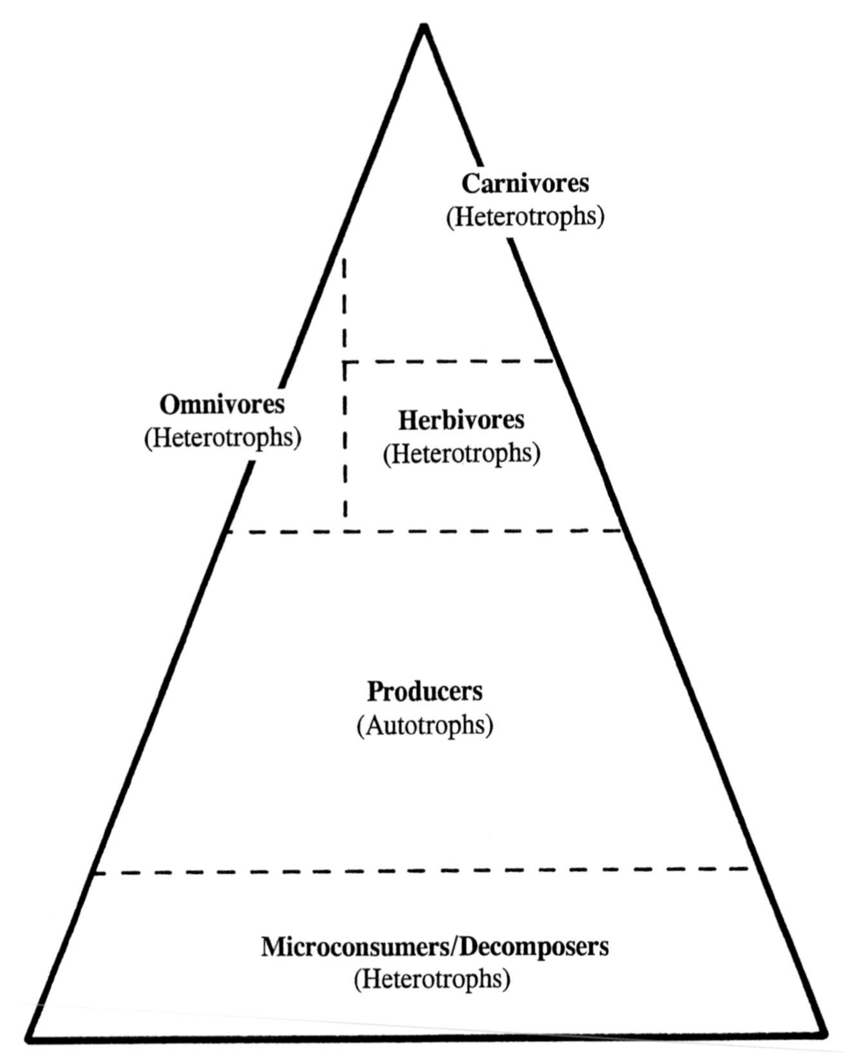

FIGURE 2.7
A generalized trophic pyramid.

(i.e., it takes 100 calories of grass to make 10 calories of cow), pigs have a 5:1 ratio, commercially grown chickens have about a 2:1 ratio (they do not move around much), and catfish that are fed on smaller fish rank an efficient 1.3:1 ratio. So 1,000 calories of solar energy are necessary to produce one calorie of beef for human consumption (i.e., 1,000 solar calories make 10 calories of grass, which make the one calorie of cow).

It is obvious that humans need to eat as low as possible in the trophic pyramid to make the most efficient use of the environment. The prudent human diet should consist mostly of plant foods, which is precisely what anthropologists currently find in surveying diets around the world (e.g., Eaton and Eaton 1999). Exceptions occur in areas where almost all the plant material is tied up in cellulose (indigestible by humans), such as in the Canadian forests and parts of the Chaco of Paraguay, or where little plant material grows, such as in the Arctic. In such cases, people must live primarily on game or fish.

The members of the different trophic levels form a food chain: organisms that consume other organisms. For example, humans eat sharks that eat tuna that eat mackerel that eat squid that eat minnows that eat zooplankton that eat phytoplankton that capture solar energy. Food webs are a series of food chains linked together by some common thread and can be exceedingly complex. Disruptions of any one link of the food chain can result in disruptions in other links, ultimately affecting people.

In living systems, nutrients, energy, and information circulate through the food chain (fig. 2.8) and are ultimately returned to the environment as the living entity dies and decomposes. Unlike energy, nutrients are recycled in circular

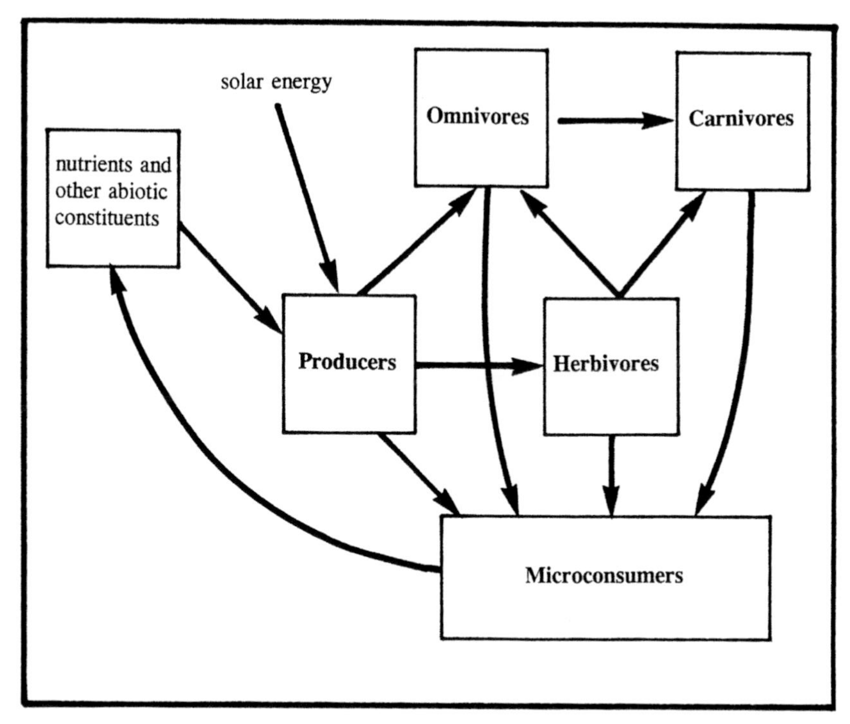

FIGURE 2.8
A generalized food chain, showing trophic levels and circulation of nutrients.

paths and can be used over and over. Nutrients required by the human body are discussed in chapter 3.

Nutrients and other resources needed and/or used by human groups are unevenly distributed in the landscape and must be obtained, transported, and traded. Thus, human social structure and economics influence the paths of materials and are integral parts of material circulation.

Diet

Within the broad category of macroconsumers (animals) are several more detailed distinctions based on diet. A herbivore is an animal, such as a cow, that consumes mostly plants. Herbivores may have either generalized or specialized diets, and some species may be very specialized, such as fruit-eaters (frugivores), depending on conditions (niche). An omnivore is an animal, such as humans, that eats both plants and other animals. As such, they can have quite a broad diet, although there are specialized omnivores. A carnivore is an animal, such as a shark, that primarily eats other animals, although it may occasionally consume plant materials. Because carnivores often eat entire animals, they consume sufficient nutrients to insure a balanced diet. This is not true for those who primarily eat muscle tissue.

Humans are omnivores; in general, our current diet averages between 80 percent and 90 percent plant foods and between 10 percent and 20 percent animal foods. This ratio makes perfect sense from a trophic and conversion perspective; there is much more plant material available to us. However, people cannot digest long-chain polysaccharides—cellulose and lignin, for example—and thus cannot consume the vast majority of plant material. We might be better off eating ten pounds of hay, but we cannot; we have to cycle the hay through a cow to get any value from it at all. The vast amount of plant tissue in wood is not edible by humans, but termites can eat it (thanks to symbiotic gut protozoa that digest cellulose), and termites are a major delicacy in parts of Africa. While it is possible for a human to be a vegetarian, it is difficult to do so without extensive knowledge or supplements.

Species Interactions

Species interact in a variety of ways, and humans will adopt one or all, depending on the situation. Some species directly compete for resources, with the loser suffering some negative consequence, such as extinction. A predator–prey relationship exists between many species; the predator kills and consumes the prey. While this relationship is of obvious benefit to the predator, it may also benefit the prey, such as by keeping populations down or removing unfit members from the breeding pool of the prey. Such a relationship is called **symbiotic**.

These relationships are quite different from parasitism, in which one organism (the parasite) exploits another (the host) to the benefit of the parasite and the detriment of the host. For example, many disease organisms are parasites, infecting and reproducing within the host and using the host to spread themselves and so ensure their survival. However, it is to the advantage of the parasite not to kill the host, at least not too quickly, so that the parasite can successfully propagate. If the host dies too quickly, the disease will not spread, and the parasite will not be as successful. In essence, then, parasites have a tenuous relationship with their host, and they tend to coevolve, the parasites becoming less detrimental and the host developing resistance.

In some cases, species will develop a relationship that is mutually beneficial. An example of this **mutualism** is the general relationship between dogs and humans. People provide dogs with food and habitat, and dogs provide people with companionship, labor, and other services.

CHAPTER SUMMARY

Ecology is the study of the interaction of organisms with their environment, defined as the surroundings in which the organism interacts, commonly called an ecosystem. Environments have both abiotic and biotic components in which all organisms interact. Biomes and ecozones, geographic regions defined by particular biotic communities, as well as the ecotone between ecozones, are measured by productivity and diversity and commonly form the basis for initial analysis of ecological questions. Understanding change in ecozones is also important for analysis.

All organisms occupy a niche and live in a habitat, with humans having a very broad niche and occupying most habitats. However, one could view humans as occupying different niches, such as hunter-gatherer, herder, or farmer. Resources, things actually used by organisms, are important factors in defining and understanding niches and habitats.

The number of organisms an environment can support is called its carrying capacity and is dependent on the availability of necessary resources, with population limited by the least-available resource. With humans, culture is also a key element, and cultural practices can significantly alter carrying capacity.

Energy is utilized and converted from one form to another, always with some loss. Materials move through the trophic pyramid of plants converting solar energy into biomass, herbivores eating plants and converting biomass into tissue, carnivores eating herbivores and converting that tissue into new tissue, decomposers eventually eating everything, and the material resulting from decomposition being used by plants to make new biomass. All of these species exist in some relationship with each other, be it mutually beneficial, as predators and prey, or as parasites. Humans exist as all of these, depending on circumstance.

KEY TERMS

abiotic
biodiversity
biomass
biome
biosphere
biotic
carrying capacity
ecosystem
ecotone
ecozone
environment
habitat
Liebig's Law of the Minimum
mutualism
niche
resource
succession
symbiotic
tether
trophic pyramid

3

Human Biological Ecology

Human biological ecology (HBE) is the study of the biological requirements of humans for survival, reproduction, and genetic success. Because culture plays a central role in fulfilling the biological aspects of human existence, there is considerable overlap between HBE and cultural ecology. As the focus of this book is on cultural adaptations, this chapter presents only a brief outline of the strict biological components of human existence, including nutrition, physical adaptations, and biological models of genetic success. An extensive discussion of HBE is presented by Kormondy and Brown (1998).

HUMANS AS ANIMALS

Biologically, all humans are almost identical. Cultures vary enormously, and sometimes the anthropologist almost despairs of finding any common threads between them. Yet humans are all one biological group, the genus and species *Homo sapiens*, and are very closely related genetically. Indeed, living humans all belong to one subspecies, *H. sapiens sapiens.*

It is important to consider human needs in the context of human biology. Different human populations do vary in a few important ways. For example, some groups of people need less food than others, and some groups can resist a particular disease better than others. On the whole, however, humans are remarkably uniform creatures. No differences in intellectual or behavioral abilities and tendencies have been conclusively demonstrated between human populations, and this homogeneity is true of most species. Domesticated species, such as horses and dogs, are interesting exceptions, as humans have purposefully bred physical and behavioral differences into the various types of domesticates.

Like all animals, humans are subject to **stress**, conditions that require either a short-term or a long-term response. We have biological requirements and tolerances and must operate within certain environmental parameters. We require adequate temperature, oxygen, health, nutrition, and a few other things, all of which must fall within certain minimums and maximums (the law of tolerance). The body needs to maintain itself at about 98 degrees Fahrenheit and has a rather narrow range of tolerance, between about 85 degrees and 105 degrees. Outside this range, the brain will not operate properly, and the individual will eventually die. Humans must also operate within a certain range of oxygen concentration, and as we cool ourselves by sweating, we have to have relatively more water than most other animals. Finally, we must maintain some level of health and nutrition in order to function and reproduce.

BIOLOGICAL ADAPTATIONS

Like all other life forms, humans adapt biologically to changing environmental conditions. These responses fall into two broad categories, **physiological adaptations** and **anatomical adaptations,** and species will manifest aspects of both types of adaptation to changing conditions (table 3.1). However, with the advent of culture and the increasing complexity of technology, cultural adaptations are becoming increasingly important.

Physiological Adaptations

Physiological adaptations are relatively short-term changes in the body in response to rapid changes in environment. Genetics and selection determine the presence of these mechanisms, but environmental conditions determine whether they are used. For example, the presence of sweat glands in the skin is genetic, but they do not produce sweat unless the body is warm. If the environment gets hotter and the body is constantly warm, there may be selective pressure for a greater number of sweat glands in the skin of subsequent generations. Thus, there is an interplay between physiological and anatomical adaptations.

Some of the most common stressors that require physiological adaptations include excesses in sunlight, altitude, cold, heat, malnutrition, and disease (see Kormondy and Brown 1998:162–226). Two basic physiological responses are defined here: primary and secondary. Primary responses, often called acclimatization, are immediate, occurring within minutes or days. These may include alterations in metabolism, respiration, blood distribution, sweating, and many other responses. Secondary responses may take months or even years and can include changes in fertility, work abilities, and even population distributions.

Table 3.1. Biological Adaptations to Some Stressors

Stressor	*Primary Physiological Adaptations*	*Secondary Physiological Adaptations*	*Anatomical Adaptations*
Sunlight	production of melanin in the skin, resulting in a "tan"	continued and increasing melanin production, deepening of "tan"	genetic selection for darker skin
Altitude	increased respiration, blood pressure, and heart rate	increase in red blood cell production, increase in number of mitochondria, expansion of vascular system	enlargement of lung and chest size, increase in size and thickness of placenta
Heat	sweating; increased respiration, blood pressure, and heart rate	general peripheral vasodilation (GPVD)	elongation of body and extremities, increase in number of sweat glands
Cold	shivering, cold-induced vasoconstriction (CIVC)	systemic, whole-body physiological responses	"rounding" of body, shortening of extremities, addition of fat layer
Malnutrition	reduction in stomach tension, mobilization of stored sugars, then fats, then proteins	decrease in ability to work, decrease in muscle and body mass, reduction in metabolic rate, growth arrestment and resorption of vital nutrients from body tissues	genetic changes in rates of metabolism, propensity to develop problems such as diabetes, reduction of body size to adjust to fewer nutrients
Disease	autoimmune response, macrophagic activity	decrease in ability to work, reduction in metabolic rate	changes in genotype to reflect the development of genetic resistance through selection

An example of people moving from a low to a high altitude illustrates both adaptations. Upon their arrival at the high elevation, people will experience oxygen deprivation (hypoxia) due to lower atmospheric pressure by which lessens the ability of gasses to pass through the membranes of the lung. The primary physiological response will be increased respiration, blood pressure, and heart rate, as well as the dilation of blood vessels. These measures suffice until secondary responses can be initiated. Secondary responses take longer and include an increase in the production of red blood cells, an increase in cellular mitochondria (which transform oxygen and sugars to energy at the cellular level), and an expansion of the vascular network.

A series of cultural adaptations may also be made. The ability of people to work will be lower, and adjustments to work schedules may be needed. Lower oxygen pressures may also cause increases in miscarriages, and in some cases, females are temporarily moved to lower altitudes to give birth. Interestingly, European females generally have difficulty reproducing at high altitudes; therefore, during the period of European colonization in the Andes, few European settlements were established there. An additional cultural adaptation to high altitude in the Andes is the use of drugs (e.g., coca leaves) to mitigate the effects of lower oxygen availability.

Physiological adjustments to cold climates include the variation of blood flow to exposed skin surfaces such that heat loss is reduced and tissues are kept alive. Another adaptation is an increase in metabolic rate so that energy burns at a higher rate, a condition that requires a greater daily caloric intake. Cultural adaptations to cold include the use of fire and clothing (skins or other materials). In addition, alcohol is used in some cultures as a means of keeping warm for short periods because it increases blood flow to the extremities.

Anatomical Adaptations

Anatomical adaptations are long-term genetic changes in genotype (and so phenotype) due to selective pressures. Unlike physiological adaptations, anatomical adaptations reflect changes that are passed to subsequent generations. If the people discussed above in the example of physiological adaptations to high altitude stayed there long enough, those with larger lung capacities would be more successful, have more children with larger lung capacities, and eventually outcompete those with smaller, less efficient lung capacities. In time, increased lung capacity and a resultant enlargement of the chest cavity would be selected for and become the dominant phenotype. Indeed, people adapted to high altitudes, such as in the Andes of South America, have larger torsos and lung capacities than people at lower altitudes.

Humans will also adapt anatomically through alteration of basic body structure. To help regulate temperature, the shape of the body will evolve to have either more or less surface area. For example, in cold environments, a shorter, rounder, more compact body is more efficient at preserving heat. The rounder the body, the less surface area through which heat can escape; thus, being short and round is adaptive in cold environments, a phenomenon referred to as **Bergmann's Rule.** In warmer climates, a tall, lean body is more efficient for temperature regulation. Other examples of anatomical adaptation to cold include the reduction of extremities (called **Allen's Rule**), hence short noses and ears to prevent frostbite, and the addition of a fat layer, for both insulation and energy reserves.

HUMAN POPULATION REGULATION

All species face the problem of maintaining their populations within the carrying capacity of the environment. Human populations grew slowly (at an annual rate of under 0.001 percent) for millions of years, reaching perhaps about ten million at the end of the Pleistocene (the last ice age) (see Hassan 1980; Kormondy and Brown 1998; Molnar and Molnar 2000). With the advent of agriculture some ten thousand years ago, human population began its soar to the six billion of today, an annual growth rate of about 2 percent. Many factors influence population growth and size, primarily the natural aspects of infant mortality, predators, disease, food supply, and habitat size. Humans, of course, add the dimension of culture (see Cowgill 1975; Campbell and Wood 1994; Molnar and Molnar 2000).

Life expectancy is closely related to population growth and is high and rising in most nations, mostly due to declining infant mortality. People are living well into their seventies in all developed countries and into their eighties in Japan and Iceland. Only the poorest nations have life expectancies below age sixty. Life expectancy is not actually a measure of how long you can expect to live. In the United States, it is measured as the average age at death of all the people who died in a particular year. If two people are born at the same time, and one dies at one day old and the other lives to be seventy years old, the average life expectancy is thirty-five ($0 + 70 \div 2 = 35$). Such a figure can be misleading and may actually reflect a high infant mortality rate rather than the age most people have reached when they die. In most cultures, if a person survives past four or five years of age, he or she can usually expect to live a fairly long life.

Natural Mechanisms

Predators

Predators are a major problem for most species, but not for humans. No animal habitually hunts humans as a major portion of its diet. Parasites, perhaps a form of predator, do impact human populations. They rarely cause death but are vectors of disease.

Disease

Human populations have most typically been regulated by disease—apparently a rather rare (but far from unknown) regulator of population size among other species. Infectious diseases have been prevalent throughout most of history, but since the conquest of most of them in the last two centuries, the noninfectious killers—heart disease, cancer, diabetes, and so on—have expanded to fill the gap (see Kormondy and Brown 1998:227–252; Molnar and Molnar 2000:210–247). Tobacco and alcohol—both voluntary choices—have become the biggest killers in the developed world. Accidents and deliberate violence are far ahead of infectious diseases in the developed nations, especially in the United States, where guns are so common. Environmental pollution is becoming a serious factor and can only worsen. Meanwhile, starvation and infectious diseases remain pervasive in poor nations.

Cultural Mechanisms

The natural aspects of predation, disease, food supply, and habitat size control the populations of most species, but humans employ culture (including technology) in an attempt to bypass these factors. Nevertheless, humans are still faced with population control issues. Thomas Malthus (1960, orig. 1798) presented an argument that people, like mice, would automatically breed up to, and beyond, the food supply (carrying capacity), causing a food shortage and thus a population crash (boom and bust). Malthus argued that the Four Horsemen of the Apocalypse (a common Christian image of the end of the world), traditionally identified as war, famine, plague, and death, set the human population limit. However, human population growth is not automatic but is regulated by conscious and unconscious human factors. Humans manipulate the environment and food supply and employ a variety of cultural rules designed to prevent a too-rapid increase.

Medical intervention is a cultural mechanism and has had a great effect on population growth. Both population size and life expectancy are related to declining infant mortality rates, a direct result of improving medical services. More effective and more available medicine also influences the distribution and impact of disease and malnutrition on the population. For example, the major effort to eliminate polio around the world through inoculation was successful until everyone thought the disease was eradicated. The inoculations were discontinued, the disease returned, and inoculations had to be resumed.

Food Supply

Unlike other animals, humans manipulate and manage their food supplies, can improve and/or intensify existing food supplies (commonly referred to as the "Green Revolution"), expand their habitat (e.g., create new agricultural lands), and even expand their niche by exploiting new foods. In contrast to the Malthus model, Ester Boserup (1965) and Mark Cohen (1977) argue that population growth had the effect of forcing people to intensify their food production to meet the challenge. On the other hand, groups faced with population problems are prone to engage in warfare, to emigrate, or to break up into smaller groups rather than intensifying food production. Intensification does occur, but only when other means of coping with the situation are unavailable or undesirable. Peasant cultivators need labor on their farms, so high birth rates are encouraged. Often they are prevented by their governments from emigrating or fighting.

Famine is a leading cause of moderate or severe malnutrition and is typically associated with agricultural groups. The causes of famine are varied. They are sometimes natural, such as floods or extended drought, but are more often due to cultural factors, such as war or embargo. In some instances, the "natural" cause (e.g., flooding) is actually cultural in origin (e.g., deforestation).

Malnutrition affects much of the population of the world and is the largest single cause of death in undeveloped or underdeveloped countries. Malnutrition comes in several levels of severity. Many populations exist in a slightly malnourished state, a condition in which the people can work and reproduce, but not optimally. Moderate malnutrition involves the disruption of the functional abilities of the individual or population, and severe malnutrition is a life-threatening situation, often leading to death.

Birth Control

Every group of people on earth employs techniques to control the number and timing of births. This type of control includes planning, sexual abstinence, and the use of birth control devices or drugs. Abortion, often a result of poor planning, is also employed by some cultures as a "birth control" measure.

Some groups breastfeed for fairly extended periods, up to four years. The fertility of a lactating woman is reduced by at least 50 percent. So, other things being equal, the longer a woman lactates, the less likely she is to become pregnant. Thus, long-term breastfeeding is one technique to space births and is purposefully employed by some groups, including many hunter-gatherers, to help control population (see Lee 1982; Kelly 1995).

The Western medical system has been introduced in much of the world. For religious and ideological reasons, which all too often are no more than a desire to keep women repressed, birth control has not been introduced as aggressively as other public health measures have. Lowering the death rate without controlling the birth rate simply leads to more and more misery in most of the world.

Where birth control information and technology have been provided, birth rates have fallen and, often, rural health and security have risen significantly. Introducing the full range of birth control techniques can also reduce abortion and infanticide.

Infanticide

Once born, infants may be killed for various reasons. Infanticide rates of 30 percent or even 50 percent (Smith 1977; Hrdy 1999) have been documented for extremely densely populated regions. Cultures that frown on outright infanticide may allow selective neglect, virtually a sentence of death in disease-ridden, food-short areas. Infanticide in dense agricultural communities usually targets females and thus is particularly effective at controlling population (Divale and Harris 1976) and in generating enough males to labor in the fields. In spite of the availability of birth control technology, female infanticide is still common in many areas of the world.

For example, for much of their past, the Eskimo (or Inuit) lived a difficult life, and hunting was very dangerous, resulting in a high adult male mortality rate and a constant shortage of males. In times of severe and chronic food shortages, a family may not be able to support a newborn without endangering the entire group. In such cases, infanticide might be practiced, but it was always a traumatic decision. Due to the shortage of males, they were more valued and more likely to be kept as infants, leading to a higher incidence of female infanticide. Infanticide was not considered to be murder since a baby was not a person until named and would be reincarnated in more favorable times (see, e.g., Condon et al. 1996). Today, life is still difficult for the Eskimo, but infanticide is no longer practiced.

Geronticide

Geronticide, the killing of old people, is rare. Such people are no longer able to reproduce, and their loss has little direct impact on population numbers. However, they do consume resources that might be better allocated to younger people during times of stress, and some groups place a high social value on old people committing suicide during famines. On the other hand, the retention of the knowledge of the older people could be a distinct adaptive advantage.

Warfare

Warfare is commonly seen as a major factor in population regulation. In reality, though, the death rate of males in small-scale warfare is very low, and such losses have virtually no influence on population size, although there are some notable exceptions, such as the Pawnees in North America (Wishart

1979) and the Waorani in South America (Robarchek and Robarchek 1998). Today, conventional warfare, even with much larger numbers of deaths, is still only a relatively minor factor in population regulation. Nuclear war, however, would probably have a profound effect on population size and growth due to the massive number of deaths, especially in regions like south Asia (e.g., India and Pakistan).

Divale and Harris (1976; also see Harris 1984) looked at the issue of small-scale warfare from a different angle. They argued that in small-scale societies, warfare functioned to control population not through combat deaths of males but through the presence of a "male supremacist complex" whereby the males instituted a female infanticide program, not only favoring males (warriors), but also reducing the growth rate of the population.

The Future of the Human Population

The human carrying capacity of Earth is unknown, but estimates have ranged from one billion (a figure already exceeded) to one thousand billion (Postel 1994; Cohen 1995; Molnar and Molnar 2000; Smil 2000). From the six billion people of 2002, it is estimated that, with current agricultural technology, the population of Earth will grow to between eight billion and ten billion people by 2050. New and unexpected innovations in agriculture, storage technology, and transport could substantially increase this figure. On the other hand, pollution and the continued loss of topsoil, forests, safe water, and other vital resources could substantially reduce it. In addition, administrative problems beset the application of the best agricultural technology. Many have speculated that the human population will level off at between ten billion and twelve billion. The question is how much of the leveling off will be due to the demographic transition from high birthrates to lower ones (as in many industrialized countries) and how much to Malthus's Four Horsemen. Current population growth rates do not inspire much hope, although some experts are optimistic (Smil 2000).

All industrial countries today have low birth rates and low death rates, and many are close to "zero population growth," or just over two children per woman. In contrast, women in frontier agrarian societies, where labor is especially valued, often have fifteen or more children.

NUTRITION

Nutrition is a measure of the ability of the diet to maintain the body. It is the raw materials, consumed in the form of "food," needed by the body to function properly (see Johnston [1987] for an anthropological perspective). These raw materials include protein, carbohydrates, fats, vitamins, and minerals (plus fiber). As body conditions change, such as during illness, seasonal variations in food availability, or pregnancy, so do nutritional requirements. These requirements are poorly known (Leslie et al. 1984) and are generally estimated from studies of the general average daily nutritional requirements of people currently living in the United States.

Such recommended daily allowance (RDA) figures probably are not directly applicable to non-Western populations. Requirements needed for growth, reproduction, and physical activity will vary depending on conditions (e.g., Froment 2001; Jenike 2001). For example, individuals existing in a chronic state of malnutrition and disease would have very different minimal requirements from individual Eskimo living in low temperatures. The daily caloric intake of extant hunter-gatherers in various regions is known to vary within a range of 1,600 to 3,827 calories (Jenike 2001:table 8.2), but seasonal changes in need or food availability can drastically alter these numbers (Jenike 2001:table 8.3).

However, the RDA figures are still worth considering as a baseline estimate. For example, young persons (eleven to twenty-two years old) require more calories than older ones, but protein requirements typically increase with age and size. Also noteworthy is that pregnant and lactating females require more nutrition (see Thomson et al. 1970; Nerlove 1974; Garber 1987:353, 357, and references therein). Although the specifics are unclear, it does seem that caloric, vitamin, mineral, and protein requirements increase to some extent for women under these conditions. Thomson et al. (1970:571; also see MacLean 1984) estimated that during lactation, the energy requirements of the human mother increase by about six hundred calories, while Duhring (1984:table 1) estimated an increase of about five hundred calories.

No one food can supply all of the necessary nutrients for optimal human health. Even mother's milk, ideal for infants, has too little iron and vitamin C for toddlers. Many important foods have ingredients that are dangerous in high quantities; for example, acorns must be processed to remove tannic acid. Meat and animal fat contain cholesterol, and eating too much of these foods can lead to an excessive cholesterol buildup. Thus, every culture must contain rules and guidelines for a varied, nutrient-rich, adequate, and safe diet. Simply assuring that enough calories and protein are consumed is not sufficient.

Garber (1987:353, 357) noted that most mammalian species, including primates, do not generally store fat prior to conception but often will change their feeding habits during pregnancy to "accommodate the nutritional costs of lactation [by increasing] their feeding time and/or exploit[ing] higher-quality food resources." Chimpanzees, on the other hand, are believed to practice preconception nutritional preparation (see McGrew 1981:41): Females expand their diets during estrus. Humans (and other animals) experience "cravings" to eat certain foods at various times. This phenomenon may be explained as the body "telling" the individual that it needs certain nutrients, but it may also be simply a bad habit.

The diets of people in most cultures are sufficiently broad to provide these various nutrients. In Western societies, with narrow diets and highly processed foods, supplements such as daily vitamins, protein supplements, enriched bread, vitamin D–enriched milk, and iodized salt are frequently consumed (exactly how much supplementation is actually needed is not clear). While it is true that most traditional societies do not know the details about specific nutrients or quantify their intake as Western cultures do, they certainly are aware of the consequences of not having enough of certain foods. The same was most certainly true of past cultures as well.

Calories often are used as the measurement or currency of "proper" nutrition (see below). However, those using calories as the prime currency ignore the requirements of other nutrients and assume that the subjects are in good health and have no special needs. In addition, caloric requirements are generally overestimated. However, when caloric (and other needs) are not met on a consistent basis, people will die.

Another response to inadequate food supplies is to exist and survive in a malnourished condition. In such circumstances, people are not "healthy" by Western standards; some are undersized or unable to work, have a low reproduction rate, suffer sickness, die young, or develop improperly. Yet as a biological population, they survive. This last point is worth noting: Western societies believe that perfect health is normal and that it is abnormal to be in less than perfect health. The opposite is more likely true: Most people throughout history have had a rough time when it comes to diet.

Calories

Calories are units of energy; one calorie is calculated as the energy needed to raise the temperature of one gram of water one degree centigrade. Some 70 percent of calories consumed by humans are utilized to maintain body temperature at about 98 degrees Fahrenheit, the temperature at which the brain and metabolic functions work best.

Caloric requirements have been calculated (but only generally) for people in industrialized societies (table 3.2), but the caloric needs of humans living in non-Western societies or in the past are not known. However, the data do provide a general estimate of human needs, recognizing that the requirements vary by condition, as noted above for pregnant women.

Carbohydrates

Carbohydrates are a class of compounds that include starches, sugars, and cellulose. Simple sugars, such as glucose and fructose, are produced in plants and can be combined and stored as starches or double sugars, such as sucrose or lactose. When these substances are consumed, they are broken down into simple sugars, which provide energy for the animal that ate them. Cellulose is not digestible by humans, but it is an important source of fiber. Glucose is a very important sugar: It is the only energy supply for the brain.

Most cultures obtain the majority of their calories from starch, such as rice or potatoes. A few societies consume large quantities of sugar, such as the Rotinese in eastern Indonesia, who eat a great deal of palm sap (Fox 1977). In this case, the various products of the palm (sap and leaves, the latter made into a bewildering array of products, from houses to fishing lines) form the basis of the entire economy, which has resulted in a stable economic system that operates without much environmental damage. In nearby regions where palms are not so abundant, people employ a swidden, or shifting cultivation, system.

Protein and Fat

Proteins are complex combinations of amino acids. The protein is broken down, "liberating" the amino acids, which are then used as building blocks for proteins in the consumer or converted into fuel. Twenty-two amino acids are necessary in human nutrition. Some of these can be manufactured by the body in adequate quantities, others can be manufactured but only slowly, and a number cannot be manufactured by the body at all and must be obtained from foods. The latter two categories are called essential amino acids (lysine, leucine,

Table 3.2. General Daily Caloric Needs

Age	*Children (Both Sexes)*	*Males*	*Females*
1–3	1,300		
4–6	1,800		
7–10	2,000		
11–14		2,500	2,200
15–18		3,000	2,200
19–24		2,900	2,200
25–50		2,900	2,200
51+		2,300	1,900

Adapted from Whitney and Rolfes (1996:table G-3).

isoleucine, methionine, phenylalanine, threonine, tryptophan, and valine) and must be included in the diet. Proteins contain different amounts of the various amino acids. A deficiency in the intake of protein will result in the body "reusing" some of its old protein; that is, proteins from the body will be broken down to provide amino acids for new protein synthesis. This circumstance will cause the body to "waste away" as it digests itself, beginning with muscle tissues. At the extreme, if the body must digest the heart muscle, death ensues. Protein is expensive for plants to produce and so is rarely found in them in high concentrations; it occurs in high concentrations in animals that eat plants.

Protein containing all of the essential amino acids can be found in a combination of various plants, and it is possible for a person to achieve optimal health by consuming only plants. However, being a strict vegetarian requires extensive knowledge of which combinations of plants must be eaten, in what sequence, and with what frequency to maintain proper nutrition. In any case, it is necessary to have vitamin B_{12} in the diet, and this vitamin must be obtained either from animal sources or from artificial supplements. Being a strict vegetarian is very difficult, and most people lack the requisite knowledge. Most simply consume some animal flesh to meet their overall protein needs.

One example of how cultures may deal with protein deficiencies is the bean/corn (maize) complex of Mesoamerica. Beans are generally high in protein but low in lysine, while corn contains more lysine. Eating beans and corn (squash may also contribute) in combination provides a more balanced protein source. A recent problem is that the new hybrid corn varieties are rather low in lysine, and nutritional problems are beginning to appear where hybrid varieties have replaced native varieties.

Some cultures, such as the Inuit, have a diet high in animal protein. In diets high in meat, it is necessary to ensure that sufficient fat and/or carbohydrates are included because lean meat is low in calories and stresses the kidneys. In addition, the lack of fat makes it difficult to utilize the fat-soluble vitamins, forcing the body to used stored fat for that purpose. Thus, in a diet high in lean meat, it may actually be that digesting the food uses up more calories than the food provides, resulting in a state of malnutrition (Speth and Spielman 1983).

Fat (lipids) occurs in several forms in the body, the most abundant being storage fat. Storage fat contains about double the energy content of carbohydrates and serves as a reserve source of energy. In general, fats contain about twice as many calories per gram as either protein or carbohydrates. In addition to being important as an energy source, fats are necessary to activate fat-soluble vitamins.

Vitamins and Minerals

Vitamins are organic compounds needed to maintain certain body functions. Most cannot be manufactured in the body and so must be obtained in the diet. The body requires a variety of vitamins, some in fairly large quantities and some in very small amounts. Some vitamins are fat-soluble and can be stored in the body for long periods. Others are water-soluble and are easily lost in the process of cooking and by the body during urination. The major vitamins include A, the B complex, C, D, E, and K, although many others also are required (see table 3.3).

Minerals are inorganic substances (elements or compounds), and many are needed by the body for a number of functions, including bone formation. While the overall content of minerals in the body is small, such elements are vital (e.g., iron is absolutely necessary for the blood) and are integral parts of many organic compounds. Sodium (salt), potassium, iodine, and manganese are some of the most essential minerals (see table 3.4).

EVOLUTIONARY ECOLOGY

A relatively new way to investigate human behavior is through the use of **evolutionary ecology**, an approach that applies natural selection theory to the choices

Table 3.3. Vitamins Important for Proper Nutrition

Vitamin	*Solubility*	*Sources*	*Required For*
Retinol (A)	fat	liver, milk, carotene in plants	vision, maintenance of skin, respiratory tissues, bone growth
Thiamin (B_1)	water	meats, grains, vegetables	maintenance of nervous and cardiovascular systems
Riboflavin (B_2)	water	milk, green vegetables, grains	vision, energy metabolism, and skin maintenance
Niacin	water	meats, cereals, eggs	metabolism of energy, synthesis of fatty acids
Pyridoxine (B_6)	water	meats, vegetables, grains	metabolism of amino acids
Cobalamin (B_{12})	water	many animal foods	synthesis of fat
Ascorbic acid (C)	water	fresh fruits and vegetables	many metabolic functions
Dehydroxy vitamin D	fat	milk, fish oil (or exposure to sunlight)	absorption of calcium and phosphorus in the intestines
Tocopherol (E)	fat	vegetable oils	protects vitamin A and some fatty acids
K	fat	vegetables, milk	variety of functions, including blood clotting

Table 3.4. Minerals Important for Proper Nutrition

Mineral	*Sources*	*Required For*
Calcium	dairy foods, leafy vegetables	bones and teeth, blood clotting, movement of ions
Chloride	many foods	stomach acid
Magnesium	nuts, legumes, cereals, seafood	protein synthesis
Potassium	fruits, potatoes	nerve transmission, muscle contraction
Phosphorus	animal protein	bones and teeth, cell membranes
Sodium	many foods, flavoring	nerve transmission, muscle contraction
Sulfur	many foods, meats	proteins, B vitamins
Copper	cereals, legumes, shellfish, fruits, vegetables	hemoglobin, cell respiration
Iodine	seafood	thyroid hormone
Iron	organ meats, nuts, leafy vegetables	blood, enzymes
Manganese	cereals, legumes	bone development
Zinc	whole grains, fish, milk, eggs	enzymes, insulin, wound healing, DNA synthesis

made by people (see Smith and Winterhalder 1992; Winterhalder and Smith 1992). Originally developed by biologists, evolutionary ecology can be used to apply general biological principles to humans, primarily in modeling resource use. An advantage in using such models is their testability and reliance on quantitative methods (e.g., statistics), allowing a more objective and comparable analysis. Archaeologists also employ evolutionary ecology in the study of past groups (see the special issue of *World Archaeology* vol. 34, no. 1, 2002).

Evolutionary ecology begins with the general premise that specific human behaviors are the functional equivalent of biological traits. Different behaviors are seen as trait variation subjected to selective pressure. Selection then acts on these traits, with the outcome depending on the particular situation (Richerson and Boyd 1992; Fog 1999; also see Dunbar et al. 1999).

If people are doing things that do not work well in a given environment, or if they fail to change their practices as conditions change, they will suffer some consequence. They may have to move, be incorporated into another culture, or, in the extreme, die. If other people with different traits enter an area, the two groups may compete if they occupy the same niche, with the best adapted replacing or absorbing the other. This is analogous to biological evolution, in which one group of organisms competes with another for space and resources. Remember that flexibility, the ability to change with changing conditions, is itself an important and highly adaptive trait.

Optimization Models

The primary approach used in evolutionary ecology is the study of **optimization** through the application of **optimization models.** Optimization models are used to explain some aspects of behavior related to the utilization of resources (Jochim 1983:157), usually on a least-cost basis. Originally developed by economists, optimization models were borrowed by biologists to model and predict the behavior of animals in relation to their diet and feeding strategy. They were then borrowed from biology by anthropologists and applied to humans (see Winterhalder 1981; Smith 1983). The central theoretical tenet of these models in anthropology is that humans will adopt "optimal" behaviors to maximize their reproductive success. The optimal behaviors are generally limited to diet and appear to be based on the premise that, in the long run, one must "make" more than one "spends" to survive, a cost-benefit formula that is obviously true. In essence, then, resources are ranked based on net return, and a model is constructed that predicts which resources will be used in the diet.

All optimization models operate with a number of working assumptions. The most basic of these assumptions is that humans will behave as other life forms do and that behavior is designed solely to maximize evolutionary (reproductive) success. In making these initial assumptions, the models rigidly adhere to rules of cost and benefit, mostly ignoring cultural factors that may override such behaviors. Further, following rational choice theory, optimization models assume that people are rational choosers, a big assumption (as discussed in chapter 1). Importantly, optimization models do not assume that people and cultures make perfect decisions. In fact, optimization models function as ideals against which actual behavior can be compared.

Optimization models are now widely applied by human ecologists, attracted by the models' relative simplicity and their ability to generate empirical data and testable models (see, e.g., Thomas 1986:254). Optimization models describe small-scale events and itemize conditions and behavior over a limited period of time and space. The sum of these behaviors can then be used to explore large-scale trends in adaptation (e.g., Pyke et al. 1977:140). Archaeologists find optimization models very useful as a first estimation in the attempt to reconstruct past societies (Winterhalder and Smith 1981; Jochim 1983:158). It has been said that optimization models are a "60 percent solution" to human behavior (a quote attributed to the late Glynn Isaac) because the models account for only biological, not cultural, behavior.

All optimization models have four basic components (Gardner 1992:18). Each requires (1) an actor to choose among the different alternatives, (2) a currency by which the payoff of the decisions can be measured, (3) some range of resources available to the actor, and (4) a set of constraints, factors that limit the alternatives and payoffs. In the study of human systems, the actors in optimization models are the people of the culture being studied.

Currency

All optimization models use a **currency**, some measure of cost and benefit that can be quantified. Possible currencies (see Jochim 1998:20) include energy, nutrients, information, goals of the actor(s), constraints in the environment, risk, options or choices (Smith 1983:626), or even material items such as hides (e.g., Keene 1979:390). Energy, as measured in calories, is the most preferred currency (Pyke et al. 1977:138; O'Connell and Hawkes 1984:530), as calories are easily quantified and measurable in foods and as calories expended in work. Other currencies are more difficult to measure and calculate, and some are so intertwined in culture and difficult to deal with that they are often "factored out of the analysis or reduced to residual behavior" (Keene 1983:141).

Energy is usually measured in calories, with net return often being quantified in calories per hour. However, the use of energy alone may result in a poor estimation of the potential of any given area because it may be that other nutrients, such as vitamins (Reidhead 1979:558), fat (Speth and Spielman 1983:18–19), or protein (see Jochim 1998:20), are more important. These problems are understood but are generally ignored because, at least for now, optimization models have to be simple in order for investigators to begin to gain even a general understanding of complex human behavior (e.g., Hawkes and O'Connell 1985:401; Jochim 1998:21).

The Resource Universe

To build any optimization model, it is necessary to know the **resource universe**, those resources that were available and utilized by the group under study. Resource universes are constructed by first using contemporary and/or paleobiological data to determine the presence and distribution of resources. Next, ethnographic data are employed to determine which resources were used. In archaeological studies, ethnographic data are used as an analogy to model the behavior of a prehistoric group.

In optimization models, resources are ranked relative to each other, generally by their overall return rate. If the resource universe is incomplete, that is, if resources are missing from the list, there will be an error of unknown magnitude in modeling the diet. Considerable further complication could result if the data on resource distribution are faulty or if the ethnographic data on the known resources are incomplete or erroneous. Lastly, some dietary elements are poorly preserved in the archaeological record, and if they are not recovered, an error in interpretation could occur.

Constraints

A number of other factors bedevil the construction, analysis, and interpretation of optimization models, including an incomplete understanding of human nutritional requirements, an uncertainty of the various costs associated with resource exploitation, and a working assumption of complete information. In addition, human culture tends to interfere with the simple cost-benefit formula of optimization models in that social obligations, preferences, and the use of storage can greatly influence actual behavior.

Nutritional Needs. The application of optimization models requires some understanding of the nutritional needs of the people under study. However, as noted in chapter 3, few such data have been compiled, and only a general estimate of the nutritional needs of people in contemporary Western culture has been made. Thus, the nutritional needs of traditional or past peoples can only be estimated based on those data.

It is also important that the nutritional content of the various resources be determined. Fortunately, this is now a rather simple process conducted by sending some of the material to a food laboratory and having a machine determine the numbers. However, for the people under study to have made decisions regarding which resources to use, they would have had to be aware of the nutritional content of the resources. Most researchers agree that such information was generally available to traditional peoples, who drew on thousands of years of experimentation and practical knowledge instead of machines that analyze calories per gram.

Exploitation Costs. The most difficult aspect of building an optimization model is the computation of the costs involved in resource exploitation. To determine procurement cost, one must calculate a whole series of figures based on the calories expended to complete each of the tasks associated with finding resources and getting them ready to eat. The major factors in this regard are (1) prey selection—the cost of deciding which resource to obtain, dependent on ranking, reliability, and location; (2) search time—the cost of locating and encountering the resource; (3) transport costs—the cost of moving the resource from where it was obtained to where it will be used; (4) processing—the cost of preparing the resource for consumption (e.g., butchering an animal or grinding seeds); and (5) storage—the cost of storing a resource. Some of these activities require the use of some technology, and the costs of obtaining raw materials for tool construction, making the tools, and then maintaining them also have to be computed (see Ugan et al. 2003). All of these costs are then combined to figure the total cost, which is then subtracted from the value of the resource, with the "net" return being used to rank the resources. An example of how this is done is presented in table 3.5.

Obtaining accurate information on all these costs is difficult, and lacking specific information, researchers often use general estimates. However, some experimental work has been done to generate real numbers for some resources in the Great Basin (e.g., Simms 1984; Madsen 1986; Metcalfe and Barlow 1992). Although these data are the best available, and much better than just estimates, they are still somewhat problematic because the studies were carried out by relatively unskilled workers (anthropologists are not skilled native people) with replicated technology leading to an unknown and perhaps variable error. However, the data are useful and could be interpreted as reflecting minimum return rates and so might be adequate for ranking resources on a relative basis.

Associated with accurately figuring costs and ranking is an understanding of the resource under consideration, including its reliability, abundance, and seasonal availability. Most resources are more abundant during spring and summer than during winter, and in many environments, food is harder to find in winter. Some resources, such as trees, are stationary, but others, such as some animals, migrate. If animal "X" is hunted but migrates out of an area for a portion of the year, the people in that area will not be able to hunt it at that time. To mitigate this changing availability, other mechanisms to ensure the availability of food

Table 3.5. Hypothetical Resource Modeling for Culture X

Resource (per 100 grams)	*Deer*	*Ducks*	*Grasshoppers*	*Acorns*	*Grass Seeds*
Gross calories	600	400	2,000	1,000	350
Calories expended searching for resource	75	25	50	25	25
Calories expended obtaining resource	25	50	50	50	25
Calories expended in processing resource	25	10	75	100	75
Calories expended in storing resource	5	5	5	15	10
Calories needed to make equipment to obtain, process, and store resource	25	50	50	50	25
Net caloric return	445	260	1,770	760	190
Rank in total diet	3	4	1	2	5

must be developed, including expanding the diet to include lower-ranked resources, moving to other locations, and storing food. If the behavior of a particular species is not understood, or if data from the wrong species are used, a serious mistake in the construction of the model can result.

For example, if one were to model return rates for pinyon in the Great Basin, it would be essential to know its distribution, cropping frequency, predictability, seed size, and a variety of other data, including the nutritional content of its seeds. One might be tempted to think that all pinyon species are the same, but in reality, a considerable difference in behavior exists between species. If the wrong species of pinyon were used in a model, the results could be vastly distorted (see Sutton 1984).

Information. Optimization models assume that the actor(s) will make optimal decisions regarding resource selection and use and that such decisions will be based on complete information regarding resource distribution, seasonality, and concentration. In reality, however, information is never complete and may even be incorrect, improperly interpreted, or inaccurately transmitted. Poor-quality information can dramatically affect the efficiency of resource procurement (e.g., Moore 1981, 1983). The price of getting information is often high, but often worthwhile.

Social Obligations. From the viewpoint of a strict optimal diet, one would expect a group to schedule its activities and movements so as to exploit the highest-ranked resources in the area. Thus, optimization models would predict how certain settlement types were located to follow the distributions of the highest-ranked resources. However, the actual activities and movements of a group are heavily influenced by social needs, such as networking, ritual activities, political obligations, mate exchange, and family maintenance. In some cases, meeting these needs takes precedence over economic needs.

Preference. People rarely utilize all possible resources and so make some selection of which to use and which not to use. Optimization models predict that such selection will be based on the currency used in the model, such as caloric return, and it could be argued that some resources are not used due to their low ranking. However, even highly ranked foods may not be consumed due to cultural preference, a factor not considered in the strict application of optimization models. There are many such preference factors, including religious taboos and the desire to have different (e.g., gourmet) foods to provide variety in the diet. Recent European history documents that people will sometimes go to extraordinary lengths for "exotic" goods, such as spices.

Moreover, humans can rapidly change their preferences. Before World War II, pizza was unknown in the United States outside of Italian American neighborhoods. During the early 1950s, pizza and related Italian food became so popular that the consumption of oregano (the favored herb of Italy) in the United States increased some 5,000 percent in just a few years. For better or worse, pizza has changed American food habits appreciably.

Another good example of cultural preference is the avoidance of eating the flesh of other humans, that is, avoidance of cannibalism. Most cultures consider cannibalism to be quite unacceptable, even abhorrent. However, from a strict optimization standpoint, why not eat human flesh? It is as nutritious as other meats, does not contain toxins, and is readily available. Thus, it should be a highly ranked resource.

Cannibalism is an emotional issue, with some researchers confirming and examining the practice (e.g., Volhard 1939; Hogg 1966; Sanday 1986; Askenasy 1994) while others (e.g., Arens 1979) deny that cannibalism exists as a cultural institution. It seems clear that the practice does exist. In a cultural context, endocannibalism (the eating of people within one's social group) is almost always prohibited, except for the ritual consumption of ancestors. Exocannibalism, eating people from other social units (strangers and enemies), is more common, and in some cases even expected (Turner and Turner 1999:1).

Three general categories of cannibalism can be defined: culinary, ritual, and emergency. A fourth category, criminal, will not be discussed (see Askenasy 1994). Culinary cannibalism is the eating of other humans as a normal, if infrequent, part of the diet. Such cannibalism seems to have been very rare, but it has been argued that the Anasazi practiced it (White 1992; Turner and Turner 1999) and that the Mexica (frequently called the Aztec) consumed some of their sacrificial victims (Harner 1977; Ortiz de Montellano 1978; Winkelman 1998; Turner and Turner 1999:415–421). Ritual cannibalism, the eating of small portions of people to honor or gain power from them, is much more common. In these cases, very little nutritional value is gained. An interesting case of such a practice was the ritual eating of a small portion of the brains of deceased relatives (endocannibalism) by certain groups in New Guinea. It was discovered that some of the brains contained a virus that causes kuru, a disease that results in madness (somewhat similar to "mad cow" disease). When the brain was eaten, the virus was transmitted to the consumer, who contracted the disease. The government stepped in to stop ritual cannibalism in order to stop the spread of the disease (Zigas 1990).

The last type of cannibalism, emergency, is the type most commonly known to most Westerners. People on the verge of starving to death will sometimes resort to eating the dead in order to stay alive until they can be rescued or find other food. Examples of such practices and events include the Inuit (e.g., Saladin d'Anglure 1984), the Donner Party (Hardesty 1997), and the soccer team whose airplane crashed in the Andes in 1973 (Read 1974; in 1993, a Hollywood movie, *Alive*, was made about this incident). Unlike ritual forms, emergency cannibalism is practiced to obtain nutrition, but for only a limited time. One could argue that emergency cannibalism is a practice retained in the dietary inventory of many groups, held in reserve for extraordinary circumstances and thus an expansion of diet breadth.

So why not regularly eat other people? It almost seems wasteful to discard such a large package of nutrition. Emotional attachment to the deceased seems to be a major factor, as well as a fear that the same could happen to you.

Power and Access to Foods. Optimization models do not distinguish which people, or groups of people, in a culture are consuming resources. However, there is considerable variation in food consumption based on age and sex. In most cultures, children consume at least some foods that are different from those adults eat. Males and females may also consume different foods, and in a number of cultures, males have greater access to meats derived from the procurement of most animals.

For example, among the Yanomamo of northwestern Brazil, Lizot (1977:table 6) reported that males consumed about 93 percent of all of the protein derived from animal sources, with only about 7 percent being consumed by females. A similar pattern of differential access to protein among the Tukanoan Indians in the northwest Amazon was reported by Dufour (1987). Males generally consumed "high-status" resources (e.g., hunted mammals) while females derived most of their protein from less prestigious sources, such as insects.

Model Refinement. Finally, it is important to consider the fact that, as these models are employed, they are continually refined. We learn new things, recalculate costs, expand the resource universe, determine new rankings, correct mistakes, and improve theory. As a result, it is necessary to revisit earlier results and reexamine past conclusions. This process is integral to scientific inquiry.

The following example illustrates the importance of model refinement. Archaeological excavations at Lakeside Cave, Utah, resulted in the recovery of a few fragmentary grasshopper parts in the site deposit (Madsen and Kirkman 1988). When a smaller mesh screen was used on some other site soils, a larger number of grasshopper parts were found, leading the researchers to consider the possibility that grasshoppers were a resource used at the site. A still finer mesh resulted in the recovery of a still larger number of grasshopper parts, forcing a reconsideration of the role of those animals in the overall subsistence system. It was discovered that grasshoppers could be obtained from the shore of the Great Salt Lake, very near to the cave, where they had been washed up on the shore, dried, and salted. The return rate in collecting the grasshoppers was determined to be about 270,000 calories per hour, a rate that would result in grasshoppers being by far the highest ranked resource in the region. This would require that the diet models for the region be reevaluated.

It was later determined (Jones and Madsen 1989) that if the grasshoppers were transported to other localities, their caloric return rate would be much lower; still high but much closer to other resources. However, the new calculations were based on the transportation of whole grasshoppers and did not account for the fact that at least some of the grasshoppers may have been processed prior to transport, allowing a person to carry many more consumable grasshopper parts than whole grasshoppers.

In sum, we do not know the ranking of grasshoppers in the Great Basin. They appear to have been very high, at least where they could be collected in abundance (e.g., Madsen and Schmitt 1998). A determination of whether the animals were processed prior to transport and whether they were stored, among other factors, must be made before the return rate, or rates, depending on conditions, can be figured. In the meantime, the ranking of all of the other resources used in that region remains suspect, pending the outcome of the grasshopper case.

Four Optimization Models

A number of optimization models have been employed. The two most commonly used models, diet breadth and patch choice, are derived from optimal foraging theory (OFT), which emphasizes net efficiency and minimization of risk as its guiding principles (Jochim 1976, 1998:14–20; Pyke et al. 1977; Durham 1981; Pulliam 1981; Thomas 1986:251–258; Winterhalder 1986). The two other most commonly used models, central place foraging and linear programming, employ similar criteria.

The Diet Breadth Model

The **diet breadth** (or optimal diet) model is the oldest and most widely used (see Simms [1984] for a detailed discussion) in optimization studies. In addition to the basic assumptions of any optimization model, the diet breadth assumes that resources have a fine-grained distribution, that is, that they are evenly or "randomly" distributed throughout the environment. This assumption takes a very complex actual situation and simplifies it for analysis. In reality, resources are never randomly distributed in an environment.

In the diet breadth model, resources are ranked by their return rate, with the highest-ranked resource predicted to form the primary basis of the diet. If the highest-ranked resource did not contribute enough calories to the diet, the second-highest-ranked resource would be added to the diet. If the diet were still lacking, the third-highest-ranked resource would be added. The breadth of the diet would keep increasing in this fashion until the caloric requirement was met. Thus, the model predicts "*the order* in which resources will be added to or deleted from the diet" (Simms 1984:33, original emphasis).

The composition of the diet will fluctuate depending on environmental conditions. As all environments are dynamic, the breadth of the diet will always be expanding or contracting. In times of dietary stress, defined as fluctuations in the availability of higher-ranked resources, lower-ranked resources would be added to the diet to make up any caloric shortfalls.

The Patch Choice Model

The **patch choice model** is similar to the diet breadth model in that resources are ranked based on return rates. However, instead of resources being ranked individually, patches or suites of geographically associated resources are ranked, and decisions regarding diet are based on the average return rate of patches, not of specific resources. The optimal strategy is "to locate oneself in the patch with the highest rate of return and remain there until conditions change" (Smith 1983:631). The distinguishing feature of the patch choice model is that patches of resources are ranked as a unit, and the patch with the highest return rate is selected first, with lower-ranked patches being added to the system in the order of their ranking, which has to include travel time between patches. Once a patch is occupied, the people should remain there until the return rate falls below that of the next-best patch. If the return rate does not decrease until the resources are completely exhausted, this strategy could result in local extinctions (Smith 1983:632).

The ranking of patches, rather than of individual resources, requires making four basic decisions: (1) which patch to visit; (2) how long to stay and when to leave; (3) which resources in the patch to use; and (4) foraging path, or which patch to visit next (Pyke et al. 1977:140). In human societies, the first and fourth of these decisions are largely tied to information systems, since a decision cannot be made prior to the patch's being "encountered" (Pyke et al. 1977:144), as well as to seasonality and cultural considerations. The second decision is related to the total resource potential of the patch, and the third to the rank of resources within the patch.

The cost-benefit criterion in the diet breadth model is the return rate of individual resources. However, with the patch choice model, the criterion is the average return of a suite of resources within a patch, the patches being ranked on their total resource return rate. Patch A, containing a lower quantity of a high-ranked resource, may be chosen over patch B, with a higher quantity of that resource, if patch A contained more of a resource of lower rank. Thus, even the best pinyon stand may be bypassed in favor of a less-productive one where deer are also present.

Factors in patch selection include the amount of time that can be spent in a patch before the group must move to another patch and the labor available to exploit a patch. If a group is too small or if time, including travel, is too limited to exploit a patch, a less productive one may be chosen. Once a patch is chosen and a group has located itself within that patch, the application of diet breadth principles within the patch should be useful because diet breadth does not make any predictions about abundance, only the order in which resources will be used.

The patch choice model has some of the same shortcomings as the diet breadth model in that it makes assumptions that do not always hold true for human groups. Cultural aspects are usually factored out of the model, and like diet breadth, a single currency (energy) usually is used. Costs are computed in the same manner as with diet breadth, except that the additional costs of traveling between patches and of gathering and analyzing the information for ranking the patches must be added.

Central Place Foraging

Central place foraging employs the same basic tenets as diet breadth and patch choice but assumes that the group stays in one central place, with specialized task teams traveling to the resources, obtaining them, and returning to the central place (Orians and Pearson 1979; also see Bettinger 1991:93–97). The central place model analysis is based on the energy expended in travel time (the round trip from the central place to the patch) and time spent in the patch, versus the return rate of the resources.

The major problems people using central place foraging must solve are (1) where to locate the central place (i.e., base camp) in relation to nearby resource patches and (2) once in a central place, which prey to seek, which patch to use, and the load size, that is, how much of the resource can be taken back to the central place (see Orians and Pearson 1979:156). Each of these problems is influenced by patch quality, risk, and competition, and considerable information is required to make good decisions on these issues (Orians and Pearson 1979:156).

In sum, the central place forager should seek larger prey, up to what can be carried back to the central place, as travel distance increases. The further the patch from the central place, the more that the time spent at a patch and the size of the prey should increase (Bettinger 1991:94). As prey abundance increases, the requisite quantity of resource is obtained more quickly, and the time spent at the patch should decrease.

Linear Programming

Linear programming (Reidhead 1979; Keene 1981, 1985a, 1985b; Shapiro 1984; Belovsky 1987, 1988) is "a mathematical technique of calculating the optimal allocation of resources towards a defined goal in the face of multiple constraints" (Gardner 1992:1). Unlike other optimization models, linear programming can use multiple currencies, which can be an advantage over the single currency of other optimization models. Linear programming is more complex than other models in that it requires calculation of return rates in each of the currencies used in the analysis.

Linear programming models have three components (Shapiro 1984:13–15; Gardner 1992:25): (1) the goal to which resources are allocated; (2) the set of variables; and (3) constraints, factors that limit what can be done. Each of the variables (currencies) is linear, that is, a change in one variable must produce a proportional change in the other variables. A sequence of events, that is, the allocation of resources, is specified and programmed (Gardner 1992:24).

A graphic example illustrates the method (condensed from Gardner 1992:25–28, fig. 2.1). Say that a family needs 8,000 calories and 180 grams of protein per day and wishes to minimize its time spent getting food (the goal). Available food resources consist of nuts and fish. It takes one hour to get one pound of nuts, and they return 1,600 calories and 20 grams of protein per pound. It takes one and one-half hours to get one pound of fish, with a return of 800 calories and 60 grams of protein. The resources are limited in the environment, and the family cannot obtain more than six pounds of nuts or five pounds of fish per day. Four constraints on the solution are present: (1) the requirement of calories, (2) the requirement of protein, (3) the availability of nuts, and (4) the availability of fish.

To obtain the requisite number of calories, the family could eat five pounds of nuts, 10 pounds of fish, or any combination that falls along the calorie line shown in figure 3.1. To satisfy the protein requirement, the family could eat nine pounds of nuts, or three pounds of fish, or any combination that falls along the

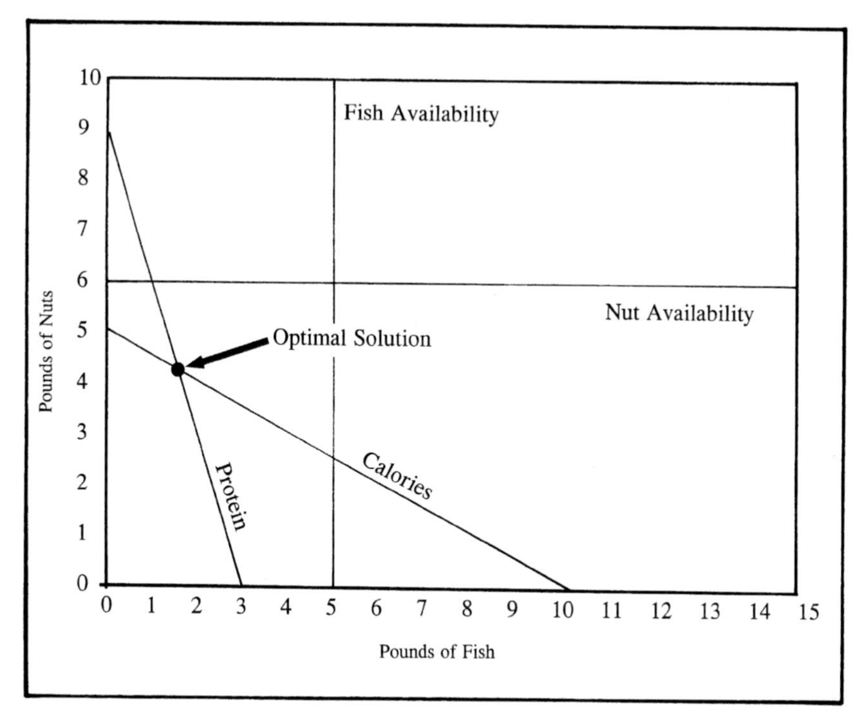

FIGURE 3.1
The graphic solution to diet using a linear programming model under the parameters detailed in the text. (Adapted from Gardner 1992:Fig. 2.1)

protein line (fig. 3.1). The intersection of the calorie and protein lines is the optimal solution: 4.2 pounds of nuts and 1.6 pounds of fish, a resource mix that requires 6.6 hours of time.

In the above example, the optimal diet solution fell within the region of the chart where the availability of the two resources (nuts and fish) was not a factor, and so those constraints are considered "slack." Calories and protein are the only factors that are relevant, and so those constraints are considered "binding." If the binding constraints changed, or if the slack constraints changed enough so that they became binding, the solution to the problem would also change.

Among the benefits of the linear programming model is its ability to employ multiple currencies in the analysis. In addition, the researcher can alter the various constraints to see changes in the solution and thereby gain an understanding of the range of acceptable solutions and the scale of change necessary in the various constraints to alter the system in a fundamental way. The weaknesses of the model are similar to those of the other models, including the assumption of perfect information, the assumption of rationality, a complete understanding of the resource universe, and an accurate modeling of return rates.

Discussion

Optimal foraging models attempt to explain aspects of *human biological ecology*; by themselves, they cannot explain *cultural ecology*, as cultural factors alter and sometimes override decisions that would be "optimal" from a dietary perspective. This is not to say that OFT models are not useful. In fact, what makes them most interesting is when they *fail* to predict human behavior; when that happens, we can look for cultural explanations.

Researchers using optimization models seem to prefer hunter-gatherers as subjects because they often assume (even if implicitly) that hunter-gatherers are somehow "closer to nature" than agriculturalists and that their activities are somehow more amenable to analysis. Hunter-gatherers are supposed to behave like other animals, foraging for their food and wandering about the landscape (Ingold 1987:11). On the other hand, agriculturalists are food-producing landholders and are viewed as somehow set apart from nature, which makes optimization models difficult to apply (but see Gregg 1988).

In addition to this bias, three major problems with the current application of OFT are apparent. First, and most fundamental to the application of any model, is the working (and implicit) premise that we understand the full range of available and utilized resources—an assumption that is invalid. In this situation, conclusions drawn from any of these models likely are invalid as well. This is not to say that such studies cannot be valuable. In fact, they are critical to test and fine-tune any model of adaptation.

Second is the use of the diet breadth model. While this model is a useful tool in gaining some understanding of resource utilization, its assumption of a fine-grained environment makes its power to predict settlement and subsistence patterns weak in many environments. However, its application within a patch and in the ranking of patches can be quite useful.

Finally, simple optimal foraging models adopted directly from biology cannot account for the diversity of cultural behavior and factors influencing economic decision making (see Jochim 1998:23–26). While admittedly difficult to quantify, multiple currencies derived from a variety of social factors could provide a more realistic view of economic systems. The lack of food (calories, vitamins, etc.) is a stressor that requires some response, and many of the responses are cultural. Such nutritional adaptations (Stinson 1992) can be central to an overall adaptation.

However, for a given small group at a given time and place, OFT can be very useful (Winterhalder and Smith 1981; Smith 1983). The study of the Aché of Paraguay (Hawkes et al. 1982; also see Hill and Hawkes 1983; Hurtado et al. 1985; and Hill and Hurtado 1996) has been particularly valuable in clarifying this point. Hawkes et al. (1982) found that the Aché do indeed forage more or less optimally except for a clear preference for fat meat. Other important studies of hunting and gathering groups have been carried out in places as diverse as the Australian coast, the Kalahari Desert, and the Arctic. These studies show that people are indeed highly sophisticated at making a living from the environment but that they make their decisions on the basis of cultural preference and social choice as well as of considerations of simple protein and calorie return. Archaeologists find OFT reconstruction extremely useful as a first pass at reconstructing past societies (Winterhalder and Smith 1981). It is now an important area of study for human ecologists.

Fairly recently, the term "human behavioral ecology" has been applied to the component of human ecology that seeks to explore "the link between ecological factors and adaptive behavior" (Smith 2000:29) through the application of strict empirical and testable biological evolutionary models for humans, such as studies of optimal foraging, population regulation, and evolutionary ecology (Cronk 1991, 1999; Cronk et al. 2000). Such biological models tend to be simplified abstractions of economic models developed for human society, and for that reason their application is limited. The problem is that using such biological models eliminates consideration of conscious choice. While doing so is useful in dealing with most animals, it is an unfortunate limit to impose on humans. Nevertheless, a great deal of new knowledge about the evolution of human behavior has been generated (Ehrlich 2000; Irons and Cronk 2000:21; also see Ehrlich and Feldman 2003), and human behavioral ecology is a useful approach.

Let us conclude with this thought: people and cultures must make a "profit" to survive. No hard and fast rules govern how much of a profit, whether an adaptation is actually optimal, or whether the decisions made are the best possible ones. The adaptation must be only good enough. If one culture is competing with another, its adaptation must be better than the other's to survive; there is no rule that it must be exceptional, just better than the competition. We believe that, in general, most humans make enough of a caloric profit on the biological side of human ecology that they can afford to do whatever they want on the cultural side.

CHAPTER SUMMARY

Humans are animals and can be studied using biological and ecological models. This approach within human ecology is called human biological ecology and emphasizes how humans adapt biologically. This is done in two general ways, short-term physiological responses and long-term anatomical changes.

The overall human population is growing rapidly, and individuals are living longer lives, but it is unclear what the carrying capacity of the planet is. Human population is regulated in a number of ways, including disease, food supply, birth control, infanticide, and, to some extent, warfare.

Individuals require daily nutrition to function properly. Nutrients are obtained in foods, and nutritional requirements vary by age, sex, and physical condition. Calories are a commonly used measure of diet. However, adequate quantities of specific nutrients, including carbohydrates, proteins, fats, vitamins, and minerals, are required for proper nutrition.

A relatively new approach in the study of human adaptation is the use of evolutionary ecological models, those that apply the concepts of selection to human behavior. Such models generally study optimization, primarily as related to diet. Using available information and, usually, calories as the measure, researchers develop a quantitative model of optimal diet, that is, what people would be eating if they were behaving optimally, and test that model against what people actually eat. In this way, one can determine how actual behavior deviates from the biologically predicted behaviors and so gain an understanding of cultural systems. The most common optimization models are diet breadth, patch choice, central place foraging, and linear programming. The successful application of these models requires an understanding of resource distribution, resource value, acquisition and processing costs, and information flow. While very useful, evolutionary ecology addresses only issues in human biological ecology. Other issues are addressed by cultural ecology, and culture has traditionally been the central and pivotal concern of anthropology.

KEY TERMS

anatomical adaptations
Bergmann's Rule
calorie
carbohydrates
central place foraging
currency

diet breadth model
evolutionary ecology
fat
life expectancy
linear programming
minerals
nutrition
optimization
optimization models
patch choice model
physiological adaptations
protein
resource universe
stress
vitamins

4

Cultural Ecology

Biological evolution and natural selection are the forces that shape organisms. Beginning some time in the distant past, culture began to influence human development, changing the relationship of humans to their environment from one of strict biology to a mixture of biology and culture. Over the millennia, culture has become more complex and influential in human affairs, and the role of biology has diminished.

To be sure, humans still require a certain level of nutrition, have physical limits to their physiological adaptations, and are still subject to the rules of biological evolution. Today, however, biology plays only a minor role in human adaptation, and now most of the problems posed by the environment have to be solved through the mechanism of culture.

Much of the ecological work relating to humans has centered on diet and subsistence, and while these are valuable studies, they focus on human biological ecology. **Subsistence** is not simply a list of foods but a complex system that includes resources, technology, social and political organizations, settlement patterns, and all of the other aspects of making a living. Subsistence is one of the vast complexities of human behavior largely related to culture. Because so many studies of subsistence have focused on food, much of the behavior of people has been excluded from many ecological studies (Jochim 1981:ix). Once we get past the emphasis on food, however, we can begin to look at the influence that other behaviors have on the adaptations of human cultures (e.g., Goodman et al. 2000). The field of cultural ecology focuses on discovering cultural adaptations.

HUMAN CAPABILITIES

The development of humans and cultures capable of building cities and creating art is tied to the development of the human brain. The increase in brain size and complexity through time is the most visible indicator of increasing capabilities, but the important outcomes of these developments are the ability of humans to act thoughtfully, rather than by instinct, and the increases in human sociability and intelligence.

Instinct and Learned Behavior

All animal behavior is guided to some degree by instinct, and there is a considerable debate in anthropology and biology about the extent to which it guides human behavior. The old tabula rasa view (see Locke 1975, orig. 1685) held that humans learned most of their culture. Locke is often mistakenly quoted as saying that humans learn everything—that our minds, at birth, are literally "blank tablets" waiting to be inscribed. On the contrary, he recognized that humans had certain inborn capacities. Individuals differ in learning ability, for instance. Locke proposed that within limits set by innate capacities, people could learn a potentially infinite range of customs.

Humans apparently do have a few instincts, such as self-preservation, but most other things are clearly learned, and it seems that humans have evolved the capability of learning certain things easily; others are more difficult.

Language is obviously learned. The human vocal apparatus is structurally capable of making a wide range of sounds, but English, for example, uses only some of the possibilities. Newborns can be raised to learn any language, but it is not clear how much of the capacity for language is hardwired and how much is learned. These issues are currently under investigation (see Barkow et al. [1992] for recent debates).

Similar questions arise about aggression. Clearly, we share with all vertebrates a tendency to fight back when attacked—to deal with perceived threats by using violence. Yet there are people, even entire groups, in whom violence is essentially unknown. These data eliminate simple theories that postulate universal, uncontrollable aggression as a human trait (Robarchek 1987; Robarchek and Robarchek 1998).

Even art has been given biological treatment (Dissanayake 1992). Humans like regularities, such as rhythm in music, regularity in geometric design, and prosody in poetry. Much of this tendency is undoubtedly inborn; the human love of rhythm has something to do with heartbeat, breathing, and other rhythmic bodily functions. However, much of it is certainly learned; Chinese and Europeans generally do not identify with each other's music. A simple biological explanation fails to account for the complex interaction of learning and genetic programming, although it provides a beginning hypothesis.

Human use of the environment has innate components as well, ranging from our love of sweets to our ability to calculate—roughly but quickly—the payoffs of alternate foraging strategies (Smith 1991). But it also has learned components. Only by studying the ways in which genes influence the cognitive ability of humans can we gain ultimate understanding of human biological ecology (Barkow et al. 1992). This understanding will not come easily. The considerable differences between cultures and the total ease with which adopted children can learn any culture if they are introduced into it early enough in life are but two examples that demonstrate a far greater capacity for learning and plasticity than many currently allow.

Sociability

No animal on earth, with the possible exception of some insects such as ants and termites, is more social than humans. Human babies cannot survive without years of care, and children must learn a phenomenal amount of cultural knowledge before they can even begin to fend for themselves. Instinct does not guide us adequately in the food quest. Human adults can live alone, but we very rarely choose to do so, and still more rarely do we thrive as hermits. Psychological health, even more than physical, depends on sociability. Experiments with monkey and ape infants suggest—and observations of neglected children confirm—that young humans will simply die if they are not involved in warm, close social relationships, even if all their physical needs are met.

On the other hand, humans differ from ants and termites in that they value autonomy. We have a psychological need to feel in control of our lives, and this feeling can be so strong that we may deteriorate and die if we feel hopeless and directionless (Schulz 1976; Langer 1983; see also Jilek 1982). Thus, we are caught in something of a paradox. The existential human dilemma is that we need autonomy yet cannot get along without sociability. This basic tension lies behind many tragedies. It is well to ask how we got this way. Why are humans such strange animals? Why do we have such exaggerated brains, such curiosity, such a need for both society and freedom? Surely, designing a better world must begin with the knowledge of just who and what we are. The study of human evolution is providing some of the answers. The broad outlines of human evolution are now well known, although there are countless unresolved questions about the details.

Intelligence

Humans are, on the average, intelligent animals. Intelligence has some obvious benefits, but it also has huge costs. The brain and nervous system make up only 3 percent of the weight of an average adult human but uses about 15 percent of the body's total energy budget, enough to fuel a good-size dog.

Moreover, the brain is spared when starvation occurs. During starvation, the body begins to digest itself to provide the basal metabolic energy it needs. First, fat is converted to energy; then muscle tissue is utilized, until finally the heart and digestive muscles are reduced to the point that death ensues. The brain is sheltered from this process. Chronic malnutrition in utero and in the first few years of life can stunt brain growth and intelligence, but long famines do not appear to have the same effect on adults. Studies of the Dutch who were subjected to starvation during the Nazi occupation, and of people during the Korean War, showed no intellectual deficiencies attributable to starvation.

Given the world history of famines and malnutrition, it is obvious that the brain must provide a huge advantage in survival. While the advantages inherent in human intelligence are obvious today, what of the early hominids? How did intelligence evolve? What were early hominids doing that made increases in intelligence so valuable to them? At least part of the answer is, broadly speaking, that early humans were almost certainly social omnivores who shared food (Isaac 1978a, 1978b). They probably foraged on anything they could find and operated in sizable social groups. Everyone would have looked for food and, when it was discovered, alerted others, who shared in the benefits. Food would have probably been brought back to the children and to the adults who were tending them.

This sort of foraging pattern is typical of several other social animals quite unrelated to us, like crows and coyotes. These animals all exhibit a similar pattern of brain development; their brain is larger than that of their less social relatives. Indeed, the brains of crows seem as highly developed beyond solitary birds as human brains are beyond the brains of solitary primates.

Intelligence evolved for two reasons. First, as omnivores, humans could never rely on specific instincts or simple rules of thumb to obtain food. Even now, one sometimes reads about the "hunting instinct" that people have because of our "hunting past." Such an instinct seems highly unlikely. Only a few small groups of humans with a highly specialized and highly developed technology have lived solely by hunting—usually in areas with few edible plants. All other known human groups live mostly on plants and/or fish and shellfish. Today, the main source of calories, and the favorite food of most peoples worldwide, is seeds. Roots and tubers, fruits and their sugar sources, oily seeds, and tender young leaves are also consumed in large quantities. However, this reliance on plants has taken shape since the development of agriculture (which began some ten thousand years ago). Before that time, the case is not as clear.

The evidence supporting a long history of omnivory lies in anatomical data. We have small, unspecialized teeth—no fangs like a carnivore or grinders like an herbivore. We lack claws, pouncing muscles, and other carnivore equipment. We also have an unspecialized digestive system. Physiology also tells an omnivore story. We need high levels of vitamin C and other nutrients that are scarce in meat, as well as vitamin B_{12}, which is scarce in plants.

The second reason for the evolution of intelligence is that as social creatures, we had to manage the incredible complexities of social life. This involved, among other things, the development of sophisticated communication. It is sometimes said that language was "invented," and rather recently at that. This may be true of language as we know it—long sentences, complex questions, embedded clauses, subtle puns, and slick lies. However, some sort of complex communication must have existed long ago. Once again, the evidence lies in anatomy. Language centers exist in the brain, usually in the left temporal lobe. Lip, tongue, glottis, and vocal-cord anatomy in humans is exceedingly complex and fine-tuned—incomparably more than that of chimpanzees, who cannot manage human vocal sounds and have to be taught sign language when we attempt to teach them to "talk." The evolution of such vocal structures must have been long and gradual; such a fine-tuned accommodation by hundreds of muscles and cords could not have evolved rapidly or suddenly.

Along with language, facial expressions and gestures are forms of communication in human societies. The whole human face is very expressive; we can display Mona Lisa smiles, broad grins, evil smirks, and false "warmth." One hypothesis is that language evolved to communicate about tools. The use of tools is difficult to accomplish with language alone, however, as we require experience to learn to use them. We are adept, though, at fine-tuning social situations with speech.

This observation brings up an important point: Humans, while intelligent, do not excel at everything. A dog lives in a world of smells that we cannot even imagine. Some fish locate prey via electric fields. Birds find their way during migration by means of magnetic sensor cell systems in their brains. What we call human intelligence is a quite specific bundle of abilities (Barkow et al. 1992), and social skills are by far the best developed. The second-most-developed skill is the ability to know almost everything about the environment we live in—every weed, every rock, every trickle of water—and to describe it to others. We have a phenomenal ability to learn about such things. Furthermore, we learn for the purpose of social rewards. Approval and prestige are the greatest of reinforcers.

Basic Human Needs

Cultures exist to satisfy human wants and needs, but how a culture does this is highly variable. Certain needs are inescapable. First, we all must eat and drink; that is, we must ingest water, calories, and certain nutrients, and so some sort of an economic organization is required. Second, we must stay healthy enough to function, so we need some sort of medical system. Third, for a culture to survive, people have to reproduce. Reproduction not only involves bearing live and healthy children but includes all the behaviors associated with ensuring that the children live long enough to reproduce—these behaviors include such things as feeding, protecting, and educating them. Fourth, people usually need some kind of protection from the elements. In areas like the tropics, this protection can be minimal, but in areas such as the Arctic, protection from the elements is critical.

Fifth, people need some stimulation and variation. We need to experience arousal, excitement, boredom, calm, wild enthusiasm, active interest, and sleep. Without variation, people may become lethargic and unproductive. Sixth, people need a social life, as people can actually die from negligence or boredom, as neglected children and older people in nursing homes have been known to do. Seventh, people have to have some sort of control over their lives. Like neglect, stress and helplessness can cause death (Peterson et al. 1993). Humans must also deal with emotion effectively. People tend to make major decisions on the basis of emotion, without considering the environmental consequences. The most extreme case is outright warfare, but any political conflict is apt to inflict environmental costs.

This highly complex social world requires two other critical elements: methods of communication and of decision making. Communication invariably goes well beyond language. Gestures, facial expressions, art, music, dance, clothing, and verbal performance are highly developed in all cultures. In small groups, decision making may be undertaken by all the people getting together and discussing issues. Larger groups must have some sort of leadership and management organization, such as a council of elders. At the other end of the scale are the huge, convoluted, powerful bureaucracies of the contemporary state. Regardless of the size of the group, however, decisions have to be made.

Being generalists in a very wide niche, humans have a vast range of valid choices about how to meet their biological needs. It is *culture* that determines which set of solutions will be utilized. In addition to meeting biological needs, culture imposes other needs, such as religious rites or taboos, and the members must also meet these needs for the culture to be successful. The subdivision of cultural ecology investigates *how* and *why* people or cultures choose one response over another.

CULTURE AS AN ADAPTIVE MECHANISM

Today, the primary mechanism by which humans adapt to their environment is culture, probably "the most potent method of adaptation" available to humans (Dobzhansky 1972:422; also see Cohen 1974; Kirch 1980). Cultural responses include technology and organization, such as the structure of economic, political, and social systems. Compared to biology, culture is an extremely flexible and rapid adaptive mechanism because "behavioral responses to external environmental forces can be acquired, transmitted, and modified within the lifetimes of individuals" (Henry 1995:1).

All people belong to a specific culture, a group of people who share the same basic but unique pattern of learned behavior. As such, each culture has a distinct ecological adaptation. One might view a culture as a "population" in that the people occupy a distinct geographic area and have a defined way of making a living (they occupy a specific niche). Cultures interact with both the natural and the cultural environment. A culture must first meet the biological needs of its members. Then it must meet the cultural needs of its members through religion, social regulation, and other mechanisms. The combination of the biological and cultural ecological interplay is quite complex.

Individuals in cultures are born into a system operating within a given environment (Dobzhansky 1972:427). In traditional societies, the cultural system one is born into tends to be more influenced by the "natural" environment than industrialized cultures are. In industrialized cultures, the environment tends to be much more socioeconomic, with class and income (access to resources) being the major environmental differences between individuals (Dobzhansky 1972:427). Thus, in industrialized cultures, selection processes operate more on socioeconomic factors, and these factors influence the genetic makeup of populations.

As the environment (abiotic, biotic, and cultural) changes, humans adapt both biologically and culturally. As all environments are dynamic (even if the changes are small ones), a culture must make constant adjustments just to maintain some sort of equilibrium, and there is a constant interplay between cultural practices and biological adaptations; for instance, a people can be anatomically adapted to cold but still wear coats.

A variety of cultural practices can "mitigate" the impact of environmental change and so "level" environmental differences. Such differences might be seasonal, such as differences in food availability between the spring and winter, or longer term, such as climatic fluctuations. Culture chooses from a variety of solutions to various problems, and as some solutions become unavailable, others present themselves. Technological change also will alter the equation, likely increasing the potential set of options available to a culture.

Each culture must somehow solve the problems faced by the culture and its individual members. To do this, each culture has institutions—rules, principles, laws, social contracts—and organizations to keep things working and to maintain a balance between the various needs. The specific solutions to problems must mesh with the institutions and organizations. If the solutions are valid (adequate), the culture survives. It is important to understand that there may be multiple valid solutions. One culture may choose one, and another culture may choose a different, but also valid, solution.

Organization

Culture is a system that is organized into a variety of components, including economic, political, religious, and social. Each of these components also has an organization. Such organizations can be relatively simple (e.g., family) to very complex (e.g., the U. S. government). Different aspects of culture have different organizations. A political system may have a hierarchy, with its members having power and responsibility at assorted levels. The religious system may or may not be interrelated but will also have a hierarchy. The completion of a simple economic task may require a fairly complex organization, including divisions of labor based on sex and/or age, political or social requirements for access to resources, and/or religious involvement. The better we understand the organizations behind such aspects, the better we can understand the aspects themselves.

Those interested in general human ecology, particularly archaeologists, concentrate on the economic part of culture in order to determine how people make a living and thus infer their interaction with the environment. While anthropologists tend to deal with these components separately, they all are interrelated. Religion is in fact intertwined with economics and thus could be considered part of the environmental adaptation.

For example, if a group obtains its food through agriculture, there are certain environmental variables that are important for its success. One such variable is rainfall. If the rains do not come, the crops will fail, the people will starve, and the culture may even die. Rainfall patterns (as part of climate) are part of the abiotic environment but are related to the biotic environment (plants need water) and therefore tied to the cultural adaptation (agriculture). To guarantee success of the entire process, the group may choose to influence the amount and/or the timing of rainfall. To ensure rain, they may institute certain religious practices, such as the rain ceremonials of the Hopi of the American Southwest. Such customs, including the effort required to organize the group to conduct these activities, could be viewed as directly related to economics.

Westerners might believe that such a practice is unnecessary, arguing that the rain would come anyway and is not influenced by "dances." Several flaws exist in this line of reasoning. First, it is arrogant to assume that their practices are automatically irrelevant and have nothing to do with rain; there is too much we do not know. Second, how can one argue with success? At any rate, as anthropologists, we are interested in understanding, not criticizing, their practices.

Social Networks

Social responses to environmental stress are varied. One simple technique to alleviate uneven resource distribution at the immediate level is sharing, giving materials to someone else in need. Sharing may carry an expectation of return (reciprocity) and can serve as somewhat of an insurance policy against future shortages. A classic example of this type of behavior is the stereotypic view of band-level hunter-gatherer males sharing meat they obtain with the other members of their group (e.g., Gould 1982). Hawkes (1993) argued that the benefit of sharing resources may lie not in calories but in the development of social ties.

A larger scale of social networking may be seen in bilateral kinship systems. Unilineal systems establish membership based on relationships through only one side of the family, either the father's side (patrilineal) or the mother's side (matrilineal). Thus, in a matrilineal system, for example, mother's brother may be an important relative while father's brother may not. A unilinear system may be preferred for a variety of reasons, but it limits one's functioning relatives to fewer than the number possible. A bilateral kinship system includes both sides of the family, and one can therefore have many more relatives than in a unilinear system. It has been argued (Helm 1965) that a bilateral system could develop in areas prone to resource shortages, giving people many more relatives from whom to seek assistance.

Settlement Patterns

Different cultures will use the same space in a different manner, reflecting the organization of the culture, and this use will be reflected in the distribution of settlements across space and time. People and their activities, residences, work localities, facilities, and sacred places are located across a landscape in a culturally significant way, called settlement pattern. Part of this pattern is the way in which a particular group conceptualizes and utilizes its space.

Settlement pattern depends on a variety of factors, beginning with the basic economic system used by a group. For example, a hunter-gatherer group would utilize a valley floor quite differently from a group of farmers, with very dissimilar components, management practices, residential localities, and support facilities. The basic economic system would also influence the types and scale of facilities and technologies employed and the kinds of resources utilized. In general, hunter-gatherers would not have large iron mines, oil fields, or reservoirs, facilities utilized by industrialized groups. Some things might remain similar, such as the use of transportation corridors (e.g., freeways tend to follow ancient trade routes).

Technology

Technology, including the ability to make and use tools, is a major factor that separates humans from other animals, although some other animals do make and use simple tools. It is mostly through technology, rather than biology, that humans have adapted to virtually every ecosystem on earth. Technology can be very general (a hammer can be used for many tasks) or designed and used for very specific tasks (a space suit). Through an analysis of technology, one can gain insights into the functions of the tools and the relationship between the user (the culture) and the environment.

For example, suppose that an archaeologist finds a rock that is battered on its edges. The archaeologist concludes that the rock was probably used as a hammer, a simple tool. But to hammer what? If we could answer that question, we would have a better understanding of the users, their culture, and their adaptation. If the hammer were used in the production of flaked stone tools, a more complex technology would be inferred (evidence of which most likely would have been found with the hammer). If the hammer were used to pound in tent stakes, the use of tents would be inferred. The more we know about technology, the more we can learn about how a culture adapts to its environment. This is a materialist and empirical approach and is useful, but it does not deal with all aspects of cultural adaptation.

Technological Change

All human cultures have technology, some more complex than others. Technology is the result of need, available materials, innovation, and influence from other cultures. If one of these conditions changes, the technology will also change, and the environment and culture will be affected.

A quick look at the development of weapons technology is instructive. When thrusting spears were replaced by throwing spears, the number and type of animals that could be successfully hunted changed, thus altering economic systems as well as both human and animal populations. Another dramatic change occurred with the introduction of the bow and arrow. For example, in western North America, one theory (Grant et al. 1968) held that the bow and arrow was so much more efficient than spears at killing mountain sheep that the human population quickly expanded, decimated the sheep population, and then had to move due to the lack of game. Another example is the impact on human and bison populations that the introduction of the horse, and then the gun (not to mention the Euro-Americans), had on the Plains of North America.

Although native peoples had been cutting trees with stone axes in eastern North America for millennia, the rate of cutting was small enough that the forest could recover. However, metal axes are much more efficient, and their introduction in North America altered the rate at which trees could be felled. As a result, Euro-Americans, with a larger population and agricultural and land-use systems different from those of the Native Americans, deforested large tracts of North America (a process that is ongoing). The same thing happened in Europe (with stone axes) and is now occurring in Amazonia and Africa (with chain saws).

Complex technology has allowed for the colonization of areas that could not previously support human life, such as Antarctica, the ocean floor, and space. Technology provided some cultures an advantage (e.g., guns over spears) that they used to conquer others. In addition, technology now allows us to eliminate other species and to alter the environment on a global scale.

Storage

Storage means taking some resource and saving it for later use. All plants and animals store energy within their bodies, in the form of fat and carbohydrates, and some, such as hibernating bears, use stored fat as energy over extended periods. A number of species collect biomass from other species for later use, a practice called practical storage (Ingold 1987:202). An example of this form of storage is a squirrel collecting acorns for the winter. Humans do the same thing, storing acorns, corn, meat, or whatever. Humans can also "store" living resources, such as domesticated animals, for later use.

Human storage practices usually differ from those of other animals in two major ways: scale and technology. Most human groups, especially agriculturalists, utilize storage on a massive scale. For humans, technology plays an important role in storage as resources will often be processed by grinding (e.g., wheat), drying (e.g., jerky), smoking (e.g., many fish and other meats), and roasting or parching (e.g., many seeds—so they will not sprout during storage). Such treatment can make some resources storable for long periods of time, sometimes years. In addition, humans will often construct special facilities to store resources, such as granaries, cairns, silos, and warehouses. Another way to "store" resources is by controlling an area where resources are found and not allowing others access, a social storage (Ingold 1987:207).

Some have suggested (e.g., Testart 1982; Binford 1990) that food-gathering economies can be differentiated by their degree of storage, which is related to cultural complexity. Thus, an economy with relatively little storage would be fairly simple, while another with large-scale storage would be much more complex. The latter society therefore would develop a sedentary lifestyle and higher population density. The environment, specifically the length of growing season and the amount of precipitation, may also play a role in the development of storage behaviors leading to cultural complexity (Binford 1990).

TRADITIONAL KNOWLEDGE SYSTEMS

Throughout time, cultures have obtained and categorized knowledge about their environments. The vast majority of this knowledge was unwritten, passed verbally from generation to generation. The amount of knowledge is staggering, and many individuals in traditional cultures know a great deal about the environment because they work in it everyday. Others hold specialized knowledge relating to medicine, religion, or other fields. The practical applications of traditional knowledge and wisdom have attracted ecological scientists as well as anthropologists (see Ford and Martinez 2000). However, intellectual property rights remains an issue in regard to this knowledge (Posey 2001; also see the special issue of *Cultural Survival Quarterly*, vol. 24, no. 4, 2001).

Each culture has a system by which knowledge is obtained, that is, a science. As noted in chapter 1, Western science is strictly empirical and requires that a specific method be used in scientific inquiry. All other cultures also use empirical science, and all recognize objective realities. Relatively few use the strict formal method of Western science, but considerable experimentation occurs in traditional science, although written records are generally lacking. It is clear that traditional peoples have learned a great deal about the physical world. If one were to de-emphasize methods and concentrate on results, the contribution would be rightly viewed as staggering. A traditional doctor might not be able to explain the specific chemical properties of the substances used, but its use as a remedy makes it clear that the results are understood.

An example of such scientific understanding may be found with the Navajo (see Grady 1993; Schwarz 1995). In 1991, healthy Navajo people in the southwestern United States were dying from a mysterious disease. After considerable scientific effort, the culprit was found to be a hantavirus carried by deer mice (*Peromyscus maniculatus*). The virus spread to humans through exposure to mouse urine and saliva. An examination of traditional Navajo beliefs, much of which was conducted by Navajo trained in Western medicine, showed that they recognized mice as disease carriers and took special precautions to protect themselves from mouse urine and saliva. This knowledge appears to be centuries old, indicating that Navajo science had identified the vector (mice) and had developed precautions to avoid getting the disease. Navajo elders blamed the recent outbreak on the movement away from traditional beliefs.

Many cultures also include a nonempirical aspect in their science. Nonempirical data are those that are neither physical nor objective, cannot be reproduced, and are not subject to verification by experimentation. Generally, nonempirical knowledge is gained by specific individuals under "special" circumstances. The use of a hallucinogenic substance is a common method by which to gain such knowledge. In some cultures, such knowledge has an equal, and sometimes privileged, status compared with empirical knowledge. Much religious behavior and belief are nonempirical, being based on faith. Interestingly, despite the pervasiveness of empirical science in Western society, many Westerners include a great deal of nonempirical belief in their lives, as indicated by the popularity of psychic readings and astrology.

Few traditional cultures adhere to the rigorous requirements of Western science, and many Western scientists consider the scientific practices and knowledge of traditional cultures to be inferior (much as the other components of their cultures are viewed by Westerners) and so ignore their results. However, much of the traditional knowledge base has been utilized by Western science, and now a serious concern exists regarding the intellectual property rights of the traditional peoples who amassed the knowledge (Brush and Stabinsky 1996). In particular, considerable traditional medical lore has been appropriated by multinational drug companies without compensating the traditional culture in which it originated. Worldwide, legal authorities are debating ways to extend something like copyright protection to unwritten traditions.

As traditional cultures disappear, much of their knowledge is being lost. Thus, recording traditional knowledge is a critical concern in anthropology (e.g., Nazarea 1998, 1999) and in science in general. Even if traditional knowledge does not "fit" the data of Western science, it is still of great interest to anthropologists trying to understand how a culture operated. As cultural ecologists, we want to know how a culture interacts with its environment, and so it is necessary to understand how a culture knows its environment.

Ethnoscience

In the late nineteenth and early twentieth centuries, ethnographers recorded as much information about native cultures as they could, believing that such cultures were rapidly disappearing. Classifications of plants and animals very different from the Linnaean system were soon recorded. Worldview, cosmology, astronomical beliefs, hunting theories, agricultural lore, and other ecological knowledge were recorded in wonderful detail. Most cultures code much of their ecological knowledge as part of religion (which may be a major function of religion [Rappaport 1971, 1999]) or at least in religious discourse; the Bible, for example, has proven to be a source of considerable information regarding plants (Moldenke and Moldenke 1952; Zohary 1982) and animals (Hunn 1979) of its time. A new component of cultural ecology had been invented: the study of what local people know about their environment, how they classify that information, and how they use it; this approach is now called **ethnocscience** or **ethnoecology** (Gragson and Blount 1999:vii; Nazarea 1999).

The first study of ethnoscience after cultural ecology came into its own was *Native Astronomy in the Central Carolines* (Goodenough 1953; cf. 1964), an island group in Micronesia (the term "ethnoscience" had not been coined yet, so "native" was the term used). The people studied are famous for the long ocean voyages they undertake, navigating by stars and wave patterns. They are perhaps the most intrepid voyagers the world has ever known—setting out over thousands of miles of open water in small canoes with no compass or other technical aids. Other work followed, including studies of the knowledge of soils, forests, plants, and farming of the Hanunóo people in the Philippines (Conklin 1957) and the religious, medical, and nutritional knowledge of the Subanun people, also in the Philippines (Frake 1962).

These works opened the floodgates, and the prefix "ethno-" was attached to all manner of words, from ethnoichthyology (fish) to ethnoconchology (shells). People had been writing about ethnobiology and ethnobotany for decades, but those terms are now defined to mean the biological and botanical views of the groups in question rather than a redefinition of those views in Western terminology. Ethnoecological information is the classification and knowledge of the environment possessed by a culture (Toledo 1992:6).

It was realized that while the study of ethnoscience was important for understanding the relationships of traditional cultures with their environments, it was also important to translate the information into Western scientific terms and make it more accessible to agricultural scientists, development workers, and archaeologists. There was a need for in-depth botanical and zoological study of the plants and animals in question and for serious evaluation of their reputed properties (e.g., see Etkin 1988 on ethnomedicine).

However, there are several possible problems in this approach. An obvious problem is taking the knowledge of other people, encoded in their classificatory system, and translating it into our own system. It is possible to "lose something in the translation" and so miss the point. Another problem is that some researchers, in an effort to be less ethnocentric or more politically correct, place an overemphasis on a traditional view of the environment. In such cases, any questioning of the traditional view would be seen as "unsympathetic," ethnocentric, or racist, making it difficult to evaluate claims. We need to take more care in studying traditional knowledge systems.

One truth that emerges is that the Western world has a great deal to learn from other cultures. The wealth of knowledge encoded in even the most (to Western eyes) "simple" culture has proved to extend beyond anyone's wildest dreams. Countless new medicines, foods, and industrial crops continue to be developed from traditional plant and animal sources. Agricultural practices are being reformed in many areas on the basis of the special knowledge of local people. Thus, it is important to find ways to integrate traditional knowledge into Western empirical science (DeWalt 1994; Nazarea 1998), and vice versa.

Classification

All cultures construct a system to classify the elements in their environments, including plants, animals, soils, rocks and minerals, climate and weather, earth surfaces, and astronomical phenomena. Each of these various systems is based on a particular starting point. For example, one culture may classify animals based on morphology whereas another culture might classify them based on habitat. In the first system, a whale would be classified as being very different from a tuna (mammals and fish), while in the second, the two animals would be seen as similar (living in the ocean).

Any classification is, at least in part, a cultural construction of reality and biased by the particular worldview and experience of the culture. Some anthropologists argue that the resulting view of the world is so far from reality as to be totally arbitrary and idiosyncratic. However, the truth is that people are constantly testing their beliefs against reality and culturally encoding the right (or wrong but plausible) ones. Many individuals may not know that poison hemlock is deadly, but it is safe to say that all cultures that occupy areas infested with that plant have encoded its deadliness in their information banks.

Studies of cultural classification systems have dealt with the ways in which people cluster the things in the world, what is included with what, and what is overlooked. People tend to finely split domains that interest them and to split less salient domains much less carefully. Hong Kong fishermen who specialize in catching rockfish have names for all the dozens of species there, while fishermen who catch rockfish only incidentally would simply call them all "green," "red," or "spotted."

Early students of classification felt that anthropologists should study knowledge and that the proper task for cultural ecologists was to study traditional systems of ecological knowledge. This led to a debate among cultural ecologists, some of whom held that we should study practice and treat knowledge systems as secondary (Harris 1968). This controversy soon passed as almost everyone came to understand that knowledge and practice cannot be separated; they are interrelated and must be studied together. Today, no one seriously attempts to ignore the constant feedback between knowledge and practice—the way that our endless testing of the world is constantly altering our representations of it.

Knowledge of the Biotic Environment

The study of the classification, use, and knowledge of the biotic environment (past and present) is called **ethnobiology** (see Toledo 1992; also see Medin and Atran 1999). Ethnobiology is a major component of cultural ecology and includes studies of human diet, classificatory systems, ritual, and the knowledge and use of plants and animals. Data on such questions are obtained through a variety of means, including standard baseline ethnographic data, analysis of oral tradition, research by other specialists (e.g., by a botanist rather than an anthropologist), and archaeology.

Ethnobotany

The study of the native classification and use of plants is called **ethnobotany** (see Ford 1994, 2001; Minnis 2000). Plants are used for a great variety of purposes, including food, building materials, tools, textiles, and decoration, among others. If one is dependent on plants for these purposes, one's knowledge of plants would have to be considerable. A detailed understanding of plant locations, season of availability, general chemistry, durability, and biology is required for successful exploitation. The knowledge that traditional people have about botany is considerable, and many plants currently unknown to Western science (including many that would be very useful) are being used on a routine basis in other cultures (e.g., Balée 1994).

Horticulturalists and agriculturalists require a more detailed knowledge of certain specific (domesticated) plants in order to understand growth cycles, nutritional requirements, pest control, and fertility. Inadequate knowledge could result in poor timing and decisions, resulting in crop failure. In situations where a swidden agricultural system (see chapter 7) is practiced, considerable knowledge of the forest environment would be needed.

Ethnozoology

Ethnozoology is the study of "the knowledge of, use of, and significance of animals in indigenous and folk societies" (Overal 1990:127). Such knowledge includes the biology, seasonality, reproduction, edibility, and utilization of animals. Some species may be used for food, some for skins, some for bone, some for poison, and some for many other things. The more intimate one's knowledge about a variety of animals, the greater flexibility one has in using them.

When most people think of animals, they generally think only of vertebrates, often mammals. However, most animals actually are invertebrates, primarily insects. While Western folk classifications generally ignore insects, this is not the case with many cultures. For example, the Navajo of the American Southwest have a very sophisticated classification of some seven hundred insects (Wyman and Bailey 1964). The hunter-gatherers of the Kalahari Desert in southern Africa identify numerous invertebrates, including about seventy species of insects (Lee 1984:25).

Ethnomedicine

The study of the traditional knowledge used for medical purposes is called **ethnomedicine**. Some of this knowledge includes the setting of broken bones and the like, but it mostly involves **ethnopharmacology**, the classification and use of plants, animals, and other substances for medical purposes. The field also includes the knowledge and use of substances to alter one's reality (e.g., hallucinogenic drugs). In traditional societies, the people who specialize in medicine, commonly called shamans, are often the same people who specialize in religion.

Considerable research into ethnopharmacology has been under way for many years. Hundreds of thousands of years of experience with medicinal uses of plants is available if we look for it. A recent compendium of medicinal plants used by the Indians of North America contains more than two thousand species (Moerman 1986). Even our own culture has an ethnopharmacology separate from the commercial drug industry. We call this practice "home" or "folk" remedies, and it includes many cures not recognized or confirmed by science.

The production of drugs on a mass scale, as demanded by the consuming public, has had a profound effect on our ecology. Considerable resources are expended in drug production that could be spent on food production, and impacts on the land resulting from drug crop production are enormous. For example, the acreage devoted to cultivating poppies in some countries, such as Afghanistan and Colombia, for the production of cocaine and heroin is truly impressive. In addition, the acreage used to grow tobacco in the United States (not to mention the government subsidies) detracts from food production.

Knowledge of the Abiotic Environment

In addition to the biotic environment, all cultures obtain and classify information on the abiotic environment, including both terrestrial and nonterrestrial elements. Among the important components of the abiotic environment are geography, soils, meteorology, and astronomy.

All groups occupy and interact within landscapes. Western urbanites tend to view landscapes as Cartesian space, geographic localities and distances often devoid of real meaning. Other groups view landscapes from an experiential perspective, seeing geography as places where important beings interacted and where important events occurred rather than just as x,y coordinates on a map (see Brück and Goodman 1999:9). Such landscapes may have great ritual significance, such as in aboriginal Australia or even the contemporary Middle East.

Landscapes themselves can contain a variety of elements. Forman and Godron (1986:passim) identified three major landscape elements, patches, corridors, and surrounding matrix (fig. 4.1). Patches consist of small, specific ecozones separate from each other. Corridors are strips of territory, such as roads, rivers, and trails, that connect patches. The matrix is the remainder of the landscape, which surrounds the patches and corridors. Thus, one could have patches of forest connected by trails in a matrix of grassland or patches of farmland connected by roads in a matrix of forest. Culturally, it may be useful to view patches as places where people live and work: towns, farms, and the like; corridors as arteries for transportation; and the matrix as outlying areas, such as forests or preserves, which may be used, but less intensively than patches or corridors. The degree of management of these three landscape elements would vary greatly.

All cultures recognize and maintain specific geographic places in the landscape. Physical places exist that were the locations of important events, such as Mount Olympus, where ancient Greeks believed the gods lived; Gettysburg, the location of a major battle in the American Civil War; and Uluru (Ayers Rock) in central Australia, where many of the deities of the Pitjandjara tribe reside. Other important lo-

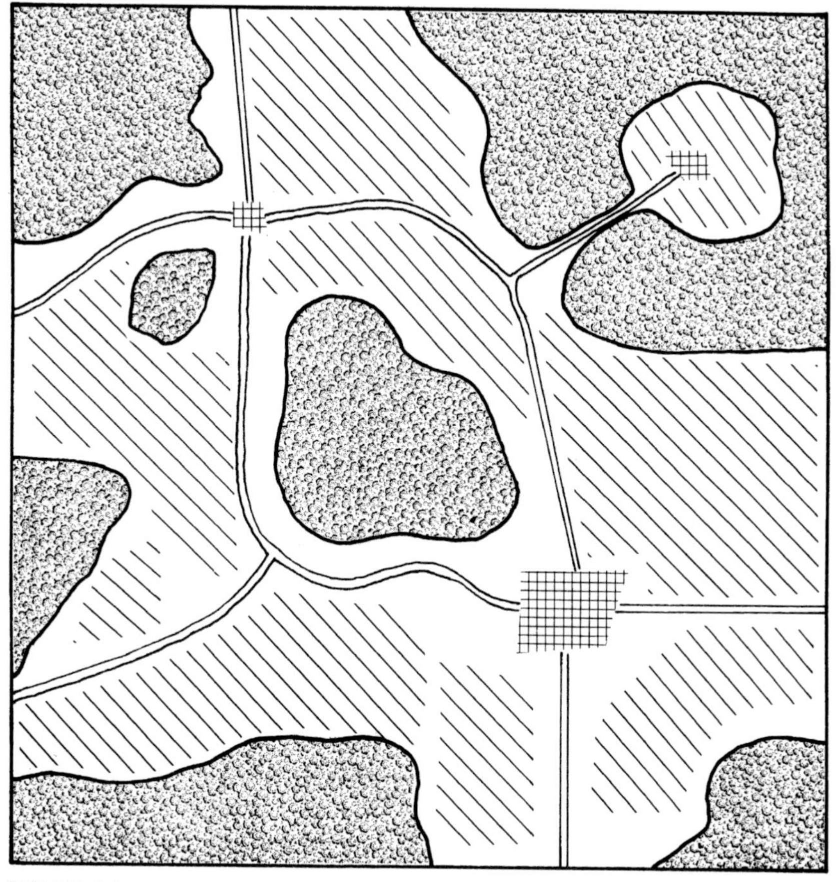

FIGURE 4.1
The major elements of a landscape: patches such as cities and towns (cross-hatched) and farmland (diagonal lines), corridors (double lines), and surrounding matrix (shaded).

calities include sources of power, sources of materials (such as major stone quarries or fishing grounds), and others, including combinations of any of the above.

A knowledge of soils and soil types can be critical, especially for agriculturalists. Soil type can also indicate the presence of other resources, such as plants, animals, water, or minerals, that can be important to a culture. For those people manufacturing ceramics, the soil (clay) itself may be a resource. The same is true in industrialized societies, which use a variety of soils for industrial purposes.

Knowing, predicting, and controlling the weather and climate are important in all societies, and most apparently in agricultural societies (see McIntosh et al. 2000). In deserts (e.g., western Australia, southern Africa, or northern Africa) the ability to track rain is essential to knowing where water can be obtained and where animals are likely to be. The proper timing of burns is also tied to weather patterns; if they are too early or too late, problems may ensue.

Agricultural societies (including our own) place an even greater emphasis on weather and its control. The effort expended by the Hopi to make rain is a good example. The Maya apparently conducted even more elaborate ceremonies and sacrificed people with the same goal in mind.

People have always watched the sky. All cultures have some classification and explanation of the changing sky, some quite sophisticated. Astronomical observations by traditional peoples have revealed regular patterns of celestial movement, such as of the sun and moon, and some not so regular occurrences, such as comets. Every culture, present and past, has some explanation for these patterns and occurrences, and many incorporate their explanations into belief systems regulating world renewal, agricultural cycles, and other important cultural phenomena. The study of the astronomy of past groups is called **archaeoastronomy**.

Art and Environment

For at least tens of thousands of years, art has been the way in which people have demonstrated how they feel about the environment. Yet cultural ecologists have shown an astonishing indifference to art. Art has portrayed not only animals and plants but also cosmological schemes, and it has symbolized religious philosophies. An ecological theory of art is needed, but as yet the nearest thing we have is the theorizing of Claude Levi-Strauss (see Strauss 1962).

Consider the art of the northern Northwest Coast Indians, specifically that of the Haida, Tlingit, and Tsimshian groups (see Jonaitis 1986; Anderson 1996; Wardwell 1996). The vast majority of images in this art is of animals, with most of the rest being of humans or supernatural beings. Animals are important to Northwest Coast peoples not only as food and clothing sources but also as social and cosmological symbols. The major descent groups are named for particular animals, including bears, wolves, eagles, and whales. Many other animals, as well as a few plants, and even meteorological items like clouds, are used as family crests.

The art of the northern Northwest Coast defines parallels between the realms of humans, animals, and plants. Animals and plants are seen as persons—not human persons, but persons nonetheless. They are talked to, prayed to, and taken as sources of magical and spiritual power. Humans and animals are often seen as reincarnations of one another. Animals can sometimes change into humans or can assume human form when they hide under mountains or seas. Thus, when animals are used to portray social facts, there is no thought of using the animals as "metaphors." The animals are actually part of society, and the art says much about human–animal relations.

Western Uses of Traditional Knowledge

Most people are unaware how many ideas and products from other cultures, "ethnoknowledge," have been adapted for use in our own culture (see Linden 1991). Some are useful in the short term, such as many current drugs, while others are important for the long term, such as learning to adapt a sustainable agriculture to the rain forest, thereby preserving the forest. The plants, animals, ideas, and technology that we Westerners have borrowed from Native Americans were discussed by Weatherford (1988, 1991).

A number of moral and ethical issues are raised by the "borrowing" and adaptation of traditional knowledge. One could argue that native knowledge is being stolen without compensation to the holder of the knowledge, a sort of copyright or patent infringement. Many see this as an extension of Western colonial practices and a further exploitation of traditional peoples. Ways to deal with this problem have been suggested (Laird 2002).

Medicine

Cultural ecologists have been part of a larger effort on the part of Western science to find plants and animals that contain useful compounds. This is the most simple, direct, and obvious use of field biology. Medicinal plants are one example (see Lewis and Elvin-Lewis 1977; Etkin 1994). Of the thousands of plants used in treating disease, the vast majority has been used in ethnomedicine, and some 50 percent of the medicines we use today originally came from other cultures. Aspirin, one of the most widely used drugs, came from the willow bark used by a number of traditional groups, such as those of the Northwest Coast of North America, long before it was synthesized and mass-produced for Western society.

Large numbers of plants (and animals) have medicinal value. Daniel Moerman (1986) compiled a database of 2,147 plant species used by North American native peoples; a large percentage of these is effective, and several have become sources of important medicines. A very large number of plants and animals are still unknown to Western science and may contain compounds of great value. That these species are being driven to extinction faster than we are able to find and test them (as depicted in the movie *Medicine Man* with Sean Connery) is a serious problem, and major efforts are under way to document and test such species.

Sometimes, a plant that was studied because of its value in ethnomedicine turned out to be useful in quite another way. One famous case is that of the common or Madagascar periwinkle (*Catharanthus roseus*). Its traditional uses did not include treatment for cancer, yet it became the source of vincristine and vinblastine, the drugs that revolutionized therapy for childhood leukemia. Variants of these two chemicals are now widely used and are credited with saving thousands of lives.

Traditional peoples discovered most of the common poisonous plants of the world, and in many cases they found the antidotes as well. In this way they saved Western investigators a great deal of time and danger. In such investigations, the role of the ecological anthropologist is usually limited to recording data on local uses of plants and collecting the plants in question. Specialists in botany and biochemistry must identify them and study their potential for medicine. It seems that much traditional medical knowledge evolved as people sought to understand the properties of plants (see Johns 1990).

Food and Fiber

New sources of food, fiber, and other valued goods also appear frequently, although they currently have less media glamour than medicines do. In the United States, about one-third of the calories we ingest come from a single plant—corn. We get corn calories from many sources: Cornmeal is fed to most of the domestic animals we eat, corn oil is in many foods, and we eat corn directly. Corn is a plant that was domesticated in the Americas by Native Americans. We have borrowed it from them and adapted it to our own use. The same is true of many other plants, including potatoes, tomatoes, and tobacco.

At present, the world is dependent for food and fiber on a very few crop species that have come to dominate the world's agricultural systems due to demand, their adaptability, and the relative ease of growing them. Wheat, rice, corn, and potatoes now contribute most food calories, and the New World variety of cotton produces most of the world's natural fiber. Most of the meat consumed comes from only a few animals: cows, pigs, sheep, and chickens. Such a narrow base is an unstable situation; a few virulent plant diseases could create havoc. For example, a potato disease spread through Europe between 1846 and 1848 and hit Ireland, heavily dependent on the potato, particularly hard. A famine ensued, hundreds of thousands of Irish starved to death, and millions more left (Salaman 1949). The population of Ireland has still not recovered. Therefore, we need to know as much as possible about alternative crops, foods, and agricultural methods, as well as the myriad uses of other plants and animals.

Anthropologists and economic botanists have described local traditional uses for literally thousands of species of plants and animals. These species provide a vast and proven reserve for the future. Not only do we now know what and where many of them are; we now know how to process them and utilize them. With more than two thousand different cultural groups in the world (there used to be some seven thousand), many thousands of useful wild species have been discovered. Even species requiring complicated processing have been brought into the realm. Both Koreans and native Californians prepare acorn meal, leaching out any bothersome tannic acids; in the future, some of the human food supply could well depend on having this technology already perfected. The deserts of the Southwest, at present little used, contain many valuable species of great potential importance (see Nabhan 1985, 1989).

Already, traditional varieties of native plants are especially useful in breeding. The inhabitants of Chiloe Island, Chile, have an incredible variety of cultivated potatoes. Some of these are so disease-resistant that the farmers of Chiloe have begun to sell seed stock to international potato-breeding companies. Mexican cultivated-bean varieties have served as sources for drought- and blight-resistant beans. Teosinte (wild or partly wild corn/maize) is so valuable for its disease-resistant genes that a major biosphere reserve has been established in western Mexico to serve as a center for corn biodiversity. Contemporary agriculture in California utilizes a number of species of native plants (e.g., dates and plums) in conjunction with imported varieties.

Agricultural Techniques

Moreover, traditional techniques of agriculture are proving to be valuable, or even necessary, in many areas (Marten 1986; Wilken 1987). Ethnographers and agronomists must work together to document the vast storehouse of knowledge (Atran 1993; Nazarea 1998, 1999). In Bali, traditional water management proved so much more successful than the innovations that the new techniques had to be abandoned (Lansing 1991). Even hunting and gathering peoples had (and many still have) extremely valuable land management techniques, many of which can still be used. The native peoples of North America, often disparagingly described in older literature as "primitive" and "backward," actually had extremely sophisticated land management skills, now of great interest to contemporary land use planners (cf. Blackburn and Anderson 1993; Hammett 1997).

Thought and Philosophy

Cultural ecologists come into their own when they study the ways people think about resources. Culturally encoded beliefs about plants, animals, soils, diseases, and even time and space have been studied in various parts of the world. Usually, local people know their resources well and have rich and complex knowledge relating to them. Everyone recognizes that sparrows are closer to finches than to ducks, and closer to ducks than to fish. Everyone recognizes that water is wet and rocks are hard. However, different groups construct very different cultural and psychological environments, and such differences are of great interest. Obviously, each group may know important facts not known to another. More interesting are the different beliefs that appear "nonfactual" to the outside observer. Why do many North American native peoples believe that plants, animals, and even mountains and rocks are persons—conscious in some sense and often able to communicate with humans and affect human destinies? Why do so many groups around the globe share a belief in a sacred mountain at the center of the world? Why is disease so often ascribed to witchcraft? Why do neighboring peoples often classify plants in quite different ways? The answers to these questions directly and immediately affect us all.

In the process of dealing with these and other questions, anthropologists have greatly informed debates about human psychology and cognition. The first and most obvious conclusion, reached before the end of the nineteenth century, was that "primitive" peoples are not ignorant, nor are their lives dominated by magic; they invariably have a deep, rich, and more or less systematic knowledge of their environments. As time has gone by, we have learned how traditional peoples can maintain a religious view of the world and combine it with a hardheaded, factual view. By contrast, Western society has developed its characteristically detached and "disenchanted" view of "nature." As a result, many are now saying that the Western world has lost something.

If, indeed, religion functions to maintain knowledge and ecological adjustments in so much of the world, does the Western world need something functionally similar? Some students of environmental ethics have argued that it does (Callicott and Ames 1989; Berkes 1999). At present, the degree to which non-Western peoples managed and conserved their resources is highly controversial. Some say they failed and that Western science is our only weapon in the fight to preserve a livable environment (Lewis 1992). If so, we are in trouble, as experience shows that science alone is not persuasive.

On the other hand, if some traditional people succeeded, to some extent, in "selling" conservation through ethics or moral teachings, then there is more hope for us. We may not find their religions persuasive or their techniques infallible, but we may be able to build on their experiences to devise more powerful ethical teachings. It seems likely that we will survive only through combining Western science with a new ethical, moral, and religious attitude toward the environment (White 1997). Certainly, the beauty and poetry of traditional peoples' views of the environment have proved extremely moving and powerful to many contemporary writers, including anthropologist-turned-poet Gary Snyder (e.g., Snyder 1969).

HUMAN CONTROL OF THE ENVIRONMENT

All cultures employ practices designed to exert at least some level of control over their resources and environment. These activities include management in the form of conservation, exploitation, and manipulation. Management can occur at several scales, from individual plants to entire landscapes, and for a variety of maximizing interests, short term or long term. The access to, and exploitation of, resources are determined by some factor, such as kinship or wealth, with some resources being individually owned and others being communally owned (or "not owned," as the case may be).

Domestication and Control

People impact and alter their environment to varying degrees, particularly plant and animal populations (e.g., Simms 1992; Preston 1997; Martin and Szuter 1999; Doolittle 2000; Grayson 2001; Kay and Simmons 2002). The scale of this impact depends on the scale of human population and the technology possessed by the group. Most groups feel that they have some control over their environment, perhaps through their ability to influence the supernatural, or their ability to change the course of a river, or perhaps through their air conditioning. In fact, many groups consider their environment to be controlled, or **domesticated,** to at least some degree, although few would claim they are in total control.

The typical definition of domestication is one related to agriculture. In that context, a domesticated species is one over which humans have developed some intentional and detectable genetic control. While it is true that genetic alteration is an effective method of control, plants, animals, or ecosystems can also be controlled in other ways (Blackburn and Anderson 1993).

A broader definition of domestication could mean control in a more general sense. All cultures have methods to exercise some control over, and so domesticate, their environment. (Whether these means are effective or not is another matter.) Environments are controlled by a variety of techniques, including the manipulation of landscapes and management of individual resources. If a culture "controls" its environment, that environment could be considered to be "domesticated."

However, Western groups moving into a region inhabited by traditional peoples generally see much of the land as "pristine wilderness," landscapes somehow untouched and unaltered by humans (Denevan 1992). This view is almost always wrong. For example, when European farmers colonized the New World, they saw what they thought to be a wild and untamed landscape. In reality, however, the native populations had long been practicing a variety of management techniques and had altered the landscapes by their use. The landscape was not wild at all but was a domesticated and highly productive environment. The Europeans interpreted the matrix (following Forman and Godron 1986; also see Vale 1982) surrounding the patches and corridors intensively used by the Indians as being unused (and so available for colonization) rather than as differently used.

Humans consciously alter and manipulate their environments, generally to achieve a desired result. Such changes brought about by humans are called **anthropogenic** (anthro = human, genic = produced). The scale of change depends on a variety of factors, including the goal of the alteration. Technology is also a factor, as someone with a bulldozer can impact the environment to a greater degree than someone with a stone ax. Large-scale alteration of the environment is called **environmental manipulation**, and small-scale management of individual resources is called **resource management** (see table 4.1). Alterations of the environment include those that affect both the abiotic and the biotic components; that is, both living and nonliving aspects of the environment are altered and manipulated.

Environmental Manipulation

Large-scale change of the environment by humans is called environmental manipulation, the alteration of entire landscapes. Manipulation can be advantageous, or at least perceived to be advantageous, to humans in the short term or the long term. The clearing of the rain forest in Brazil (and other places) has a very short-term economic benefit to the farmer or rancher but a very long-term negative effect on the environment and human welfare in general. Some alteration of the environment is undirected, but much manipulation is planned and conducted for specific purposes. Manipulation tactics fall into two general categories: active and passive.

Active Environmental Manipulation

Active environmental manipulation is the purposeful, physical alteration of groups of species and of ecosystems on a large scale. This is the active manipulation of landscapes rather than of individual species.

Table 4.1. Some Methods of Environmental Control

Method	*General Principle*	*Scale*	*Examples*
		Environmental Manipulation	
Active	actual hands-on, purposeful modification of landscapes to achieve a goal	large, as in landscapes	burning tracts of land to clear brush and encourage new growth clearing large tracts of land for agricultural fields altering natural water systems for irrigation
Passive	ritual activities to effect control and change	large, as in landscapes	ceremonies for world renewal stewardship of areas to maintain their power
		Resource Management	
Active (light to moderate)	actual hands-on, purposeful alteration of a resource to achieve a result	small, generally individual resources (e.g., a species or a water source)	pruning specific plants to enhance production of some product limiting access to a spring by members of other groups so as to preserve the water
Active (intensive)	actual hands-on, purposeful alteration of a resource to achieve a result	small, but intense focus on an individual species, often to the point of genetic control	management of reindeer herds agricultural domestication of a species, such as corn or cattle, where the movements, reproduction, and lives of the individuals are controlled
Passive	ritual activities to effect control and change	small, focus on specific need	conducting fertility rituals for specific species giving thanks to a species (e.g., deer) for allowing themselves to be hunted and killed

Burning. Burning may be the most widely employed method of environmental manipulation in human prehistory and history (see Lewis 1973, 1982; Flannery 1995; Pyne 1995, 1998; Boyd 1999; Williams 2002). Fires clear away dry, dead understory and grasses, reduce competition, and eliminate thorns and animals dangerous to humans. The fires are usually timed and managed so that they do minimal damage to forests and other long-lived resources. Most of the world's plant cover has been burned repeatedly by hunter-gatherers and agriculturalists. This is a point to remember in contemporary management schemes; fire control is not always the best thing in areas where humans—to say nothing of lightning—have been causing fires for millennia.

Many plants are well adapted to fire, and some even require burning for seed germination. The ash contributes fertilizer and makes room for new growth. Natural fires periodically consume the fuel on the ground, never permitting it to build up to such a degree that any fire would be catastrophic. A recently burned area can be quite productive in its abundance of new growth.

Hunter-gatherers knew this and would often set fire to areas with the specific purposes of eliminating dead materials and encouraging new growth. In this way, a good stand of seed plants, after harvesting, would be encouraged to produce another good stand the following year. This was common practice among the California Indians (Lewis 1973, 1982; Timbrook et al. 1982; Blackburn and Anderson 1993) and was widely employed to produce materials suitable for basketry (Anderson 1999). It would also kill "weeds" and help preserve the biological "purity" of the stand.

In addition, the resultant new growth would attract certain animals to eat the new foliage. This source of food often augmented the numbers of animals available for hunting, thus raising the rate of successful hunting. In optimal foraging terms, encounter rates would be increased, resulting in a lower caloric expenditure for procurement, changing the ranking of that particular animal!

Burning also may be related to religion, such as world renewal activities. Among the *Gagadju* in north-central Australia (as seen in the documentary film *Twilight of Dreamtime*), the burning of the river plain signals the renewal of an annual cycle of life. Without the ceremonial intervention by "proper" people, as caretakers of the land, this cycle would be broken, and life would cease to exist.

Cultivation. Another good example of environmental manipulation is that of cultivation and agriculture. To plant a crop, land (often substantial amounts) must be cleared of its natural ecosystem so that an artificial ecosystem can be installed (planted). This practice results in the loss of biodiversity through the replacement of the many wild species with a few domesticated ones, an often considerable alteration of the ground surface through mechanisms such as plowing, and a disruption of adjacent ecosystems.

Constructed Landscapes. All cultures strive to regulate their landscapes, part of the abiotic environment, to some degree. Some modify landscapes in a relatively minor way while others actually construct new landscapes to suit their needs. A classic example of the latter approach is feng shui, the Chinese science of proper arrangement of elements in a landscape to ensure harmony (Burger 1993). Groups will also move materials great distances around the landscape for the construction of specific facilities. The movement of large stones used in the construction of Stonehenge in England is an example.

The most common example of constructed landscapes are those associated with agriculture, where a whole series of alterations are made to fashion the land to the use of the farmer. The construction of fields could involve just the simple clearing of forest but may also involve creating and/or leveling mounded earth, removing or altering geological features such as rock outcroppings, constructing walls, digging irrigation or drainage ditches, and constructing and using terraced field systems, such as those found in China, the Philippines, and Peru. In some cases, terrace systems can cover hundreds of square miles, as in the Colca Valley of Peru (e.g., Guillet 1987).

Irrigation systems are common elements in constructed landscapes. Such systems can be very small to quite extensive, with many miles (even thousands of miles) of canals, ditches, and other facilities being constructed. Dams such as Aswan in Egypt and Three Gorges in China can flood large areas, creating lakes and swamps and eliminating portions of rivers, streams, and valleys. In addition to the irrigation systems, the water itself can transform landscapes from arid regions to lush agricultural areas.

Other large-scale constructions can transform landscapes. Many groups have built large ritual centers, altering the local terrain and even influencing much larger regions. The construction of cities, especially the giant ones of the twentieth century, has drastically altered landscapes with housing, transportation, water, and waste management systems.

Passive Environmental Manipulation

Ritual activities designed to maintain the environment in its "domesticated" state are considered **passive environmental manipulation**. Such activities include world renewal ceremonies, such as fertility rituals and even human sacrifice. The Mexica conducted both of these practices to ensure that the sun would continue to rise.

One way to ritually control and maintain the environment is **stewardship**. In Australia, the land and its resources were formed during Dreamtime. Certain places are very special and contain power, and certain people are responsible for the maintenance of those places. The places are not owned; people are caretakers rather than owners. Failure to properly maintain these places could result in catastrophe. Such responsibilities are even set out in the Bible, where Adam was charged as the steward of the land (Gen. 2:15) and the consequences of bad land management were detailed (Isa. 34:11).

Resource Management

The management of specific resources is called resource management. Such activities are generally on a smaller scale than environmental manipulation, but there can be some overlap. For example, burning may be conducted to manage a specific resource, but it could affect a larger system. The pruning of tobacco, as practiced by a number of groups, affects a small area, notably the plant itself. Like environmental manipulation, the management of resources can be either active or passive.

Active Resource Management

Some specific resources receive **active resource management** or control to ensure productivity; that is, some physical action is taken to control the productivity of the species. (Of course, some resources are managed for their beauty rather than for food or materials.) Most species are managed such that they remain "wild," but some are managed so intensively that they may eventually become domesticated (see chapter 6).

Plants can be actively managed using a variety of techniques. Burning, often considered to be environmental manipulation, could be considered resource management if conducted as a management technique for one or two plants. Pruning is also a management technique. For example, tobacco, important to many Native American groups for ceremonial purposes, was monitored and pruned so that it would produce larger leaves. Storage of harvests, such as pine nuts, grass seeds, acorns, and the like, could also be considered a form of resource management.

In Nevada and elsewhere, Indians made bows from staves of juniper wood (see Wilke 1988). These staves were taken from special juniper trees that had unusually straight trunks free of knots and twists. Trees with such trunks were purposefully shaped, not found and then used. Planning far ahead, the Indians would select, prune, manage, and nurture trees to grow straight and knot-free trunks that could provide bow staves, a process that could take decades. Once the tree was ready, a bow stave would be cut in outline on the tree, left to dry, and removed (already cured) a few years later. It would take many years for the tree to fill in the resulting gap, and it would be monitored and managed to make sure that no branches grew in the stave area. Another bow stave could then be removed, and some trees had many staves removed. These trees were very valuable resources that were constantly monitored, maintained, and reused over hundreds of years.

Animals were also actively managed. Everyone knew that females gave birth and that it was important to maintain some level of females in animal populations. In some cases, there were rules against killing female or young animals, with adult males being preferred prey. There are such rules in the contemporary United States for hunting deer; one must have a license and a tag, and one cannot kill the females or young.

Another way to actively manage a resource, such as a water hole, is to limit access to it. Sometimes access is limited only by formality. For example, among the San of the Kalahari Desert in southern Africa, water holes are owned by bands and one must have permission to use them (Lee 1984). While permission is always granted for the asking, it is still a way in which to exercise some power over the resource. In other cases, such as a water hole being defended by force, access is more seriously limited. Such control could influence the adaptation of other groups by preventing travel through an area.

Passive Resource Management

Resources can also be managed through passive means, those that do not involve direct physical contact with the resource, and all groups employ some sort of **passive resource management**. Several approaches to passive resource management are discussed below.

Ritual Management. Rituals, as part of religious activities, can function as a form of resource management in a number of ways. All groups believe in some sort of supernatural power that has control over major elements of the environment, such as gods that control rain or the movement of the sun. Rituals, including prayer, ceremonies, and even some art, are used to influence these deities so that the sun will rise and the rain will come. These rituals, then, serve to manage the environment.

Other practices can *function* as resource management, even if that was not their original intent. One example is the ritual ownership of an important resource (or sometimes even an area or locality) such as a spring, a common custom in many groups. Such ownership may be passed through generations and would encourage stewardship. Ownership may also function as a conservation technique: a way to discourage the use of the resource by others. It is also a way in which to control resources, wealth, and power, even if ecological management is not a goal.

In many groups, individuals or social entities, such as families or clans, may have a special relationship with some entity in the natural world. These entities are often called totems. In some cultures, there are rules about their utilization by persons or groups claiming the relationship. For example, let us say that the totem of one person was deer, so that person would not normally consume deer. The totem of another person might be elk, and that person would not eat elk. This practice would serve as a passive resource management technique to reduce hunting pressure on both deer and elk because fewer people would be eating them.

Another example of passive management of game was put forward by Moore (1957:71). To determine where to go to hunt caribou, the Naskapi Indians of eastern Canada used a caribou scapula to "divine" the location of the herds, rather than using past experience and knowledge. Moore suggested that the divining would have sent the hunters off in different directions and so may have served to randomize the selection of hunting localities. This in turn would lower hunting success and so reduce the pressure on game.

Certain resources, such as meat or milk, may be avoided at specific times. In some cultures, menstruating females are not allowed to eat meat. Other resources might be consumed only during ritual activities. This type of taboo would function to lessen the demand for certain resources and could be viewed as a management technique.

Other ritual behavior might include resource renewal ceremonies. If an animal is killed for food, it may be necessary to ritually thank its spirit. Failure to do so may result in the animal's becoming angry and refusing to allow itself to be killed in the future. If this were to happen, the people would starve. In Western science, such behavior would not be considered resource management, as a connection would be denied. However, in the science of the culture practicing the custom, it may be a very real and important technique of ensuring the continued availability of the resource. This same type of behavior could be extended to general world renewal rituals.

Knowledge Retention. Knowledge itself is a resource, managed both actively (through learning and experience) and passively (encoded into ritual). Most people around the world possess a very considerable knowledge of their environment and use those skills in their everyday activities. These skills center on the use of resources currently in the environment. If the environment occupied by a certain group contained many different ecozones, the number of skills and the amount of knowledge needed to exploit those ecozones would be huge. By understanding the amount and types of knowledge a group has, one could learn a great deal about how that group adapted.

The retention of knowledge regarding the exploitation of resources that are no longer actively used may be a safeguard against bad times. People might retain knowledge of the use of certain resources not needed in good times so that they could be exploited when and if necessary. One example of such retained knowledge is the "survival techniques" taught to contemporary armed forces. Such knowledge might not be present in everyday practice but might be retained in oral tradition. The recent popularity of relearning the "traditional" practices of eighteenth- and nineteenth-century America is a related phenomenon.

DECISION MAKING

Another way to analyze environmental manipulation and resource management techniques is through studies of decision making. All groups and all individuals make decisions regarding their actions in any given situation. As all systems are dynamic, constant adjustment is necessary, usually minor, but sometimes major. In all cases, however, decisions regarding what to do and when to do it are required. We somehow assume that the process of decision making is rational and well thought out.

However, bad decisions are made, particularly (it seems) by humans. People make poor decisions for a number of reasons, such as having incomplete information, giving too much weight to emotion, or just plain erring. Some poor decisions are annoying while others, such as invading Russia, can be catastrophic (just ask the French or the Germans). If a culture makes too many bad decisions, it could end up in evolutionary failure (extinction). A decision that seems good in the short term may be very bad in the long term (and vice versa), and it is important that a balance be reached. The theory of decision making is, in itself, a field of study within anthropology, and no effort is made here to explore it in any detail (see Mithen [1990] for a discussion of hunter-gatherer decision making). However, there are several important points to consider.

Information

Decisions are based largely on information (new or old). However, no one can possibly take account of all the things in one's immediate environment. Economic theory assumes that humans act from "perfect information," but of course this is never true. Perhaps the farthest galaxies and the smallest dust particles influence us somehow, but we normally overlook them. Even much more important things, such as the long- and short-term consequences of our actions, tend to get ignored.

Anthropologists, therefore, must consider decision making with an assumption of imperfect information (following the ideas of Simon [1960]). This approach, in turn, involves studying our simplifying premises. Humans simplify, generalize, and assume that other people are more like us than they really are. In dealing with people, we routinely believe that they will do about as we would, even if their circumstances are different (Piatelli-Palmarini 1994). Two major realms of study have emerged from this more general concern with "information processing": explanation (inference, attribution) and classification. The former has been largely the domain of social psychologists (see review by Fiske and Taylor 1991), while the latter is the domain of anthropologists (see above).

To make the optimal decision, decision makers require complete and accurate information. However, these conditions are rarely, if ever, met, and a truly optimal decision cannot be made, except by accident. Poor information will likely result in deleterious decisions. It is unrealistic to believe that humans will always have perfect information, and bad decisions must be expected, even if the information is good. Information is communicated in several ways, including speech, gesture, smoke signals, and material culture (Mithen 1990:70–71). It can be used in the immediate future or stored for future use through oral tradition and the teaching of the young (about past experience).

The knowledge base of traditional peoples is usually very impressive and often greater than that of professional biologists working in the same area. This is particularly true of remote, poorly studied areas. For example, the Aché of Paraguay hunt as a main staple food flat-headed peccaries (*Platygonus* sp.), an animal zoologists had believed to be long extinct. But even well-known areas have surprises. The Sahaptin Indians of Washington State have recognized, for thousands of years, certain root-crop species that biologists only recently realized were there (Hunn 1991).

A hunter or gatherer must know not only how to recognize a species but how to use it. Consider a wild seed crop. The gatherer must know where it grows, when it sets seed, and whether it is worth seeking out in a good or bad year. For some years, one of us (ENA) monitored a patch of chia (*Salvia columbariae*), an important wild food in California. The yield of seeds from this patch varied from year to year. In most years, it was not worth the time to gather. Only with moderate, well-distributed rain did it provide much seed. A gatherer would have to decide whether it was worth the effort to utilize the patch.

Traditional peoples possess a great amount of knowledge about their environment, but the sheer quantity of information presents a major problem in systematizing, storing, and retrieving information. The solution lies partly in religion. Religion provides powerful emotional and social involvement. Knowledge is given the absolute value of a sacred text. Moreover, it is encoded in oral tradition (stories or myths)—exciting stories, often well laced with sex and violence! In areas as disparate as aboriginal Australia (Gould 1969) and southeastern California (Laird 1976, 1984), anthropologists have found that children in deserts learn the location of water sources through stories. The hero travels from water hole to water hole. At each one, dramatic adventures occur. The story is exciting and united by its coherent and orderly development. Therefore, it is much more easily remembered than a simple list of water holes.

Moreover, moral rules are learned along with basic survival knowledge. Most hunting and gathering peoples have rules about sharing resources and about taking only what is needed so as to leave some for others. These rules—along with other rules the group may have—figure prominently in the stories, and the hero or villain always suffers by failing to observe them. This interweaving provides a united worldview: Religion, ethics, and practical knowledge are not separated. Indeed, no hunting-gathering group seems to have a concept of "religion" as separate from "secular." Some have claimed that for "primitive" people, everything is sacred. The truth is that for these people, everything is both sacred and secular. Everything is religious, but almost everything is practical.

Before information can be used to make a decision, it must first be obtained, then classified, and then shared with others. Many hunter-gatherer groups are constantly out in the landscape for various reasons. While moving around, they obtain information on a variety of things, including animal spoor (tracks, dung, sounds, and/or smells), weather information, environmental parameters (e.g., the condition of one species, which may reflect the condition of another), and the activities and movements of other people, and they store it all in memory. This knowledge is used and/or shared, even with other groups, as needed. To people who make their living on such resources and who depend on good decisions being made, resource information is a very high priority.

One specific method of information procurement is **resource monitoring**, a technique that virtually everyone uses. People pay attention to the status of what is around them, the condition of resources, the presence and absence of things, and opportunities. For example, when driving anywhere, you will notice the price of gas at various stations and remember where the best bargain is. You will pay attention to what new restaurants are being built, noting for future reference where they are and what kinds of food they serve. All people monitor their environment for the resources they use.

In many cases, people will gather information on specific resources. When it comes to deciding which pinyon stand to go to in the fall, several years of data from monitoring the various stands may be utilized. In addition to the data on the actual resource, one could consider information on travel time, social opportunities, and the like. The decision about which pinyon stand to visit would be based on all these factors.

Information is gathered constantly, either as an adjunct to what one is doing or by conscious effort. Even when traveling to a specific place, hunter-gatherers will rarely travel in a straight line. By meandering, they can gather and process information regarding a variety of resources, such as the condition of plants and the movement of animals.

A route a hunter-gatherer might take from one place to another is shown in figure 4.2. The actual direct route would miss many of the areas that contain resources used by the group, so the traveler would meander across the landscape to check things out along the way. He or she would first visit Resource Patch A to determine animal density and inspect the progress of seed maturation to determine whether the seeds there would be worth collecting when they finally become ripe in the future. The person would then begin to travel to Resource Patch B, gathering information on the general condition of the land along the way to assess whether rabbits might be worth hunting there. Once at Resource Patch B, our hunter-gatherer would examine the condition of that group of resources, again to assess whether they should be gathered in the future. A visit to the water hole would be made to make sure it was productive, and if it were clogged or dirty, some maintenance might be performed. Next, a game trail might be examined to check on the animals' numbers, frequency of movement, and direction so that decisions on when and where to hunt can be made. A visit to Resource Patch C would then be made to assess the condition of the pinyon crop, the number of cones per tree, the number of trees with cones, and the probable timing of cone opening. The traveler would finally arrive at the destination, having taken much time to get there, but bringing a wealth of useful information. Such resource monitoring is constant for each of the members of the group, and the quantity of information thus obtained is staggering.

All of the data obtained through perception (e.g., vision and hearing) are subject to cognitive interpretation and filtration in the mind of the observer/player.

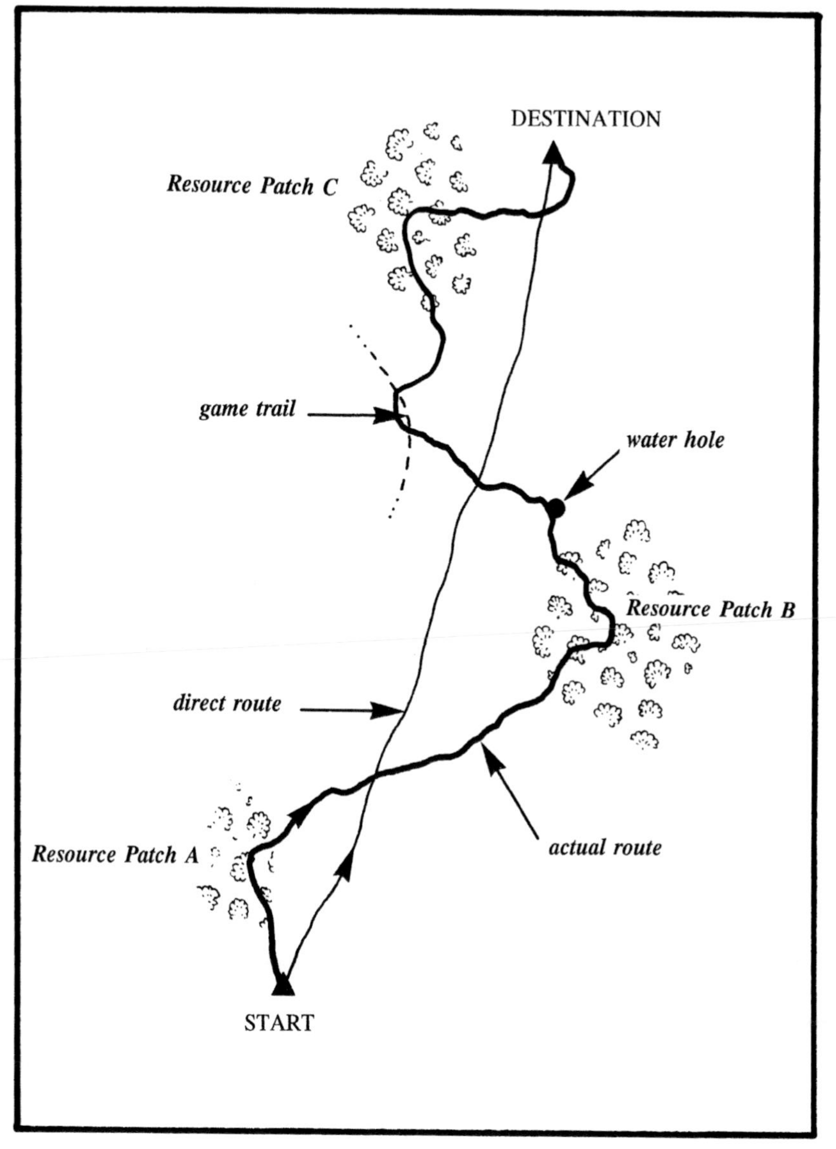

FIGURE 4.2
A generalized hunter-gatherer travel route to gather information and monitor resources.

Cognition, the way a culture views things, is an integral part of any information system and must be taken into consideration in any analysis. Classification systems are part of the cognitive doctrine of all cultures and are important analytical elements (see above).

Scheduling

Scheduling, the planning of when to move and which resources to exploit, also requires good information. To obtain the seeds of plant X, one must have knowledge of the life cycle of the plant, information on where it is, its condition, and when and where the next resource is to be used. Sometimes a schedule might be very tight and leave little room for error; other times, timing may not be so critical. If two resources are available at the same time, a decision on which one to use may have to be made. If schedules are not adhered to, serious problems could result. If it is too late to get to resource Y, a group could experience considerable stress (from its system being out of equilibrium), and drastic measures might be required.

A CONCLUDING THOUGHT ON MANAGEMENT

Cultural ecologists have shown a broad concern with various strategies for managing the environment. On the one hand, this broad view involves looking at soils, fisheries, wildlife, and pests, as well as simple food gathering and farming. On the other, it involves expanding from classical concerns of agriculture and agricultural economics to look at the social framework of food procurement: social systems and their religious and ideological representations, individual decision making and its psychological roots, and how these interact.

Broadly, the central question of resource management is: How can resources be managed to provide optimum benefit over time? This question involves issues of allocation and, above all, the trade-offs between long-term, widespread interests and short-term, narrow ones. Every culture struggles with how to manage resources, and some cultures and individuals within them make better decisions than others.

CHAPTER SUMMARY

While the study of human biological ecology can tell us much about human adaptation, the cultural aspect of adaptation must be examined through cultural ecology. The flexibility of human adaptive responses requires the study of human learning, sociability, intelligence, and basic needs to help understand how people and cultures deal with their everyday problems. Cultural ecology is about explaining how and why cultures adapt in one way and not another.

In essence, culture itself is an adaptive mechanism. Cultures contain a number of elements, such as social and political systems, settlement patterns, and technology and storage, that are adaptive in their form and evolve as environments change.

Knowledge is also part of these adaptive cultural elements. Each culture practices science in some form and each retains and classifies knowledge of the various abiotic and biotic components of its environment. These knowledge systems, collectively called ethnoscience, contain the accumulated knowledge of thousands of cultures over hundreds of thousands of years and constitute a resource of immense value to humanity. For example, much of the knowledge of drugs in contemporary Western medicine was derived from traditional pharmacologies.

Each culture expends some effort to control, or domesticate, its environment and the resources within it. Environmental manipulation entails the regulation of large-scale entities, such as landscapes, while the control of individual resources is called resource management. Both types of practices are accomplished by both active (e.g., physical alteration) and passive (e.g., religious) methods.

Decisions about what resources to use and how to manage them are made through a process that includes using the knowledge of the environment. This knowledge is obtained, classified, and stored in oral tradition and religion for future use. In addition, up-to-date information is obtained through resource monitoring. Once obtained, information has to be interpreted, a task completed by each group using its own cognitive system.

KEY TERMS

active environmental manipulation
active resource management
anthropogenic
archaeoastronomy
cognition
domestication
environmental manipulation
ethnobiology
ethnobotany
ethnoecology
ethnomedicine
ethnopharmacology
ethnoscience
ethnozoology
feng-shui

passive environmental manipulation
passive resource management
resource management
resource monitoring
stewardship
storage
subsistence

5

Hunting and Gathering

Anthropologists often classify cultures based primarily on general subsistence strategy, generally the most visible or important aspect of how they obtain their living, a definition primarily based on ecology rather than on social criteria (although ecology and social factors are related). Those cultures that make their *primary* living from obtaining and using so-called wild foods are classified as **hunters and gatherers** (often shortened to hunter-gatherers). In addition, hunter-gatherers are characterized by "the absence of direct [active] . . . control over the reproduction of exploited species" (Panter-Brick et al. 2001a:2). Until about ten thousand years ago, all the people in the world practiced a hunting and gathering economy; for most of human history, we were hunters and gatherers, and much of our culture and biology are still adapted to that basic lifestyle (e.g., Barkow et al. 1992; Eaton and Eaton 1999).

The term "hunters and gatherers" has become an anthropological cliché. It has been used so much that some anthropologists have sought to find alternatives, such as foragers, collectors, or preagricultural peoples. Some of this classificatory manipulation is intended to account for political and social complexity, that is, to distinguish small, relatively simple hunter-gatherer groups from large, complex ones. While it is true that an environment may limit the carrying capacity of some hunter-gatherers and so necessitate small groups, it is not necessarily true that small groups must have simple social and/or political systems. The term **forager** is now commonly used as a substitute for hunter-gatherer to avoid "privileging the hunting side of hunter-gatherer" (Kelly 1995:xiv), and the phrase "gatherer-hunters" (e.g., Bird-David 1990) has been used for the same reason.

The use of the single classificatory term "hunter-gatherer" presupposes that the technological and economic similarities between all hunter-gatherers unify their cultures to a sufficient degree that they can profitably be compared to each other. This is a significant assumption (see Testart 1988) because other cultural institutions, such as kinship systems, can vary widely, as seen by a comparison of the Australian Aborigines and the Inuit (such variation also occurs within other such classifications). Nevertheless, in this book we classify groups that make their principal living from wild resources as hunter-gatherers, irrespective of resource particulars (seeds or fish, large game or small) or sociopolitical details. We do not believe that we can simply substitute the terms "foraging" and/or "collecting" for hunting and gathering as we view these categories as subsets of hunting and gathering (as explained below).

Hunter-gatherers exploit wild resources rather than domesticated ones as their primary source of food and so have no direct control over the genetics of the species they exploit. As discussed above, resources often have two dimensions: time (when they are available) and space (where they are located in the landscape). Like all groups, hunter-gatherers must solve the problem of getting resources and people together. Herein lies the major complexity in the study of hunter-gatherers. Hunter-gatherers are likely to be tethered to certain resources, such as springs, due to their technology and transport capability.

Hunter-gatherers manifest a vast range of structures, forms, and adaptations (see Bettinger 1980, 1987, 1991; Price and Brown 1985; Thomas 1986; Cashdan 1990; Burch and Ellanna 1994a; Kelly 1995; Lee and Daly 1999; Binford 2001; Panter-Brick et al. 2001b). Their villages and towns may reach populations in the thousands, as they did on the Northwest Coast and in coastal California just prior to contact with Europeans, and probably in the European Mesolithic between twelve thousand and seven thousand years ago. The social networks of hunter-gatherers may extend over an entire continent, or at least a good portion of one, as may be seen in Australia. Their religions, literature, art, and music can be as complex as those in some state-level societies. Of more direct relevance to the present book, their ecological adjustments are extremely varied and fine-tuned. The nonagricultural peoples of the world most often depend on the same types of staples that agriculturalists do—seeds and root crops, greens, fruit, meat, and fish—and often in the same basic proportions: mostly seeds and roots, with meat significant and the other things less so.

As noted by Bruce Winterhalder (2001:13), four basic generalizations can be made about hunter-gatherers. First, they tend to "underproduce": to not exploit all that is exploitable and to have relatively few material possessions. Second, they routinely share food. Third, they tend to be egalitarian. Finally, they commonly employ a division of labor whereby men do much of the hunting and women do much of the gathering.

HUNTER-GATHERER CLASSIFICATION

Most hunter-gatherer cultures are classified based on a description of the basic lifestyle or settlement/subsistence system practiced by the group. A popular view of hunter-gatherers, even among some anthropologists, is that they are nomadic, wandering about the landscape aimlessly in an endless search for food. This is simply not true. No hunter-gatherers "wander aimlessly" as a normal part of their economic pattern. Individuals may wander, but cultural groups do not. Groups entering a region for the first time, such as the initial colonists entering North America or Australia, may have wandered about until they learned the landscape, but it would have been a short-lived practice. The formal term "nomad" is generally used by anthropologists to refer to mobile pastoralists, but the term has also been applied to hunter-gatherers (e.g., Krader 1959:499). Even groups like the Inuit do not wander aimlessly; they are aware of the locations (specific or general) of resources and move around to utilize them on a regular basis, though it may be that a great deal of adjustment is needed in their schedule to accommodate changes in local conditions.

It is sometimes problematic to neatly classify hunter-gatherers separately from agriculturalists because all of the latter depend on hunting and gathering for at least some resources. This is still true of our own industrialized culture; even we still use some hunted and/or gathered foods, such as deer killed by hunters on vacation or wild berries gathered on the weekend. In some rural areas, wild resources may be very important, just as they have been in times of severe stress, such as in the Great Depression of the 1930s, when some Americans even adopted hunting and gathering full time.

Resource Procurement

Defining the various procurement activities generally subsumed into hunting and gathering is not a simple task (see discussion in Ingold 1987:79–100). Most humans are omnivores, and most hunter-gatherers live primarily on seeds and roots (or tubers). This tendency is especially true in warmer parts of the world. In far northern regions, and in some other habitats, such as the thorn-scrub of the South American Chaco, very little of the plant material is edible. Here, people tend more toward the carnivore side of subsistence. If they are anywhere near water, they live primarily by fishing and sea mammal hunting. Only in the deep interior, away from abundant seeds or roots and rivers or seas, do we find actual terrestrial hunters. The best-known cases are the Aché of Paraguay, whose favorite prey is the armadillo (Hill and Hawkes 1983; also see Hill and Hurtado 1996), and the Indians of subarctic Canada.

Plant collectors include specialists, who rely heavily on one type of plant food (such as the San groups of the Kalahari Desert in southern Africa, who are dependent on mongongo nuts), and generalists, who eat a great variety of things. Fishing specialists exist along most coasts and major rivers and may live primarily on fish, shellfish, or sea mammals.

Gathering

Gathering has been defined as the collection of "wild plants, small land fauna and shellfish" (Lee 1968:41–42). Key components of the definition include the small and nonmobile nature of gathered resources and the common use of some technology for resource extraction and transport, such as digging sticks and containers (see Ingold 1987:82, 84–85). The inclusion of small fauna in gathering adds some ambiguity because some are quite mobile and unpredictable. The term "collecting" is sometimes used as a synonym for gathering. However, collecting generally implies that resources are in known and predictable locations, as are many plants and shellfish, and that little searching is required. The predictability of a good yield is high, generally much higher than in hunting. The term "collecting" has now come to define a specific type of hunter-gatherer adaptation (discussed below).

A second synonym for gathering is "foraging," in which individuals or small groups move about the landscape looking for food on a daily or regular basis. Although few groups of people actually practice such an approach, the term "foraging" has come to be applied by many to hunter-gatherers in general. However, like collecting, foraging also implies a specific type of hunter-gatherer adaptation (see below).

Hunting

The term "hunting" generally refers to humans actively looking for, killing, butchering, and consuming animals. However, a key part of the definition is that the animals are mobile species pursued and captured by some method (see Ingold 1987:80), with the hunter having no guarantee of success. Domesticated animals would not be pursued and so are not hunted. Most definitions of hunting exclude fish, seemingly because they do not have to be pursued across the landscape and are generally captured in traps or nets. Sea mammals are also sometimes excluded from hunting and included in a fishing category (Murdock 1969:154).

Richard Lee (1968:42) defines hunting as the pursuit and procurement of wild "land and sea mammals," a definition generally adapted herein. However, this definition excludes nonmammals, such as birds, reptiles, shellfish, and insects. Most would include birds (but not bird eggs) and reptiles among the prey of hunters. [Shellfish are considered as being gathered. Insects have rarely been considered at all but probably should also be classified within gathering (McGrew 1981:45; Ingold 1987:88).]

Fishing

Fishing is not an obvious separate category in hunter-gatherer subsistence. Lee (1968:42) separated fishing from both hunting and gathering and defined it as "obtaining of fish by any technique." The distinction between hunting and gathering was one of basic method: collection or pursuit. Fishing, however, seems to be distinguished by type of animal rather than by method of procurement (Ingold 1987:80). Although inconsistent, this distinction persists.

Fishers are obviously rather tightly tethered to certain aquatic ecozones. In large terrestrial biomes, fishers may be tethered to sea coastlines, lakeshores, or rivers and streams (see the case study on the Nuu-chah-nulth at the end of this chapter) and give relatively little attention to other ecozones. In island settings, there may be no other ecozones of any size, resulting in a specialization in fishing.

Scavenging

In anthropology, **scavenging** generally refers to locating and using animals that are already dead, such as a gazelle killed by a lion, rather than to hunting and killing. A number of species will scavenge the carcasses of animals that have died naturally or were killed by other animals, and some, such as hyenas and vultures, make their primary living by scavenging. Humans also scavenge, but only opportunistically. People will butcher and eat a beached whale and even chase away a large carnivore from a carcass so they can take the remaining meat. In one sense this activity could be classified as gathering because the resources are nonmobile. From another angle, it could be seen as hunting, because most scavenged carcasses are those of large game, and one would have to look (hunt) for such opportunities.

Some researchers have argued that in early hominids the strategy of scavenging preceded the strategy of hunting or at least formed an important aspect of early hominid subsistence (see Shipman 1986; Blumenschine et al. 1994; Rose and Marshall 1996; Speth 2002), but others (e.g., Wolpoff 1999:222–223) do not agree. It seems more likely that early humans both scavenged and hunted, much like chimpanzees do today (Stanford and Bunn 2001). If scavenging were an early niche, and humans changed to a hunting niche, we would have another clue to understanding the evolution of the ecology of humans. Scavenging is still important in some hunter-gatherer groups.

An Evolving Ecology

Part of the ongoing argument over these terms has to do with trying to understand the adaptational processes that led to becoming human. For a long time, researchers believed that hunting was what distinguished humans from nonhuman primates (e.g., Laughlin 1968:318–319), that is, that many animals foraged for food, but only humans hunted other animals with the aid of a technology. Having to carry tools for hunting and then carry the meat back to the family was seen as the reason hominids adopted bipedalism. The evolution was seen as being from forager-predator (without technology) to forager-hunter (with technology) (see Ingold 1987:85).

It has been argued subsequently that the initial development of food-extractive technology revolved around gathering rather than hunting (Zihlman 1981:108–109). Adrienne Zihlman (1981:109) suggests that digging sticks and plant-processing tools developed first and that hunting tools such as spears came later. In this view, the forager-predator would have first evolved into a gatherer-predator and only later into a gatherer-hunter (Ingold 1987:85). Still, foraging, predation, scavenging, and other methods of procurement remain in the inventory of the hunter-gatherer.

THE HUNTER-GATHERER STEREOTYPE

Anthropologists and the public have formed a stereotypical view of hunter-gatherers (see Ember 1978). This view includes the idea that hunter-gatherers are nomadic, wandering about the landscape on the edge of starvation; are primitive and not fully evolved; behave like many other animals; and languish at the bottom of just about any scale of culture anyone can generate. An oft-cited example of this stereotype is the Great Basin Shoshoneans, the very culture(s) studied by Julian Steward (i.e., 1938). David H. Thomas (1983:59; also see Thomas 1981) examined this stereotypical view of the Shoshoneans, noting that they neither had the simplest technology nor lived in the harshest environment. Through a series of historical and interpretational processes, Steward's "fundamental ethnographic description of a few very simple hunter-gatherers [primarily the western Shoshone] ultimately escalated into a theoretical statement of major anthropological importance" (Thomas 1983:62). In reality, the western Shoshone are not typical of the Great Basin groups, nor is "Great Basin culture" typical of hunter-gatherers.

A common misconception regarding hunter-gatherers is that they were always egalitarian, that is, that they lacked significant social distinctions and status. As hunter-gatherers manifest a diversity of adaptations, they also have a diversity of social complexity. Some groups, such as the Inuit and the San, were largely egalitarian, but others, such as the Nuu-chah-nulth (see the case study at the end of this chapter), have complex class systems in which social ranking is very important.

Early cultural ecologists, notably Steward (1955), saw the simplest hunter-gatherers as being organized in small, nomadic, and unstructured "bands." Such bands would have consisted of a core of elders, typically males, who made informal decisions. All members would have done more or less what they wanted, and group size would rarely have exceeded forty or fifty people. Little authority would have been exercised, and formal social institutions would not have existed.

Societies of this sort very possibly existed in pre-*sapiens* phases of human evolution. However, it is highly questionable whether such societies have existed in recent millennia. Steward's example of this social type, the Great Basin Shoshone, is now known to have had much more complex societies. Steward knew these people well, but he knew them after smallpox, tuberculosis, and Euro-American settlers had severely impacted them.

Other groups that initially appear to be socially simple emerge on inspection to be fragments of larger social orders. For example, some of the San of the Kalahari Desert in southern Africa, often viewed as "pure" hunter-gatherers, were trading partners with their agricultural Bantu neighbors. The Tasaday of the Philippines were almost certainly an agricultural group who lost their land—but their simplicity and isolation were romanticized, or outright misrepresented, in early studies (Headland and Reid 1989).

Hunting and gathering societies that appear to have been isolated from agriculturalists, and that had their lives recorded fairly accurately before recent cultures changed them beyond recognition, show a far different pattern. Leadership was usually informal, but hereditary chiefs with real power over large tracts of land did exist. Groups broke up into small bands for foraging purposes but aggregated into orderly settlements of hundreds or thousands when resources were concentrated and plentiful. The Great Basin Shoshone came together in such groups for pine-nut gathering and for fiestas. Networks of kinship could be even more widely flung. Thousands of people over thousands of square miles of land could trace relationships that allowed them to claim friendship, protection, and support during difficult years. Cycles of ceremonies and rituals maintained kinship and political links, and these cycles could be coordinated over vast areas. Trade links extended over these areas and often spread even more widely.

Nonetheless, the concept of the band remains important in the study of hunter-gatherers (Testart 1988:4) and has formed the basis for the common hunter-gatherer stereotype even though it is recognized that many other hunter-gatherers had more complex political organizations. A different approach might be to diverge from the band, tribe, and chiefdom categories (defined in chapter 1) and to classify hunter-gatherers on the basis of other criteria, such as being fishers or mounted hunters (Murdock 1968), or to distinguish between storage-based and non-storage-based hunting and gathering economies (Testart 1988; also see discussion below on foragers and collectors). The struggle to deal with these issues continues.

BIAS IN HUNTER-GATHERER STUDIES

Several factors contribute to a biased view of hunters and gatherers. The first is an overemphasis on the role of males and therefore on hunting rather than gathering. Second, anthropologists come from an industrialized agricultural society that tends to "look down" on hunter-gatherers. Third, most hunter-gatherer cultures are extinct or have been severely impacted by agriculturalists, and the biomes that they occupied are largely gone. As a result we have few real examples to study.

Hunter-gatherers are often viewed as the "most ancient of so-called primitive society" (Testart 1988:1), living fossils, archaic, and representative of preagricultural life (see critique in Headland and Reid 1989:49–51). This view may have some limited usefulness as general analogy in certain cases, such as the Australian Aborigines, who possess an unbroken cultural tradition at least fifty thousand years old; but in all cases, hunter-gatherers survived because they were adapted to their environment, demonstrating evolutionary *success*, not failure. True, some anthropologists construct models of ancient life using hunter-gatherers as ethnographic analogy, and a great deal can be learned from this. However, contemporary hunter-gatherers are alive *today* and cannot be identical to ancient societies (Testart 1988).

Sexual Bias

There has been a tendency for hunter-gatherer researchers to emphasize the hunting aspect of the system, to the detriment of understanding the gathering component. There are at least several reasons for this. First is the initial working assumption that hunting was a male activity and gathering a female activity. This assumption, coupled with the fact that most of the people working with hunter-gatherers have themselves been male, led to an emphasis on hunting in research. Second is the assumption that the hunting of animals is more "sexy" and attracts more attention from both the native people and the anthropologists studying them. A third reason is the anthropological view that hunting was a very important factor in the evolution of early humans.

This view is expressed in the title of a then state-of-the-art treatise on hunter-gatherers published in 1968, *Man the Hunter* (Lee and DeVore 1968). While the term "man" was meant to mean "human," it was not long before the role of females in hunter-gatherer societies became a focus of major research. The book *Woman the Gatherer* (Dahlberg 1981) represented a reaction to the male bias in hunter-gatherer studies. Beginning in the 1970s, with the recognition (or admission) that gathering was as important to the success of ethnographic hunting-and-gathering groups as hunting, if not more important, the role of women in such societies has received considerable attention (e.g., Linton 1971; Begler 1978; Leacock 1978; Dahlberg 1981; Hunn 1981; Tanner 1981; Kurz 1987; McCreedy 1994). This work continues, somewhat complicated by the emerging understanding that sex roles are not set in stone. Women do some hunting and men do some gathering.

Ethnocentrism

Anthropologists are the people who study hunter-gatherers. Virtually all anthropologists grew up in societies that are agricultural and industrialized. Few, if any, anthropologists were raised as hunter-gatherers. The result is that the view of anthropologists is biased toward agriculture. When anthropologists study a particular area, there is an inclination to judge it based on agricultural standards. For example, deserts are viewed as being "harsh," largely on the basis of the lack of greenery and the perception that agriculture would be difficult or impossible. Thus, hunter-gatherers in such environments "must" be living on the edge.

In addition, the notion that hunter-gatherers are "primitive" (a very poor choice of words) and somehow inferior is implicit in their classification based on unilinear evolutionary principles. This notion is ethnocentric and erroneous. We also tend to view the life of a hunter-gatherer as very difficult, which has led to the notion of hunter-gatherers "eking a meager living from a hostile land" or "walking around, picking things up off of the ground, and putting them in their mouths" and subsisting "on the edge of starvation."

The Marginal Environment Problem

At one time, hunter-gatherer cultures occupied the entire planet (most terrestrial ecosystems); all peoples made a living hunting and gathering. With the domestication of plants and animals, those areas that were suitable for domesticates were colonized by the agriculturalists, and the hunter-gatherer groups were generally either expelled or assimilated. Over the last ten thousand years, most of the areas formerly occupied by hunter-gatherers have been appropriated by agriculturalists. Thus, contemporary hunter-gatherer groups are limited to those areas not (yet) desired by agriculturalists—those areas currently unsuitable for agriculture (and thus "harsh" in the view of agriculturalists).

Because anthropologists today study hunter-gatherer groups that have survived in environments considered marginal or "harsh" by farmers, those hunter-gatherer groups then become the stereotype for hunter-gatherers everywhere and throughout time. We tend to forget that the most productive ecosystems were once occupied by hunter-gatherers, likely living in a very different way from the extant groups we somehow see as "typical."

Affluence

It has been argued (Sahlins 1972) that hunters and gatherers were "the original affluent society" because they had abundant resources to satisfy their relatively limited wants and needs (also see Koyama and Thomas 1981; but contrast with Shnirelman 1994). Affluence refers to the amount of free time an individual has and is based on studies that suggest hunter-gatherers "worked" only four to six hours a day to meet their subsistence needs. While it may be true that hunting or gathering took place for a limited number of hours per day, the calculation did not consider the amount of time spent back in camp processing, repairing, and manufacturing things. In reality, then, hunter-gatherers probably spent much of the day conducting a variety of tasks, although they still probably worked less than most agriculturalists. However, this may not have been the case for hunter-gatherers in more "productive" environments. For example, in Northwest Coast groups, a very considerable amount of spare time not required for basic subsistence was available for other pursuits, mostly in the winter after the salmon fishing season.

There may be a bias in the other direction as well. It may be that some view hunter-gatherers as the perfect society, the original ecologists living in harmony with the environment and having few cares or worries. In truth, like any group of people, hunter-gatherer groups have problems that have to be solved.

POPULATION

Like all cultures, hunters and gatherers must stay within the carrying capacity of their environment (see Baumhoff 1981). Unlike many agricultural groups, they cannot easily manipulate the system to create long-term food increases and must pay closer attention to population control.

Hunter-gatherers have developed (as we all have) a number of mechanisms for demographic control (Cowgill 1975). Obvious ones include migration and celibacy. Resource shortages or other factors may lead to warfare, but such small-scale warfare generally has little impact on population. Other, less obvious methods for population control are used. In San groups that move frequently, for instance, mothers nurse their babies for a long time—up to three years or more. Because nursing reduces the likelihood of conception by about 50 percent, San births are kept more or less spaced out.

Population can be regulated more directly through birth control methods, abortion, and infanticide. Inuit groups once employed all of these practices during hard times. They also had a tradition in which old people would go off by themselves to die if conditions became extremely difficult, a relatively common situation. In the absence of such methods, even disease and warfare are not effective at holding down population growth, and a hunter-gatherer population will soon outgrow its resource base. It is, then, not surprising that all well-studied hunters and gatherers have reported knowledge of abortion and birth control techniques, and most have histories of migration and land conflict.

SETTLEMENT AND SUBSISTENCE

As noted above, hunter-gatherers are so designated on the basis of their economic system. Within that general classification, groups are further divided on the basis of the particular way in which they pursue hunting and gathering. Hunter-gatherers leave their camp or village and venture into the landscape to obtain resources. These trips are most often patterned and planned.

Seasonal Round

If a group is dependent on wild foods, it must be well versed on the seasonality of the various resources it exploits. Different resources become available at different times and places. To obtain these resources, the groups (or portions of them) have to move to the resource. The same principle applies to agriculturalists as well, but most of their food resources are located in the same place, meaning that the people do not have to move very far. The system of the timing and movement of groups across the landscape is called a **seasonal round**.

A seasonal round might be regular or irregular. A very simple regular seasonal round (fig. 5.1) might be one in which the group goes to Resource Area A in the spring to use particular resources. In the summer, it moves to Resource Area B, to Resource Area C in the fall, and to Resource Area D in the winter. The next spring, it would move back to Resource Area A, the same location and resource(s) that it used the previous spring; the cycle would begin anew.

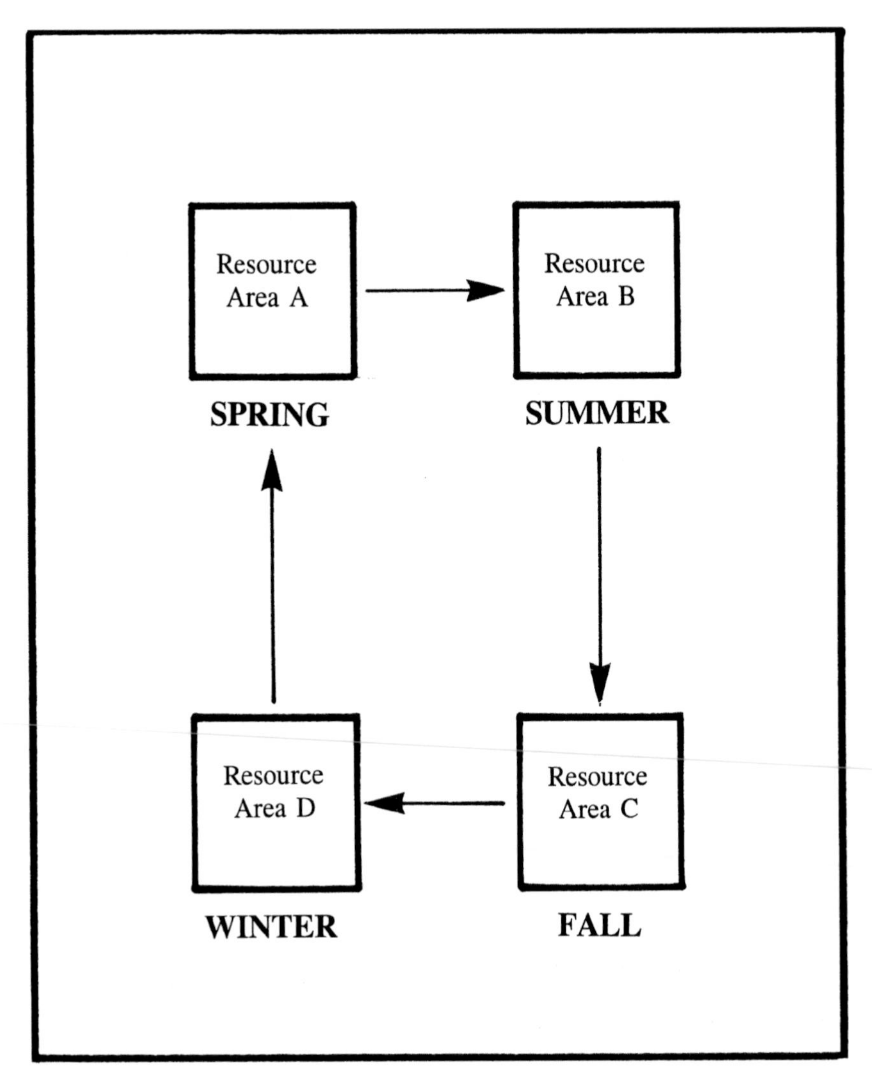

FIGURE 5.1
A very simple seasonal round (arrows indicate the direction of movement).

A more complex round would involve an increase in the number of locations and resources used and/or a decrease in the repetition of the use of specific locations. If the group were to go to a *different* location every spring, instead of the same place every year, the round would be more diverse (see fig. 5.2). This example could be stretched out infinitely, but the group would always know where it was going.

Fission-Fusion

Groups may change their size and distribution depending upon conditions. If a group splits up into several smaller groups that separate from the larger group, this occurrence is called fission (splitting). If the groups come back together again, it is called fusion (joining). Some groups practice splitting and rejoining as a regular part of their seasonal round. If the resources at a location **are** not enough to support the entire group, it may split, with some of the group going to location 5 at the same time as the rest of the people are at location 1. Later, they will all meet at location 2 to exploit resource B (fig. 5.3). This **fission-fusion** tactic may mitigate seasonal resource shortages, representing a flexible response to changing conditions.

Fission-fusion may also be used to resolve disputes between members of a social unit. For example, the various hunter-gatherer groups in the forests of central Africa (see the case study of the Mbuti) simply leave a particular social unit (village or camp) when there are too many people or if conflict becomes too great. The departing people may join other groups or may form their own group. Membership in these groups is very fluid.

Opportunism

As discussed below in greater detail, hunter-gatherers have a planned economic system; they do not do things in a haphazard way. However, they do take advantage of opportunities that present themselves. If a group exploits a particular resource and happens to encounter another useful one, it likely will take advantage of the situation and exploit the second resource as well as the first (or perhaps instead of the first).

Many hunting expeditions are conducted to obtain specific prey, for example, rabbits or bison. However, some hunting excursions do not have a particular prey in mind, and any (culturally acceptable) prey encountered will be taken. This is opportunistic in some sense, but there is still planning and knowledge of general prey locations (e.g., around water), and it is not a random operation.

Flexibility

We have already discussed information systems and resource monitoring. To hunter-gatherers, these factors are central to their success during the year. A

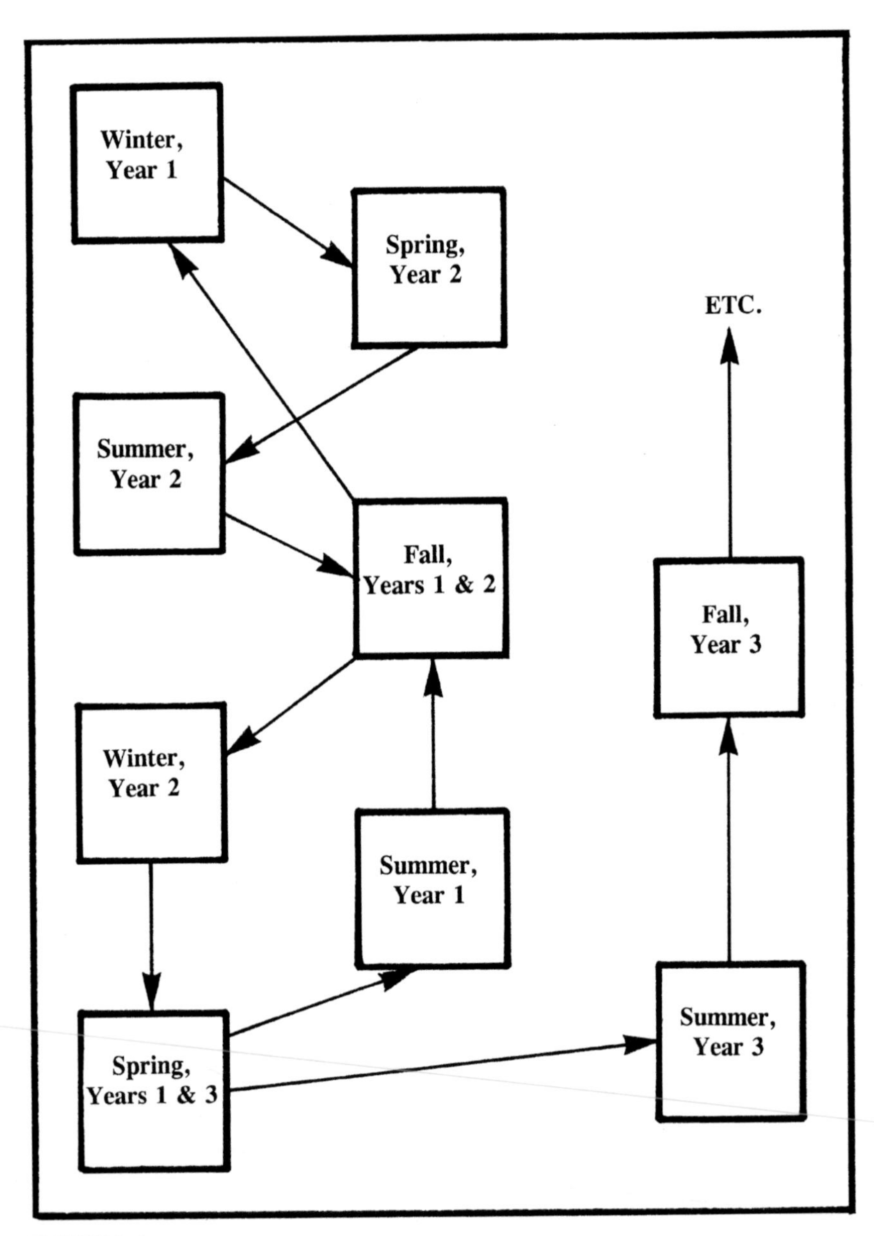

FIGURE 5.2
A fairly complex seasonal round (arrows indicate the direction of movement; note seasonal and yearly changes).

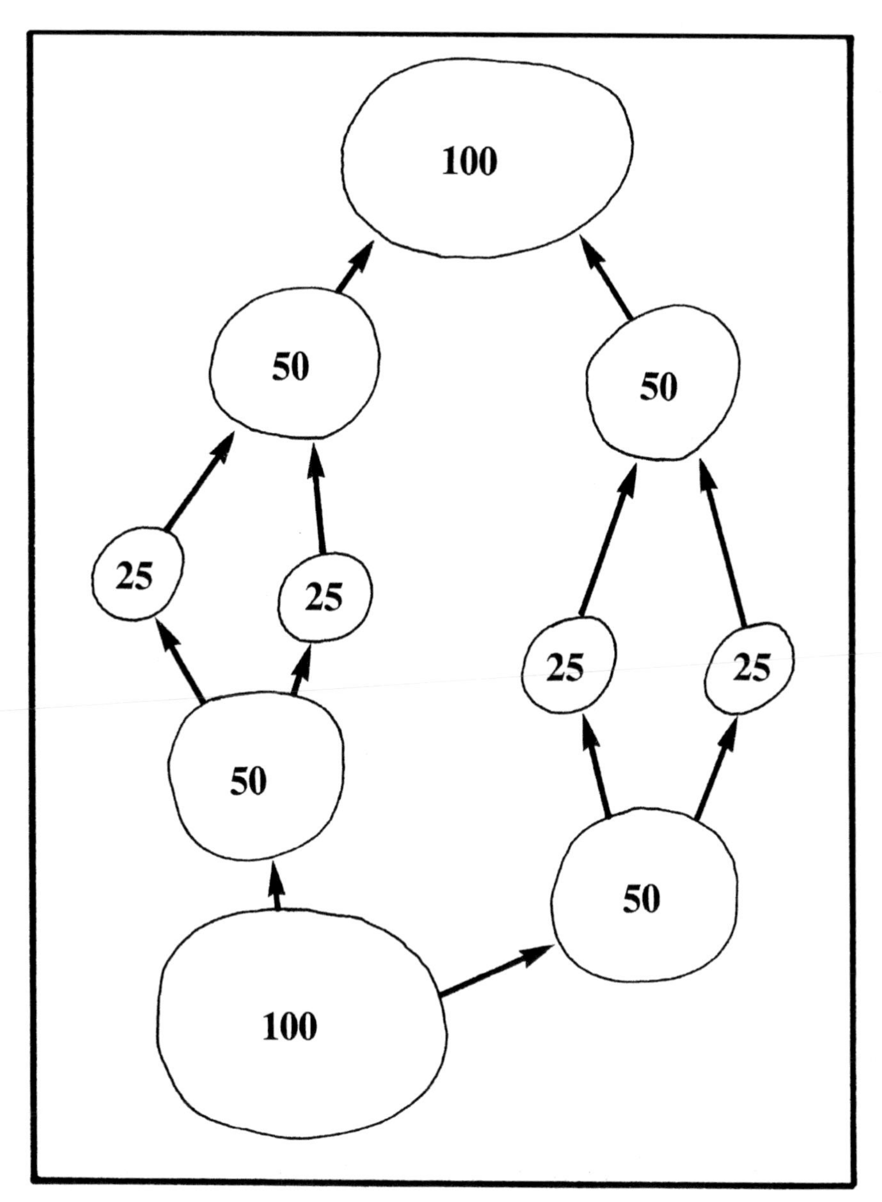

FIGURE 5.3
A very simple fission-fusion model for a culture of 100 people.

group may have a set seasonal round, but that round is dependent upon conditions in the environment. If conditions change, so must the group's response. If resource A did not ripen or has moved from location 1, a decision must be made regarding where to go and what resource to use instead. Such a decision would involve the consideration of a variety of factors, including which resources were where, the location of other social units, group boundaries, and/or kinship networks. Information derived from resource monitoring would be critical to the decision. The gathering and sharing of such information enhances the ability to be flexible in times of stress, as does the retention of knowledge, the maintenance of kinship networks, and the ability to use fission-fusion. In times of stress, the greater the flexibility, the greater the chances of success.

Foragers and Collectors

Anthropologists have generally defined two basic hunter-gatherer strategies: foraging and collecting, sometimes also called traveling and processing (Bettinger and Baumhoff 1982:487; see discussion below). Most anthropologists recognize a "forager-collector continuum," meaning that a group is more or less of one than the other. However, we tend to pigeonhole groups by their strategy, and the use of terms like "forager" and "collector" cloaks the complexity inherent in the way that hunter-gatherers make a living. Their flexibility usually is ignored, even if inadvertently.

The terms "foraging" and "collecting" have common or literal meanings even though anthropologists give them specialized meanings. However, even if it has a specialized meaning, a term still implies its literal meaning, particularly to those not aware of another definition. We tend to fall into ruts in our thinking, as it is limited by the terms we use.

Foragers

The term "forage" generally means to wander in search of food (or other resources), as in the actions taken by armies supplying themselves in the countryside (Burch and Ellanna 1994b:4). Wandering implies no specific route and no set destination. Thus, when people actually forage, they have no specific goal or resource in mind; they just wander about until some resource is encountered.

The operational anthropological definition is different. **Foragers** are seen as people who have a seasonal round, occupy a series of "camps" as they move about the landscape, and have no permanent home. Relying on monitoring, they move their residences from place to place depending upon the season and the condition of resources. (Some groups may not have great seasonal variation in resource availability and so may not move often for that reason.) Forager groups are viewed as being generally small and mobile and having a relatively meager material culture (because they carry most of their possessions with them). Thus, foragers move "people to the resource" (see fig. 5.4). In sum, "foragers generally have high resi-

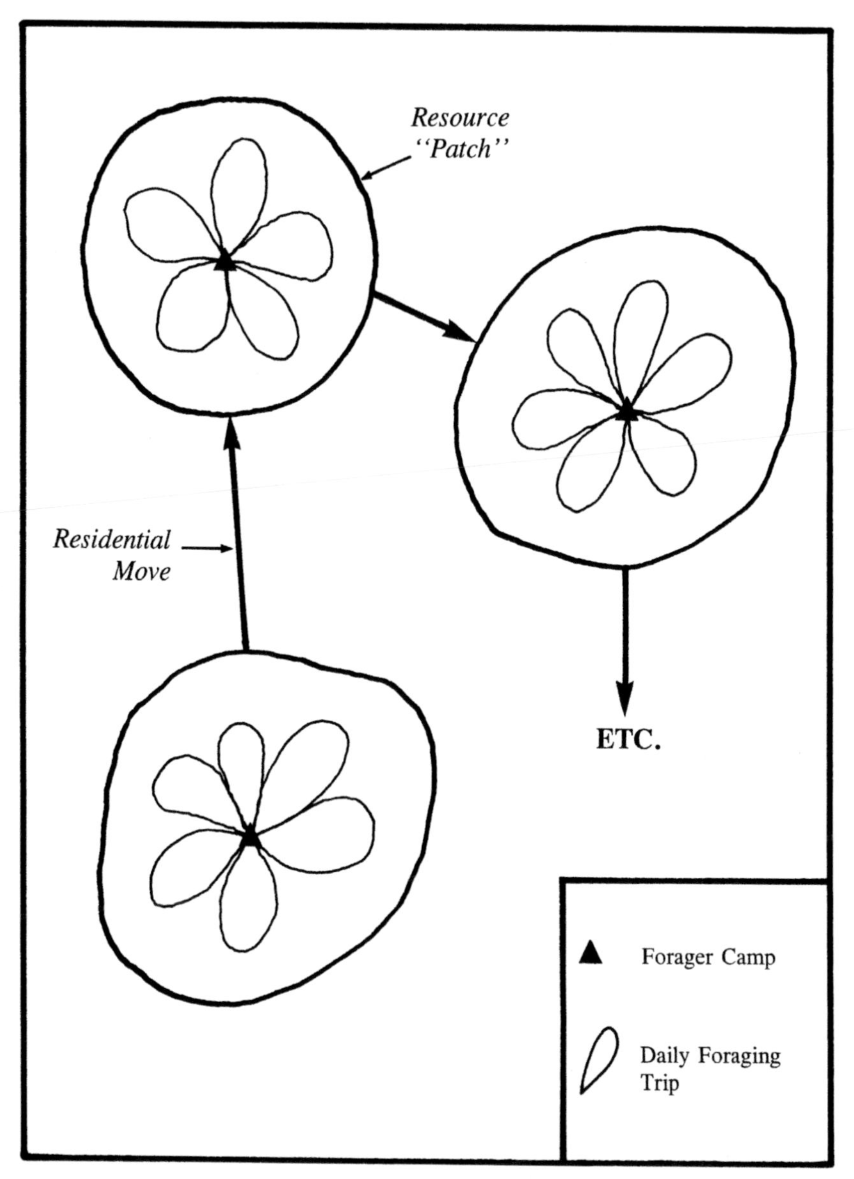

FIGURE 5.4
A very generalized forager settlement model.

dential mobility, low-bulk inputs [i.e., they gather small quantities at a time], and regular daily food-procurement" (Binford 1980:9). The daily food-gathering trips (represented in fig. 5.4 as the "daisy") result in the people's gathering information on resources as well as the resources themselves.

Collectors

Collector is a term employed (see Binford 1980:10) to describe hunter-gatherers who use specially organized task groups to exploit specific resources, often in bulk. In this strategy, the "resources are moved to the people" (different from the forager approach). Groups practicing a collector strategy would have permanent or semi-permanent residences and many smaller activity locations used briefly by specific task groups to obtain resources (fig. 5.5). Moving less frequently would allow for a larger population and a greater and more complex material culture. Because the resource would be moved to the people, storage of certain staple resources would be common. This storage would tend to limit the mobility of the groups; they would be tethered to the storage facilities. As with foragers, the short-term movements of task groups serve the purpose of information gathering as well as resource procurement.

As defined above, foragers tend to occupy smaller camps, have generally less material culture and a less specialized material culture, rely less on storage, and have a smaller population than collectors. Specialized task groups are present in a collector group but not in a forager group. Other parts of the definition of these strategies are settlement, whether villages or camps, and behavior, such as types of social systems and the use of resources. Using these criteria, cultures are classified by strategy as either foragers or collectors. The classification may be based on limited information, and once assigned, a culture is assumed to manifest all criteria, a situation that can mask diversity and confuse analysis.

Strategies and Tactics

As noted above, strategy is a commonly used term, in a classificatory sense, with hunter-gatherers (see Sutton 2000). A **strategy** is a broad, overall plan; the term refers to long-range goals (e.g., making a living), not short-term practices (e.g., one's current job). Short-term practices are called **tactics**; they are the methods used to execute or accomplish the strategy. Cultures practice a wide variety of tactics to make a living. They possess and retain an inventory of tactics, some of which are used often (even daily), and some of which are not but are still retained in the inventory (part of the retention of knowledge and flexibility discussed above).

Hunter-gatherer cultures frequently are classified as practicing one of the two primary "strategies," foraging or collecting, based on a few observed criteria, such

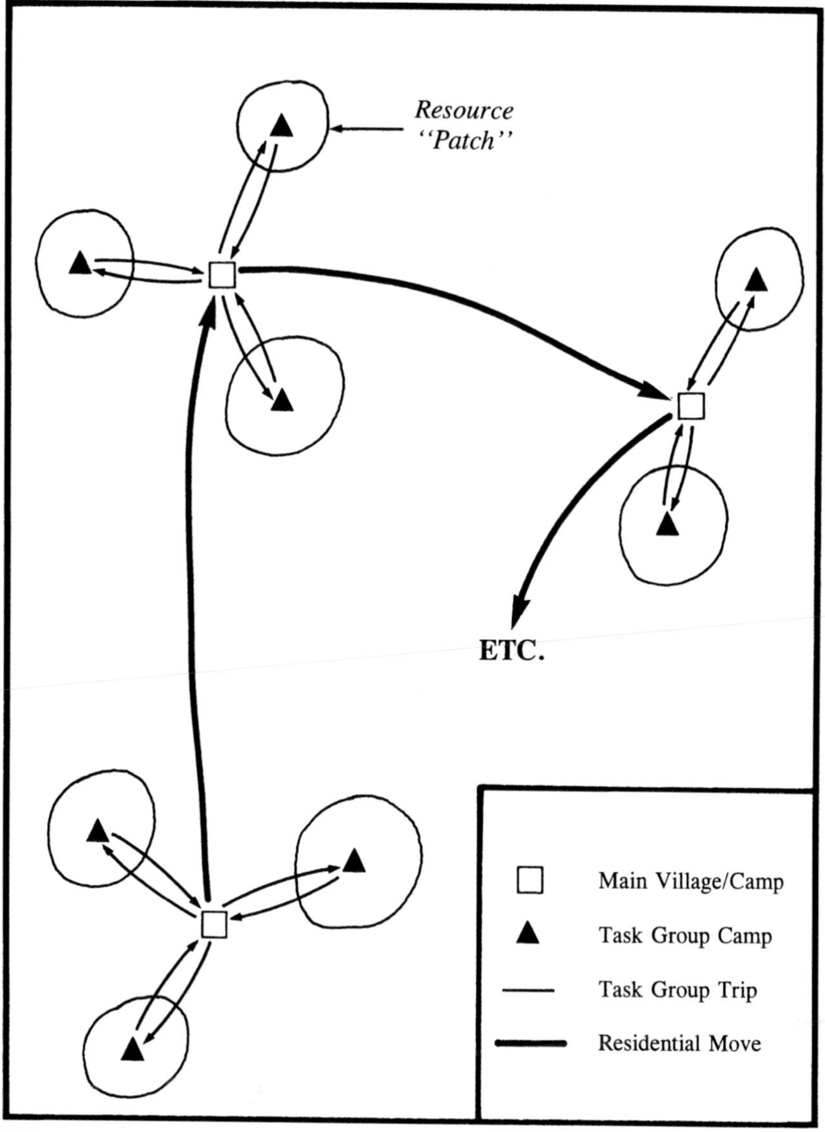

FIGURE 5.5
A very generalized collector settlement model.

as having villages or traveling to resources. Once classified, they are commonly assumed to adhere to the other, even though theoretical, criteria of that strategy as well; for example, if they have camps, they must have a small population. Once these assumptions are made, they somehow become "facts," and the variability within the adaptation of the culture is overlooked.

The actual "strategy" of a hunting and gathering group is to make a living. The methods employed to accomplish that end are "tactics." The group uses a wide variety of tactics to make a living, and these tactics vary, depending on the situation. This argument regarding strategy and tactics applies to all cultures, including agriculturalists. If foraging means going out and getting wild foods, then all cultures, including Western cultures, do some foraging, such as fishing, hunting, or picking wild berries. The same is true with collecting. No culture is totally isolated from hunter-gatherer tactics; it is just a matter of degree. One could argue that there is a hunter-gatherer–agriculturalist continuum as well as one from forager to collector.

If the focus of analysis becomes tactical inventories (or adaptations) rather than "strategies," perhaps we can better understand the variety of responses and behaviors. Each culture employs a wide range (or repertoire) of tactics, many of which will overlap with those of other groups. It is this diversity of response that is central to their adaptation. Without understanding this diversity in tactics, one cannot understand the adaptation of the culture or the culture itself. The use of "strategies" as pigeonhole categories limits our understanding.

This way of looking at hunter-gatherers is, perhaps, more important from an archaeological standpoint. In ethnographic situations, we can (in theory) observe the diversity of responses by the people and understand their settlement/subsistence system as an integral unit. In archaeological contexts, we observe (i.e., excavate) one segment of a system (an archaeological site) and use that to infer the system as a whole. One could easily find a collector camp that "looked" like a forager camp and so classify the cultural system in question as "foragers." Based on our theoretical construct of how a forager system is organized and behaves, we might "reconstruct" the ancient culture in a completely erroneous way. While this error might be corrected in the light of additional archaeological work on that system, it might be avoided altogether if we were less myopic about the strategies.

ENVIRONMENTAL MANIPULATION AND RESOURCE MANAGEMENT

Most hunter-gatherer groups practiced relatively little active environmental manipulation, as large-scale changes in the landscape were usually not required. However, when such methods were used, burning was the most commonly employed method (as discussed in chapter 4). Passive environmental manipulation through the use of ritual was a much more common practice. Such methods included ritual control of weather and places of power, such as the stewardship of Dreamtime places in Australia. Like all people, hunter-gatherers strive to exert at least some control over both abiotic and biotic elements of their environment.

The management of particular resources was common among hunter-gatherers. Active methods included taking adult male animals instead of juveniles or females where possible, pruning specific plants (e.g., tobacco) to enhance productivity, and observing some religious taboos. The intensity of management varied greatly, and in some cases resources were monitored only as to their availability and condition. By the end of the Pleistocene, some cases of the intense active management of animals and plants had led to the genetic domestication of those species, giving rise to agriculture.

Many other resources were also passively managed. A common practice was a ceremony to thank the soul of the animal that allowed itself to be killed. Failure to thank the animals would anger them, impeding future hunting success.

RELATIONS WITH OTHER GROUPS

Hunter-gatherers always had some contact with other groups, and until about ten thousand years ago, the other groups were always hunter-gatherers. Such contact often resulted in peaceful relationships, but it would be very difficult for two hunter-gatherer groups to occupy the same territory because they would fill the same basic niche in a common habitat. Sharing habitat with pastoralists, who have a different niche, would be easier as long as the seasonal rounds of the two groups did not conflict and the hunter-gatherers did not burn the grass too often.

However, it is generally difficult for hunter-gatherers to coexist with agriculturalists (e.g., Spielmann and Eder 1994; Layton 2001). The farmers generally alter the landscape in a significant manner, and in doing so they disrupt the hunter-gatherer system.

When agriculturalists enter a region, they appropriate certain lands for crop production. It may be that these lands were also highly productive for wild foods such as grass seeds or animals. The farmers will claim ownership of the land, sometimes fence it, and always defend it. This practice in effect removes that land and its former resources from the seasonal round of the hunter-gatherers, changing the geographic distribution and availability of their resource universe. The hunter-gatherers must adapt by altering their use of resources, seasonal round, and territory, or they will be either killed or absorbed by the farmers. Many times, the farmers appropriate such a large portion of the hunter-gatherer resource base that adjustment is impossible and there is no choice but to acculturate. Resistance to the farmers usually results in the hunter-gatherers being defeated since farmers generally have larger and better-supplied populations.

There have been, however, cases where hunter-gatherers have expanded against agriculturalists, usually with warfare as an important mechanism. The Apache expansion into the American Southwest during the last five hundred years is a good example. The Apache always knew where the sedentary farmers were located, raided their fields as they would exploit any productive resource patch, and generally created so much havoc that the farmers had to move. In this way the Apache took control of the region.

Mutualistic Relationships

Despite the problems noted above, some hunter-gatherers have developed a mutually beneficial relationship with their agricultural neighbors. This arrangement may include the trade of wild foods for domesticated ones or labor exchange. A good example of such a mutualistic relationship is that of the Mbuti and the Bila in the Ituri Forest of central Africa (see the case study at the end of this chapter).

CHAPTER SUMMARY

Hunters and gatherers, often referred to as foragers, are those people who make their primary living by exploiting wild plants and animals, and until fairly recently all human groups were hunter-gatherers. Even today, all cultures rely to some extent on wild foods. Hunter-gatherers display a vast array of structures, forms, and adaptations, from very small, simple groups to very large and complex ones. Gathering primarily involves the collection of plant resources while hunting primarily involves the procurement of animals. Scavenging does not fit neatly into either of these practices but may have been an important early strategy.

Hunter-gatherers are widely misunderstood, often being viewed as small and simple groups barely surviving in some hostile environment. Understanding of hunter-gatherers is clouded by an analytical overemphasis on hunting, ethnocentrism by researchers, and the fact that all contemporary hunter-gatherers live in marginal environments not yet occupied by agriculturalists, which gives rise to a perception that all hunter-gatherers lived in marginal environments.

Most hunter-gatherers move about the landscape to resource localities, a seasonal round, and may employ some sort of fission-fusion. Those who move their populations to resource localities may be classified as foragers, and those who bring resources back to their settlements may be classified as collectors. Both foragers and collectors implement their strategy through tactical actions.

Hunter-gatherers generally do not make intensive efforts at environmental manipulation or resource management, although there are important exceptions. Relationships with other groups, now mostly agriculturalists, are typically strained, and many hunter-gatherer groups are now under intense pressure to drop their livelihood.

KEY TERMS

collectors
fission-fusion

foragers
gathering
hunters and gatherers
hunting
scavenging
seasonal round
strategy
tactics

CASE STUDY

CASE STUDY: THE NUU-CHAH-NULTH OF BRITISH COLUMBIA

This case study of the Nuu-chah-nulth shows that hunter-gatherers can develop complex political and economic institutions and have dense populations, all features fairly atypical of hunter-gatherers. The Nuu-chah-nulth also illustrate some of the variability in hunter-gatherer complexity and provide an example of a group that made its living primarily from fishing.

The Nuu-chah-nulth are hunter-gatherers living in an area containing abundant resources along the Northwest Coast of North America. They live on the west coast of Vancouver Island, a large island off the west coast of British Columbia (fig. 5.6). The island is mountainous, cold, rainy, and covered with dark, dense forests (fig. 5.7), areas where hunting and gathering were difficult. The waters, however, teem with life. In the 1770s, the Nuu-chah-nulth numbered perhaps ten thousand people. By 1900, due primarily to European diseases, their population had shrunk to around five hundred; it has since recovered somewhat. Today the culture remains vigorous. The most recent summary of the Nuu-chah-nulth was presented by Eugene Arima and John DeWhirst (1990), who used the term "Nootkan" to refer to the group (see the note at the end of this case study).

THE NATURAL ENVIRONMENT

The western coast of Vancouver Island is irregular and has many small islands, inlets, and fjords. The terrain is rugged, with high mountains, thick forests, many watercourses, and narrow beaches. The climate is mild but very wet, with the rainy season extending from October through April.

CASE STUDY

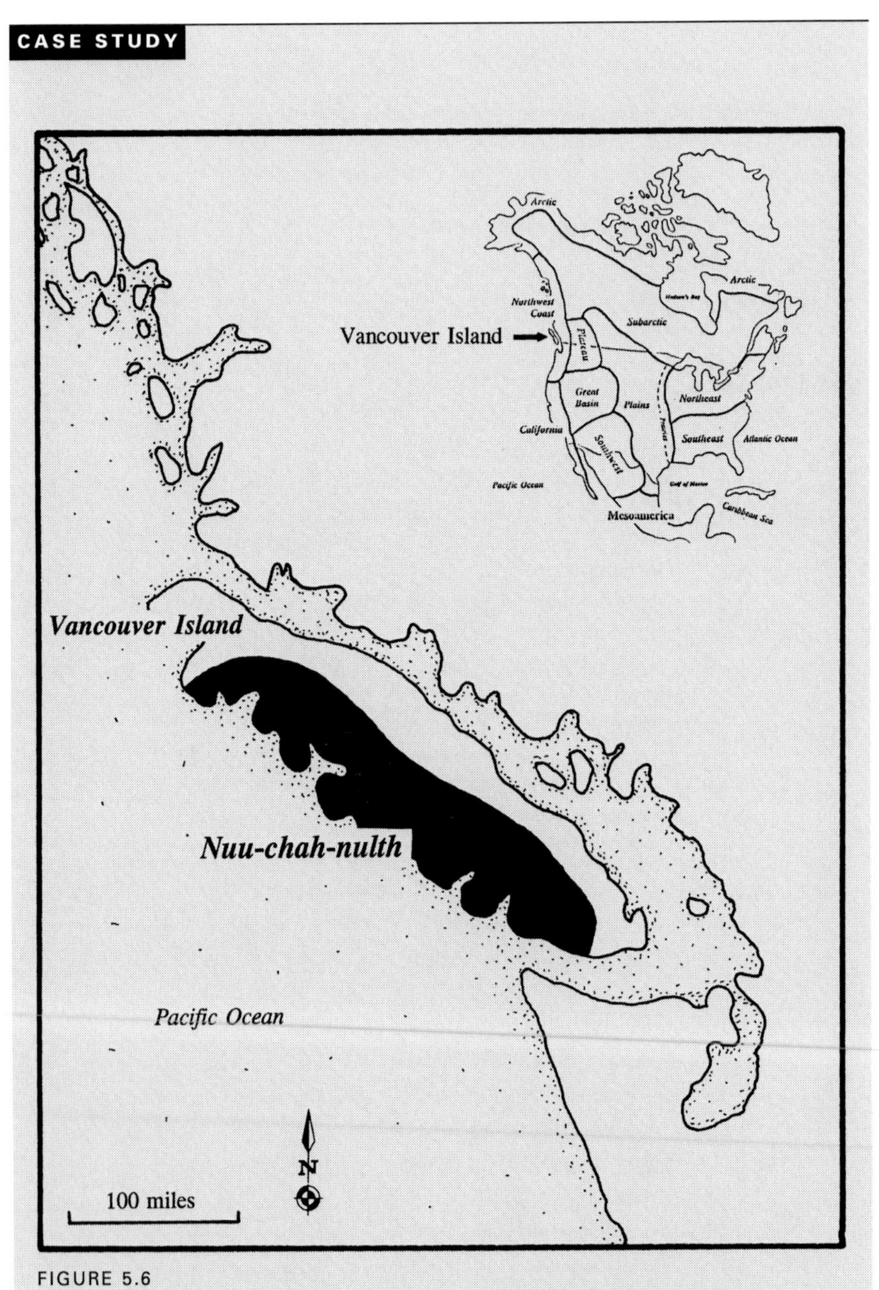

FIGURE 5.6
Location of the Nuu-chah-nulth along the Northwest Coast of North America.

CASE STUDY

FIGURE 5.7

The forest along the shore of Vancouver Island. (Photo by Mark Q. Sutton)

The Nuu-chah-nulth exploit four basic ecozones: the forest, the many riparian zones, the littoral zone, and the open ocean. The forest consists mostly of cedar and fir trees and a dense undergrowth of hemlock and ferns. The Nuu-chah-nulth utilized many animals living in the forest, including deer, bear, elk, and mountain goats, as well as a number of forest plants. The rivers and streams that ran through the forest to the sea were the primary sources of salmon, the most essential of the fish taken by the Nuu-chah-nulth.

The ocean, including the littoral zone, was the main source of food, the primary avenue of travel, and the foundation of identity for the Nuu-chah-nulth. Shellfish were the primary resource of the littoral zone, and the open ocean was home to a number of marine mammals, including seals, sea lions, sea otters, porpoises, and whales. These animals were very abundant and constituted important resources. Many species of fish, including some salmon, were

CASE STUDY

also taken from the ocean. The ocean continued to produce vast quantities of fish and sea mammals until the devastation of those resources by the contemporary fishing industry.

SOCIOPOLITICAL ORGANIZATION

The Nuu-chah-nulth once comprised a number of small, politically independent bands whose primary political unit was the permanent winter town. Groups were designated by the name of their town or sometimes by their principal food (such as Mowichaht, "people of the deer"). Life revolved around these winter towns, which might contain up to a thousand people. In the summer, the towns were almost deserted as the people broke into small groups and traveled to hunting and berry-picking areas. The various groups intermarried and recognized that they had a common language and culture. While the Nuu-chah-nulth had no formal political unity, they were united by ties of kinship, ceremonial interaction, and trade.

The Nuu-chah-nulth were stratified into three social classes—nobility, commoners, and slaves—and lineages, clans, and moieties were important. A complex series of ranks was recognized, with each person occupying an individual rank in the family, each family occupying a rank within the lineage, each lineage having a rank in the clan, each clan being ranked in the moiety, and each moiety being ranked in the town. The positions of rank were often hereditary, but the system required that the person acquiring a rank validate that position by sponsoring a potlatch (see below). The inheritance of property, including titles, rights, and power, was a critical element in maintaining rank.

Religion among the Nuu-chah-nulth focused on maintaining the social order. Myths and religious texts reinforced the social system and its key values. They also focused on maintaining the ecosystem. The social world did not stop at the borders of the human species; all things were involved. Everything, living or not, had spirits that had to be treated as one would treat a human. Thus, virtually all things in the environment were subjects of passive resource management.

CASE STUDY

ECONOMICS

The Nuu-chah-nulth were hunters and gatherers, and many resources were obtained by a variety of tactics. However, fishing was the most important pursuit. Except for perhaps some local raising of tobacco, the people practiced no agriculture until the late 1770s, when Europeans introduced the potato—virtually the only crop that thrives in the cold, wet climate.

Aquatic Resources

The principal resource was salmon. Five species of salmon (*Oncorhynchus* spp.) and two species of sea-running trout occur in the area. The five salmon species are lumped together into one category by outsiders, but the Nuu-chah-nulth have quite different names for all of them and regard it as strange that anyone could lump them together. Local English speakers have now adopted this view. They do not use the Nuu-chah-nulth names, but they do use a different name for each species: chum, pink, sockeye, and so on. The word "salmon" is never used in ordinary conversation. Thus does ecology affect language.

Each of these fish has very different characteristics. Sockeye have the richest meat, but the young mature only in lakes, so sockeye do not exist except in river-and-lake systems, where they used to occur in the millions. Chum, which spawn only in the lower reaches of rivers, have the leanest meat and are thus ideal for drying in the damp climate; they were the staple for storing and so became the staff of life in winter. Pinks run in the millions, but only every other year, and they do not store well, so their usefulness was limited.

The processing of salmon presented a logistical problem. Salmon runs tend to be brief, and a vast number of fish swim up the river in a few days. Large rivers have several separate runs. Of particular concern was the chum run because the lives of the people largely depended on how many chum a group could store for the lean season. Runs sometimes fail. The Nuu-chah-nulth coped as best they could by monitoring fish populations and artificially stocking streams (Sproat 1868). When genuine famine threatened a particular group, it dispersed to live with relatives or to comb usually neglected areas.

CASE STUDY

Whales were also a major food source. Migration routes and breeding grounds lie close inshore. The huge animals often come into the large bays of the coast to rest during their travels. Here they become quite tame. The Nuu-chah-nulth were superb whalers and could easily harpoon them in the bays and even in open sea. Sea lions, seals, seabirds, shellfish, and many other resources provided a rich supplement. Additionally, despite the relative poverty of the land, terrestrial resources such as berries and other plant foods were fairly abundant, and hundreds of species were known and used (Turner and Efrat 1982). There were even a few places in the northern and central portions of the island where rich inland pastures fed large numbers of elk and deer, permitting a few groups to live largely on venison.

These resources were not evenly distributed, however. Shellfish were concentrated on the outer coasts, where open-sea currents brought nutrients to them. There is not much else on the outer coasts, however, and residents of these areas were teased as "eaters of dead minnows on the beaches" (Drucker 1951). Whales were confined primarily to the great bays. Salmon ran best in major rivers with lakes. The prime location for a winter town, then, was a sheltered island in a bay with a major river system draining into it. Such localities were considered highly desired prizes and were often fought over.

Terrestrial Resources

Animals hunted on land were less important than those taken from the ocean. The most significant land mammal was the deer, which was not abundant on Vancouver Island. Some mammals, including bears, elk, mountain goats, and hares, were taken for food, while others, such as beaver, mink, and ermine, were taken for their fur. Some seabirds were also taken, along with their eggs. The only domestic animal was the dog.

Hundreds of plants were known and used as food, including many roots and tubers, ferns, numerous kinds of berries and fruits, and some greens, as well as several types of ocean algae. Other plants were used as material in the manufacture of basketry, cordage, and rope and for woodworking.

CASE STUDY

Seasonal Round

Although the primary residence locality is the winter town, occupied for much of the year, the residents of the town disperse in the summer to a series of smaller settlements to hunt and gather various resources distributed across the landscape in a generally patchy fashion. Thus, the Nuu-chah-nulth employ fission-fusion in their seasonal round.

The Potlatch

The potlatch is an integral component of Nuu-chah-nulth politics and economy; it is a complex cycle of gift-giving ceremonies of varying size and detail whose primary function is to validate title. Potlatching could also elevate a low chief; he could rise higher than a chief who started well but failed to potlatch often. A friend of one of us (ENA) was the eldest successor to a chieftainship of the Clayoquot. He was no power seeker, so he declined to potlatch, in favor of his younger brother, who held the potlatch and now has the position. Potlatching was highly competitive, especially in the nineteenth century, when trade with the whites was successful for the Nuu-chah-nulth.

The potlatch was not mere self-aggrandizement. A chief attracted followers by giving them potlatch gifts. He was not just paying them to join him—he was displaying his success at obtaining and amassing wealth and his generosity in sharing it. In war, these followers supported him. If he lost the physical prowess and spiritual power that were believed to give him success, he lost the followers. Nuu-chah-nulth society is not fluid or unstructured, but it is a society in which kinship links are both far-flung and close-knit. Everyone is related to everybody else. A man disappointed in his chief can easily find another chief to whom he is more or less as closely related.

In the mid–nineteenth century, several factors accelerated the potlatch cycle. First, the introduction of guns had made warfare a far more dangerous activity but far more profitable for the winner. Second, trade injected vast new wealth into the system. Third, disease had decimated the population, leading to an abundance of unclaimed land and chiefly titles. Fourth, social disorganization probably led to heightened competition. The result—among the

CASE STUDY

Nuu-chah-nulth as among their neighbors—was an increase in warfare and in potlatching. Enterprising chiefs embarked on ambitious military conquest. The Ahousaht conquered several islands. The Clayoquot (more accurately *Tla'okw'aht*) took over most of the sound now named for them. The area around Alberni was taken from speakers of an unrelated Salishan language; their descendants are now assimilated into the Nuu-chah-nulth linguistic and cultural world. In short, potlatching was not "just a ceremony," nor was it pathological waste of material goods. It was a way of mobilizing followers in a war-torn society (Drucker and Heizer 1967).

The potlatch exists in one form or another among all the Northwest Coast peoples and has frequently been a subject for study in anthropology. Initially anthropologists were influenced by local missionaries, who condemned it as a "pagan" ceremony that was nothing but a "waste" of valuable goods. Ruth Benedict, in her famous book *Patterns of Culture* (1934), saw the potlatch of the neighboring Kwakiutl as "megalomaniac paranoid." Others saw it primarily as a way of shoring up the kinship system.

Much of this literature was based on the mistaken belief that the Northwest Coast was so rich an environment that resources were always lavishly abundant. This mistake arose from two sources. First, early scholars did not realize the enormous extent of population reduction due to disease. They saw five hundred Nuu-chah-nulth and several million salmon and could not imagine a time of want. But ten thousand to fifteen thousand Nuu-chah-nulth were a very different matter! Second, these early scholars did not realize the frequency of failure of the critical salmon runs. Many Nuu-chah-nulth groups depended on chum for stored food; even if other salmon were abundant, there was no way to store them, unless exceptionally hot and dry weather permitted successful drying. If the chum run failed, the community faced severe food shortages.

Finally an ecological explanation was devised. Wayne Suttles (see Suttles 1987 for the original paper and further views on the subject) argued that potlatching among the Straits Salish (southeastern neighbors of the Nuu-chah-nulth) served to spread the food around evenly and thus reduce the risk of famine in bad years. A group that was blessed with an exceptionally rich run of salmon could potlatch.

CASE STUDY

When they hit a bad year, they could collect on their potlatch debt—they could be fed by the people they had fed before.

Stuart Piddocke (1965), a student of Suttles at the time, generalized the Suttles hypothesis to cover the entire Northwest Coast. This view was adopted by Marvin Harris (1985), but he added—correctly—the observation that social breakdown in the late nineteenth century caused the system to get out of hand. In a classic paper, Martin Orans (1975) showed that there was no good evidence presented for the Suttles hypothesis and indicated that anyone claiming it would have to show that potlatches were organized that way—that they could and did actually serve that function.

Subsequent research has shown that they did not generally do so. In the old days, a number of years might be required to organize a potlatch. It was thus impossible to predict whether the final year would be a good or bad one (if it was a bad one, the potlatch was delayed a year or so). Moreover, people who were fed in a potlatch were usually called on for more immediate services—labor, help with chiefly duties, or war. Suttles accumulated data that showed how ordinary feasts helped even out the resource picture, but potlatches were a different matter—both ecologically and socially.

Drucker and Heizer (1967) criticized Suttles's theory, reviewed the literature, and concluded that potlatches were related to warfare (also see Ferguson 1984). For the Nuu-chah-nulth, at least, Drucker and Heizer's theory is definitive. Potlatches did improve the ecological picture for the most powerful chiefs, but not directly. Potlatches did not provide just goods; they provided allies.

ENVIRONMENTAL MANIPULATION AND RESOURCE MANAGEMENT

The Nuu-chah-nulth practiced relatively little environmental manipulation, with the exception of some burning intended to open the landscape to attract deer and enhance the growth of berries. They also practiced some passive techniques aimed at controlling bad weather (one of the abiotic elements of the environment).

Several resources were actively managed, primarily salmon. Fish populations in streams and rivers were closely monitored, and some artificial stocking of streams was also done (Sproat 1868). Pas-

CASE STUDY

sive management was practiced on virtually all resources, and all things—animals, trees, and rocks—had spirits that had to be treated as one would treat a human. For example, when stripping the bark from a tree, people had to ask permission from the tree and explain that the bark was necessary to them and those for whom they were responsible (Drucker 1951; Turner and Efrat 1982). Anyone falsely claiming this necessity was subject to deadly punishment; a tree would soon fall on the person, or other disasters would ensue. Whales had to be placated by months of ceremony. The salmon were believed to be people under the sea; if they were treated politely, they would sacrifice themselves for their beloved friends on the land. When Europeans first appeared on the West Coast, some Nuu-chah-nulth thought they were the salmon arriving in their undersea houses (ships; Kirk 1986). To this day, whites are called mamalni, "floating house people."

People made sure these ethical messages were enforced because they believed that natural disasters, such as storms and failures of salmon runs, were often punishments for failing to observe these ethical standards. Exhaustion of a stream from overfishing, for instance, was caused by anger of the salmon people at humans who took more than their share. Ceremonies reinforced this view. The simplest was the brief apology and prayer uttered when taking a piece of bark or root. At the other extreme were months-long, secret, terrifying ceremonies associated with whaling.

The strong tendency to see the land as holy, natural entities as "people," and the world as guided by spirit is still very pronounced on the west coast of Vancouver Island. Not only has it not died out among the Nuu-chah-nulth; some of the belief system has been adopted by some whites.

NOTE

A word about names is necessary here. The Nuu-chah-nulth had, until recently, no name for themselves. They are known in older literature as the Nootka from a mistake made by James Cook when he landed among them in the 1770s. He asked the name of the landing place. According to the most believable story (Kirk 1986), the Indians thought he was asking if he could sail around it, and they replied

CASE STUDY

"you can go around," which sounds like *nootka*. For obvious reasons, this name is not used by the people themselves. In Vancouver Island English, they are the "Westcoast People." In the nineteenth century, they were called the Aht, from a suffix that means "people of" (like the "-ian" in "Californian"; Ahousaht means "people of the place Ahous"). The name Nuu-chah-nulth, roughly meaning "people on our side of the mountains," was coined in the twentieth century. All of these names are still in use; no one name has universal acceptance.

In 2001, the Nuu-Chah-Nulth agreed to a treaty with the governments of British Columbia and Canada that would give the groups self-rule, cash, and shared control of much land. The treaty is awaiting final approval and implementation.

CASE STUDY

CASE STUDY: THE MBUTI OF THE ITURI FOREST

This case study of the Mbuti provides an example of a mutualistic relationship between hunter-gatherers and agriculturalists interacting within ecological and cultural ecotones, with all of the difficulty in maintaining such a relationship. In addition, the Mbuti have developed a unique inventory of hunting tactics, one that effectively creates multiple human hunting niches within a single habitat.

The Mbuti are one of a number of groups of hunter-gatherers living in the Ituri Forest of central Africa (fig. 5.8). They are of short stature (generally under 4.5 feet in height) and have relatively short legs. This small body size is thought to be an adaptation to humid equatorial forests, as other people of small stature are found in other such forests worldwide. Historically, such people have been called "pygmies," although many now use the term *Twa*, a Bantu word meaning something like "short people," to refer to African pygmies. The people of the African rain forests have been known to the outside world since the time of ancient Egypt, but early European characterizations made them out to be subhuman monsters, mythical flying creatures, and the like.

CASE STUDY

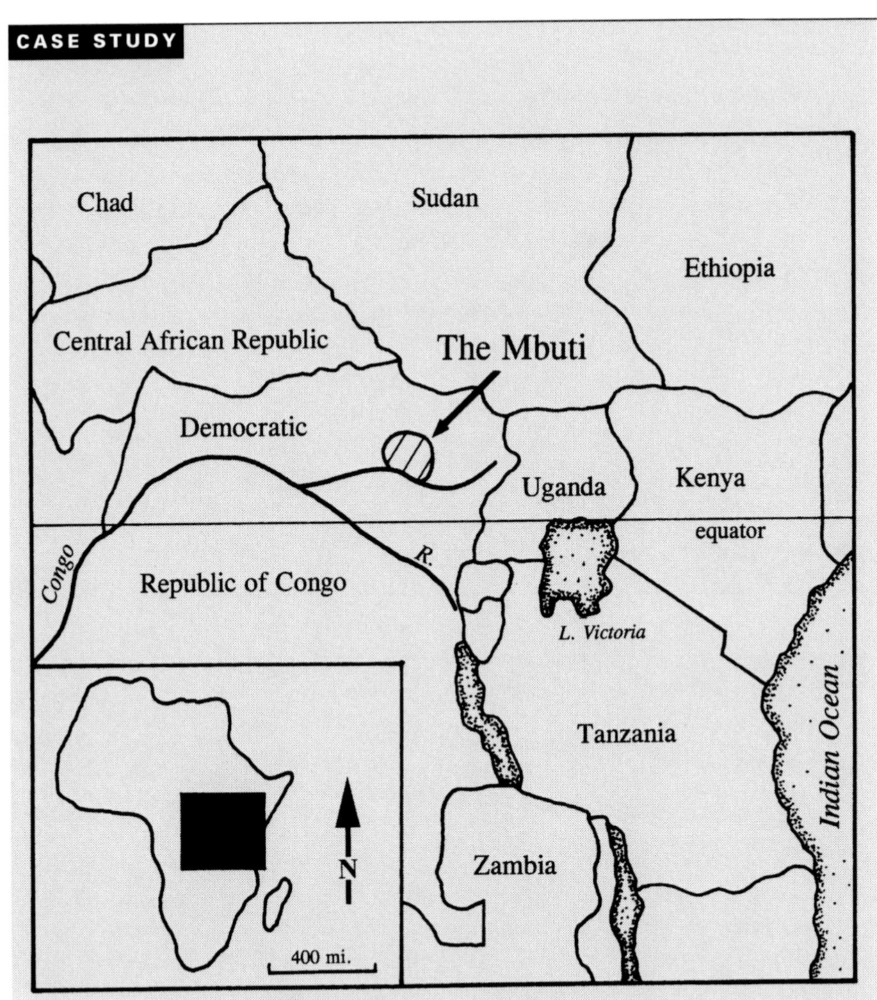

FIGURE 5.8
Location of the Mbuti in the Ituri Forest of Central Africa.

It is not clear how long people have inhabited the Ituri Forest. Some believe that the carrying capacity of the forest was fairly small until the arrival of agriculturalists about four thousand years ago (see Bailey et al. 1989). As the Bantu farmers entered the region, they cut down sections of the forest for agricultural fields. Over time, the fields were abandoned and the forest regrew, creating a mosaic of old- and secondary-growth forest. This use of the forest by the Bantu continues, and their old fields go through a series of succession stages, each of which is utilized by the Twa for different suites of resources.

CASE STUDY

A number of different Twa groups live in central Africa and have been the subject of numerous studies. Two of the best described groups are the Efe (see Bailey 1991) and the Mbuti. Each of the various groups lives in a different specific environment, and their adaptations vary. This case study considers only the Mbuti as they were in about 1900, when some thirty-five thousand Mbuti lived in the Ituri Forest and interacted with the Bila, their agricultural Bantu neighbors. Much of the information on the Mbuti discussed below was derived from the work of Turnbull (1962, 1983), Hart and Hart (1986), and Duffy (1996). Other Twa and Bantu groups share similar mutualistic relationships, each tailored to the group's specific conditions.

THE NATURAL ENVIRONMENT

The Ituri Forest is located in the Congo Basin along the equator in the approximate center of Africa (fig. 5.8). It is the forest of Henry Stanley and Dr. David Livingstone, where the American stereotype of "darkest Africa" was born, and covers about fifty thousand square miles. The climate is warm all year, and there is considerable rainfall. Being located at the equator, the Ituri Forest experiences few major seasonal environmental fluctuations, although the winter is generally dryer than the rest of the year and has an impact on adaptations. Seasonal differences exist in the availability of some resources, such as honey and insects.

The forest contains abundant game. However, it has been argued that many of the animals have relatively little body fat in some seasons and that eating lean meat alone could result in malnutrition (Hart and Hart 1986; also see Speth and Spielman 1983). It is not clear how important a factor this risk of malnutrition is in the Mbuti economy. Still, plants provide most of the food, with meat being eaten but mostly traded to farmers in exchange for agricultural products. Interestingly, many of the wild plants utilized as food by the Mbuti are obtained from secondary forests, those areas cut down by farmers and later reforested, rather than from the old-growth forests. In this sense, the Mbuti are dependent on farmers not only for traded foods but also for their cutting of old-growth forests so that secondary forests containing food plants can grow.

CASE STUDY

Prior to the arrival of farmers, the Ituri was a dense and largely unbroken expanse of old-growth forest. Farmers have cut down portions of the forest for fields, but the forest has grown back over old fields (creating a secondary forest). Many hundreds of tree species grow in the forest, and it is home to many other plants and animals.

The forest is central to Mbuti life. The forest is called "mother," and the Mbuti consider themselves to be the "children of the forest." The forest is viewed as a deity that provides food and materials to support the Mbuti and to which the Mbuti appeal for help and offer thanks in ceremony.

Both the Mbuti and the Bila believe the Mbuti to be the original inhabitants of the forest. The Mbuti are thus thought to have special rights in, and perhaps magical power over, the forest. The forest holds few fears for the Mbuti, although they rarely travel alone, and they make noise when they walk to avoid being attacked if they surprise an animal. Material possessions are of little concern to the Mbuti; well-being and happiness are much more important. The Bila are afraid of the forest and enter it only for specific reasons, such as the very productive annual caterpillar harvest and their puberty ceremonies. The Mbuti reinforce this fear to maintain their social and economic distance from the Bila.

SOCIOPOLITICAL ORGANIZATION

The Mbuti are organized into a number of bands, and there is no formal leadership or larger polity. Each band has at least one camp, usually fairly small but of at least six families. A band has a generally wedge-shaped territory with an associated Bila village on its outside, other band territories on its sides, and an area in the center defined as being off-limits to hunting, perhaps to provide a "sanctuary" (a conservation technique) for pursued game. Hunting territory is not owned, and bands hunt in each other's "territory." Bands are very mobile and practice a fission-fusion settlement system. When game is depleted in an area, the decision will be made to move the camp to a new area within its general territory. However, if an elephant is killed, the camp will move to the kill site rather than try to bring the meat back to the original camp.

CASE STUDY

The Mbuti are patrilineal and have extensive kinship networks. With everyone's having many relatives in different bands, the Mbuti bands have generally peaceful relationships with each other. Mbuti youths reach sexual maturity at a young age and get married at about the same time. People are expected to marry someone from a different band, the farther away the better to establish and maintain the extensive kinship network. The women move away to live with their husband's band, and a few men have more than one wife. An ideal marriage arrangement is the exchange of women between bands.

Small-scale disputes are settled by shouting at each other and showing contempt and ridicule for each other. Social control of small infractions is enforced in the same way. If a dispute or other problem becomes major, a band will split into two new bands, thus ending the dispute.

While there is no hard and fast division of labor, men are the ones who tell the stories and conduct the ceremonies while women have the responsibility of cooking and building houses. In food-getting activities, the work is often shared, with men helping to gather plants and women and children helping in hunting.

ECONOMICS

The primary subsistence pursuits of the Mbuti are hunting the game of the forest, gathering wild plants, and trading for agricultural products to supplement their diet. Hunting the many small and large game remains a focus of the Mbuti, but like most hunter-gatherers, they derive a minority of their food from hunting. Other important food resources include honey and caterpillars.

The Mbuti have two basic hunting adaptations; net hunting and bow hunting. The net hunters live mostly in the west and south of Mbuti territory while the bow hunters mostly live in the east, but the two adaptations overlap as related to terrain and ecozone. In the area of overlap, one could argue that there are two niches within the same habitat.

CASE STUDY

Hunting

The Mbuti are excellent and stealthy hunters and have an intimate and detailed knowledge of the forest, its travel routes, dangers, and resources. Many kinds of animals are hunted, but duikers (a small antelope) form the major prey species. Elephants are also hunted, being speared in the belly from ambush on jungle trails. Meat is shared within the group via a very complex system of reciprocity and is traded to the Bila. In sum, there is plenty of food available in the forest, and the Mbuti say that "the only hungry Mbuti is a lazy Mbuti."

Net Hunters

One hunting approach is the use of nets to capture game (see Noss 1997 for some comparative data). Net hunting is cooperative, requiring a fairly large number of participants to be effective, and the resulting "catch" generally includes a broad variety of species. Within groups of net hunters, each family will own a net, generally between one hundred and three hundred feet long. A number of families (at least six are needed) will join together and connect their individual nets into a single net that may be as long as two thousand feet. This large net will be placed in a U shape so that game can be driven into the trap (fig. 5.9). In addition to the men, women (who are considered equal partners) and children will participate, providing logistical support, holding the net, driving game (sometimes with the aid of dogs), or processing carcasses. The meat obtained is shared by all the participants.

Net hunts are conducted on an as-needed basis when a sufficient number of families with nets can be assembled and an agreement reached regarding the place and time of the hunt. The net hunters prefer to operate in areas where plant foods are available so the women can also gather. The animals obtained by net hunting are usually small to medium in size and of many species. Large game, such as buffalo, would run through the nets, ruining them, and are rarely captured using nets.

Bow Hunters

The other basic approach is hunting with the bow and arrow, an effort undertaken largely by solitary men or small groups of men.

CASE STUDY

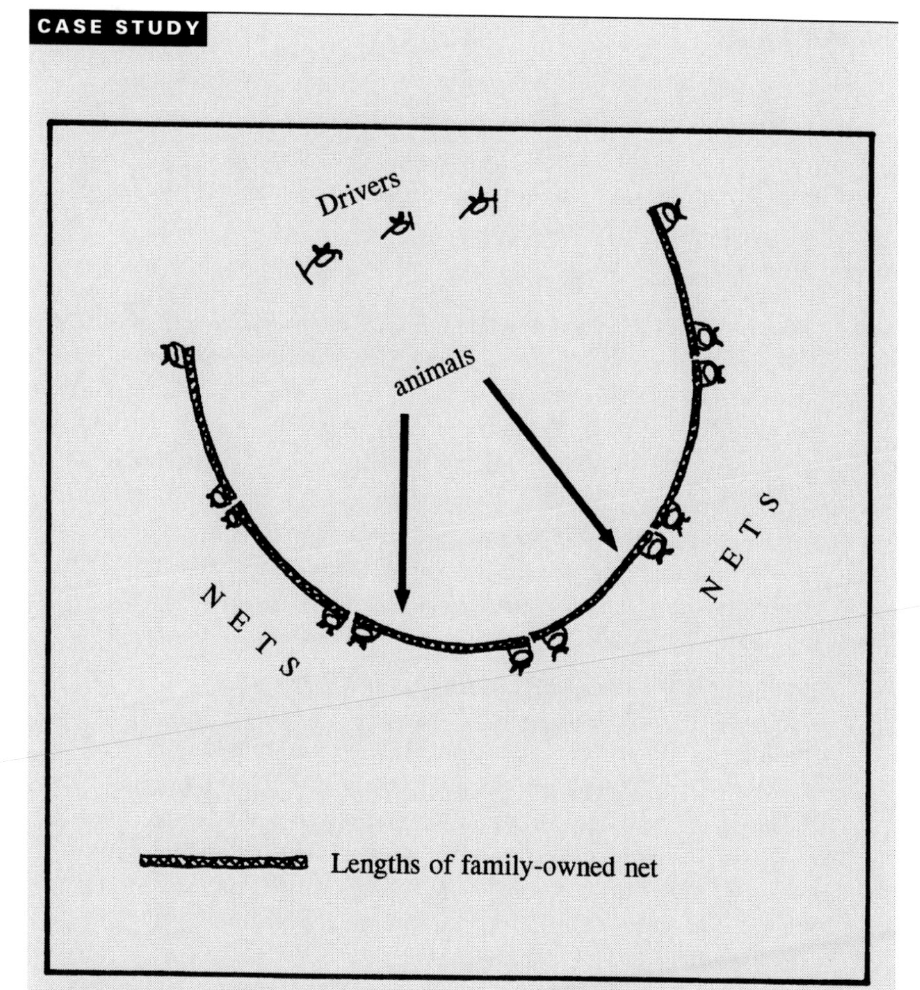

FIGURE 5.9
A schematic of the Mbuti net hunting tactic, with some people driving the animals into the large U-shaped configuration of nets.

Game is generally hunted on an opportunistic basis, with the hunters leaving in the morning and returning in the evening, although overnight hunting trips are sometimes undertaken. Any meat obtained by the hunters is shared with the community. Mbuti arrows are small, but poison is used on the tips of the arrows, making them lethal.

Bow hunters will take some of the same animals as net hunters, but larger animals, such as buffalo and elephants, are more likely to be sought using spears and traps, and many of the smaller animals

CASE STUDY

captured by the net hunters are not pursued by the bow hunters. With their different technologies and different species as prey, bow and net hunters can hunt in the same area (two niches in the same habitat, as mentioned above).

Gathering

The Mbuti, generally the women, gather a number of wild plants. However, the number is fairly small compared to the number of plants they know and classify, and wild plant foods are less important to the diet than agricultural products, although this pattern may have developed only within the last several hundred years. The old-growth forest apparently contains relatively few species used for food, so much of the plant gathering is undertaken in secondary forests. It could be argued that the Mbuti are dependent on farmers' generating secondary forests as they abandon fields.

ENVIRONMENTAL MANIPULATION AND RESOURCE MANAGEMENT

The Mbuti themselves do little to actively manipulate their environment. They try to keep the Bila out of the forest as much as possible, and that helps to maintain and manage the forest. However, the Bila utilize the forest to some extent, practicing some slash-and-burn horticulture. The Mbuti eventually take advantage of the Bila's work, monitoring the forest succession and utilizing wild resources that colonize the abandoned fields. The Mbuti manage the forest through ritual, maintaining a ceremonial relationship with the forest to ensure its continued productivity and protection.

The Mbuti employ several interesting techniques to manage the game they hunt. First, it could be argued that the bow and net hunting adaptations target different animals, preventing overexploitation of game through the reduction of capture rates by any group of hunters. The use of these techniques could be viewed as having separate niches in the same habitat. Second, the Mbuti and their Twa neighbors recognize a central area where hunting is not conducted and so provide a game refuge.

CASE STUDY

RELATIONS WITH THE BILA

A relatively large percentage of the plant foods consumed by the Mbuti are domesticated species obtained in trade with the neighboring farmers, the Bila. The Mbuti will travel to the Bila villages to trade, and most speak the Bila language. The Mbuti provide meat to the Bila in exchange for goods such as bananas, corn beer, manioc, and tobacco. If they do not have meat to trade, the Mbuti will provide labor in exchange for these products.

The Bila consider themselves as "owning" the Mbuti and want to "rescue" them from a life in the forest. Mbuti will "hunt" (cheat, trick, and/or steal) things from the Bila and feel that the Bila are "stupid" villagers, deserving to be taken advantage of. However, the Bila are usually aware of the Mbuti actions but excuse them as the acts of poor and desperate people. In addition, the Bila do not want to offend the Mbuti by making an issue of their actions.

Individual Bila farmers will develop a close trading relationship with individual Mbuti, whom they regard as poor hunter-gatherers requiring assistance. The Bila thus "adopt" Mbuti, feel responsible for them, sometimes consider them to be property, and will even "bequeath" them from generation to generation. The Bila want the Mbuti to settle down and become sedentary farmers so they can be controlled. To accomplish this, they offer the Mbuti land and women. Much to the dismay of the Bila, few Mbuti accept the offers (being ethnocentric, the Bila find it hard to believe that the Mbuti would actually prefer the forest). However, the Bila know that if they become too oppressive, the Mbuti will simply move to a new location, leaving the Bila with no trading partners.

The Bila are dependent on the Mbuti and so avoid offending them. In addition to trading the very important resources of meat and honey to the Bila, the Mbuti include Bila boys in their male initiation ceremonies. The Bila lack the ceremonial expertise to conduct such initiations and so must rely on the Mbuti. If the Mbuti did not perform this service for the Bila, Bila boys would not become men and would not be eligible for marriage.

CASE STUDY

DISCUSSION

There are four interesting points to be made here about Mbuti ecological adaptations. First, the Mbuti and Bila maintain a mutualistic relationship. The Bila need the Mbuti for meat and labor, and the Mbuti need the Bila for agricultural products. Second, the Bila need the Mbuti for religious reasons: to conduct the initiation ceremonies for Bila boys. Third, the Bila's use of some parts of the forest for farming creates the secondary growth where many of the wild plant foods used by the Mbuti grow. Lastly, the Mbuti do their best to reinforce the Bila fear of the forest, both to retain their monopoly on providing forest products and initiations to the Bila and to reduce the hunting pressure on the forest animals.

Thus, the Bila and Mbuti share a mutualistic relationship. The Mbuti get agricultural products and trade goods and the Bila get labor, meat, honey, and their boys' initiations. Both parties benefit, and neither wishes to offend the other: Both are dependent on the relationship. One of the reasons the system can work is that the two groups have mostly different niches and generally occupy different habitats.

RECENT DEVELOPMENTS

In recent times the people of the Ituri Forest, including the Mbuti, have faced a number of challenges. Over the last hundred years or so, the government has tried to resettle the Mbuti into "villages" outside of the forest so as to "free" the Mbuti from their bleak existence as hunter-gatherers. These efforts failed since the Mbuti were adapted to a mobile life and a diet of forest products rather than to a sedentary life eating agricultural products.

Today, increased logging of the forest is rendering game extinct and reducing habitat of both the game and the people. In addition, the expansion of farming and mining in the forest is cutting it down at an increasing rate, further reducing the habitat of the Mbuti as more and more farmers and miners enter the region. The Bila demand for meat from the Mbuti is steadily increasing, and the supply is beginning to be overexploited. It is possible that the mutualistic system of the Mbuti and Bila, apparently in place for thousands of years, is in peril.

CASE STUDY

Perhaps a larger problem is the civil war raging in the region since 1994 (see Hooi 2002). In this war, the Twa find themselves impressed into military service and fighting their former neighbors and trading partners. Services such as education and health care, as well as supplies, have been disrupted or stopped, and the infrastructure of the region has suffered. Trade has been suspended in many parts of the forest.

Ironically, the disruption caused by the recent civil strife has strengthened the mutualistic relationships between the Twa and the Bantu, who rely on each other even more with the erosion of outside support. The relationship between the two groups is actually quite flexible and adaptive to stressful conditions on their exteriors. ■

6

The Origins of Food Production

Beginning about ten thousand years ago, the relationship between humans and the environment changed dramatically. Until that time, all people were hunters and gatherers. At the end of the Pleistocene, or Ice Age, the climate shifted, and in some regions people began a process of intensive control over certain plant and animal species that ultimately led to their domestication. As part of this process, people became dependent upon those domesticated species for food, and the domesticates became dependent upon humans. The process further involved major changes in social and political organizations. In addition, the use of land radically changed; humans cleared fields and diverted rivers and streams. In essence, the development of farming created a new niche for humans.

The process of domestication seems to have first begun in the Middle East, possibly in the Jordan Valley or neighboring northern Syria and southeastern Turkey. The first domesticated species were dogs (which may have been domesticated earlier), sheep, goats, wheat, and barley. Neolithic villagers began to raise these species between about ten thousand and twelve thousand years ago (Zohary 1982; Zohary and Hopf 1988; Bar-Yosef and Meadow 1995). Wheat and barley grains are attached to the stalk by a thin stem, the rachis. In wild stands, the rachis is brittle and shatters when the grain is mature. The grain falls to the ground and serves as seed for the coming year. Occasionally a wild plant occurs whose rachis is tough and does not shatter. When early harvesters cut wild grain with sickles, the ordinary heads would shatter and scatter grain, but these few exceptional heads would not. The tough seed heads could be carried back to the village with no fear of loss. Probably, over time, stands of these unusual grains arose around the villages. Here people harvested them and encouraged them to grow.

Deliberate seeding became necessary because the shatterproof heads do not naturally scatter their seeds (Wilke et al. 1972), and agriculture was born.

Animals, including dogs, sheep, cattle, and goats, also were domesticated. Biologically, dogs are wolves (see Serpell 1995). The ancestor of the dog was the small southwest Asian wolf, which was something of a scavenger and presumably was common around camps and villages. Families probably kept young ones as pets—as many hunter-gatherers around the world today keep pets. The most docile animals lived the longest; fierce ones were soon killed. Probably many of these dogs escaped, living around the village in loose symbiosis with humans. Eventually, the docile ones began to breed in captivity, and "man's best friend" emerged. Domestic dogs developed smaller teeth and jaws.

Sheep and goats also grew smaller and tamer. Originally sheep did not have wool—that was not to come for thousands of years—but their fleeces probably grew softer over the generations. Cattle, domesticated later, were bred to be smaller and tamer than the fierce wild strains. Milk did not become a focus of breeding and effort until perhaps seven thousand to eight thousand years ago.

Later, beans, lentils, and tree crops were domesticated. Bitter and poisonous chemicals were bred out of some of them. From among the stony, sour, wild fruit, the least stony and sour were picked to grow, and they interbred. Their least stony and sour progeny were then selected again. It took thousands of years of this artificial selection and breeding for the almost inedible wild pears and apples of the Middle East to become the large, sweet ones of our time. By the time tree crops were emerging in the Middle East, agriculture had arisen independently in several other parts of the world.

By ten thousand years ago, villagers in northern China were raising foxtail millet (*Setaria italica*)—one of several small grains indiscriminately called "millet"—and villagers near Shanghai were raising both short- and long-grain rice (Loewe and Shaughnessy 1999:46). By six thousand years ago, agriculture was well established all over north and east China. Pigs, dogs, sheep, chickens, and Chinese cabbages had been added to the crop roster. The pigs and sheep were probably domesticated independently of the Middle Eastern center. Sheep tame themselves, as visitors to Banff National Park (Canada) well know; you have to drive them away to keep them from living with you. Chickens are native to south China and southeast Asia and must have come from the more southerly parts of the agricultural zone.

At about the same time, and certainly by seven thousand years ago, people in what is now Mexico domesticated squash (*Cucurbita* spp.), chiles (*Capsicum* spp.), and millet (*Setaria viridis*)—ironically, the American form of the plant used by the Chinese. By about five thousand years ago, domesticated corn appeared on the scene, replacing millet; millet is more nutritious, but corn has greater yields. When corn was introduced to China in the sixteenth century, the same process began, and corn has now replaced millet in most of the world.

Common beans (*Phaseolus vulgaris*) and avocados (*Persea americana*) were also domesticated in Mexico, along with many other crops (see Fedick 1995 for a recent review of New World agriculture). When the Spaniards conquered Mexico, they were astonished at the incredible variety and productivity of the indigenous agricultural scene. There was nothing like it in Europe.

In South America, meanwhile, people domesticated lima beans (*Phaseolus lunatus*)—appropriately first appearing near the city of Lima. It is probable that potatoes (*Solanum tuberosum*) and other crops were domesticated by six thousand years ago. At about the same time, the llama and the alpaca were domesticated from their wild forms. In the lowlands, people were probably cultivating manioc (*Manihot utilissima*, tapioca) and yuca (not to be confused with yucca, a different plant). Eventually, they began to grow peanuts (*Arachis hypoglauca*), sweet potatoes (*Ipomoea batatas*), and countless other crops. Like the Mexicans, the South Americans proved to be outstanding plant breeders and agricultural system developers. Their ideas and crops are only now coming to be appreciated (National Research Council 1989).

Archaeological work in New Guinea has disclosed ancient fields that may have grown taro (*Colocasia antiquorum*) or something similar. These fields date as early as six thousand years ago and reveal another independent source of agriculture, either in the highlands of New Guinea or somewhere in mainland or insular Southeast Asia.

It is possible, even probable, that agriculture was independently invented in other areas as well. For example, it began very early in eastern North America (Smith 1995), and the crops found there are markedly different from those of Mexico, where early agriculture was roughly contemporary. Other areas where agriculture may have originated independently include northern Southeast Asia and northeastern South America. For further information and discussion on agriculture, refer to E. N. Anderson (1988), David Harris and Gordon Hillman (1989), Wesley Cowan and Patty Jo Watson (1992), Jack Harlan (1992), Richard MacNeish (1992), Douglas Price and Anne Gebauer (1995a), Alasdair Whittle (1996), and Dolores Piperno and Deborah Pearsall (1998).

AGRICULTURAL DOMESTICATION

Domestication can be defined at a number of levels (see Rindos 1984; Hayden 1995). As discussed earlier, one might define domestication to mean a general control, such as over landscapes, through active or passive management. A more specific definition is an active process through which the genetic makeup of an organism is purposefully altered by humans to their advantage. Many would narrow the definition to making the species dependent on humans. In the case of agriculture, we favor the latter definition.

People have always manipulated and interacted with plants and animals. This interaction has always resulted in the genetic alteration of the species, even if the alteration was not planned. For example, when deer are hunted, the gene pool of their population is changed, which will influence their evolution to some degree. In other cases, people purposefully altered some plants, such as pruning and weeding native tobacco. However, such alterations were minor or incidental, and no serious attempt was made to control genetics (see Rindos 1984:154–158).

Agricultural domestication involves a process in which people purposefully and selectively breed for more of what they want and less of what they do not want. People have long been aware of the combination of traits, selective breeding, and other aspects of basic genetics, although the details of the specific mechanism were not discovered until the mid–twentieth century. Hunter-gatherers understood this basic biology and took advantage of it by way of environmental manipulation and resource management. Sometime about ten thousand years ago, however, people began to focus on specific plants and animals, developing sets of tactics to enhance the productivity of these species. The seeds of the largest and most succulent fruits were planted so that the next-generation fruits were even larger and more succulent. Animals that were aggressive were killed and eaten before they could reproduce, which left the tamer individuals to parent the next generation, which would be even more tame. Over many generations, grains became bigger, fruits larger, and animals more tame, and all became dependent on humans.

Domestication often involves breeding out some of the natural defenses of the species. Wild sheep are fast, tough, and armed with large horns and can defend themselves well. Tame sheep are notoriously helpless. In fact, after people had domesticated the sheep, they had to breed specialized dogs to care for them!

Taking care of crops involves labor and skill. The more the natural defenses are bred out of crops, the more the farmers have to work. Thus, two choices are available. One can retain the natural defenses. This choice means less need for care in the form of human labor, chemical pesticides, or other things. Or one can breed single-mindedly for desired traits. This choice results in more effort being devoted to chemicals, fences, guard dogs, or whatever.

These processes are exemplified by two Mexican crops: avocados and corn. Avocado trees have been bred to bear much larger fruit than any wild avocado tree could manage, but otherwise they are not very domesticated. They contain a number of interesting chemicals that deter insects, herbivores, and diseases. They thin their own fruit, prune their own branches, and sprout with glorious exuberance from their own seeds.

Corn is an extremely domesticated plant. Its tassel, cob, huge and numerous seeds, tall and succulent stalk, and numerous husks are all "unnatural" traits that do not occur in wild corn. The tough husks protect the cob, but otherwise corn has major problems taking care of itself. It is simply a free lunch counter for insects, fungi, and wild animals. It cannot seed itself because the dense husks prevent scattering of the seed. Seedlings will come up from fallen ears, but they crowd each other to death. Contemporary domesticated corn requires people to plant it, harvest it, and protect it. The domestication process resulted in a genetically narrow species susceptible to disease. In 1973, one-third of the U. S. corn crop was wiped out by a single fungus in a couple of months. We now have to breed fungus resistance from wild or primitive strains into corn.

People do not breed only for food values. Within the last few decades, it has been discovered that as early as four thousand years ago, some groups along the coast of Peru domesticated cotton to manufacture nets with which to catch fish (Moseley 1975). In this instance, early domestication was undertaken for technological purposes rather than as a direct source of food.

In the United States, sorghum (*Sorghum vulgare*) grain is used for animal feed; the stalk is relatively useless. Therefore, sorghum has been bred for short stalks. In China, the stalks are used for firewood and thatch; thus, the sorghum there is bred for long stalks. In both cases, the grain yield is kept as high as possible, but the Chinese trade off grain yield against stalk yield while the American breeders have reduced the stalk to the bare minimum necessary to keep the grain off the ground.

THE TRANSITION TO FARMING

Hunter-gatherers did not instantly transform into agriculturalists once they began to domesticate plants and animals. In fact, it was a transition (see Price and Gebauer 1995a) in which hunting and gathering became less and less important while domesticates became more and more important. One could say that domesticated species and farming cultures developed a mutualistic relationship and coevolved or coadapted. This process is still going on; virtually all contemporary cultures are dependent to some degree on agricultural products, and all still utilize at least some wild resources. This fact sometimes makes classification of certain cultures difficult because subsistence strategy is usually defined on the basis of the most visible or important aspect of the economy, but the most important aspect is sometimes not clear.

Hunter-gatherers and farmers utilize their environment in fundamentally different ways. Hunter-gatherers "largely live off the land in an *extensive* fashion, exploiting a diversity of resources over a broad area, [whereas] farmers utilize the landscape *intensively* and create a milieu that suits their needs" (Price and Gebauer 1995b:3–4, italics in original). In that sense, farmers intensify their use of the environment and resources, and in doing so, they become more narrow and less diverse and lower their adaptive flexibility. In most cases, the scale of environmental manipulation and resource management increased with farmers, and a general increase in the ritual management of resources began. Farmers have manipulated environments on such a scale that much of Europe was deforested during the Neolithic and much of North America during the last five hundred years, and an increasing proportion of Amazonia is suffering this fate today.

The evolution of food production has been a process of reducing the land area needed to feed an individual. Hunting and gathering required many acres to support one person; incipient agriculture (agriculture in which crops are raised, but only a few, and they are not significantly modified from wild-type ancestors) supported only a few more people per square mile. Every change and modification of agriculture allowed more people to flourish per square mile. Today, agricultural development and agricultural intensification are basically directed toward the same goal. This is not to say that the amount of land dedicated to agriculture has not increased; it certainly has. Today, about one-quarter of the land surface of the earth is under cultivation. Granted that agriculture only rarely increases the amount of food available to the cultivator, there are two logical reasons for intensifying agriculture. First, an elite might order the peasants to grow more to support the elite group's lifestyle. Thus, while the elite would have been better off than the average hunter/gatherer, the peasants would have been worse off. The existence of an elite class presupposes some sort of intensive agriculture because most hunting and gathering and simple agricultural economies cannot support elites. Second, there could be a need to get more food per acre even if it does not mean more food per person. For example, the population may increase, the food may be sold or traded, the land may grow scarcer, or the extra food may be desired for some special reason, such as fattening animals.

The next logical thing to recognize is that agriculture requires some inputs besides land. Labor and crop varieties are obviously vital to all agriculture. Fertilizer, pest control, and improving technology are not vital but are very useful indeed. Contemporary agriculture would not support so many people without them, but they also have their negative ecological consequences, such as pollution.

In the end, the average human cultivator is probably worse off now, in regard to food and general welfare, than the average person of ten thousand years ago, if our knowledge of contemporary hunters and gatherers is any guide. It is common to believe that agriculture is much more productive than hunting and gathering. However, most of the world's population consists of peasants and poor cultivators in the Third World, often malnourished and racked by disease. Agricultural intensification has probably never benefited the majority of mankind.

ON THE ORIGIN OF AGRICULTURE

Theories of the origin of agriculture abound (see Rindos 1984; Harris and Hillman 1989; Price and Gebauer 1995a) (table 6.1). People had been dependent on wild plants and animals for millions of years. Why, then, did hunter-gatherers abandon a relatively stable and productive adaptive strategy to take up agriculture, an economic pursuit that requires more labor and is subject to catastrophic crop failure? The process may have been partly unintentional and even accidental. However, it is also possible that people deliberately strived to improve the plants and animals on which they depended, making deliberate choices for desired traits (e.g., larger cobs on corn).

Either way, once people became dependent on domesticates, they tended to congregate and make sedentary life the rule. These transformations resulted in a whole series of other changes, including a decreasing dependence on wild resources, a much greater emphasis on land ownership, and increases in political complexity, population, and specialization. Why and how all this happened are still unclear, but many hypotheses exist. Because domestication occurred in at least several different places at the same time, it may be that each of the models, a combination of them, or one as yet unknown is correct. Factors believed to have led to agriculture fall into three basic categories: environmental change, population pressure, and changes in organization (Price and Gebauer 1995b:4). Each of these three factors is interrelated, and all may have had some influence on the process. While the origin of agriculture is not at all clear, most now agree that agriculture initially developed among relatively complex groups in areas with relatively abundant resources.

Table 6.1. Some Theories of the Development of Agriculture

General Theories	*Summary*	*References*
	Environmental Change	
Oasis theory	Environmental changes at the end of the Pleistocene forced people into a close association with certain plants and animals, leading to their domestication in some instances.	Childe 1936, 1942
Hilly flanks theory	Intensive exploitation of the native grasses along the hilly flanks of the Tigris-Euphrates river valley led to domestication of those species in some areas.	Braidwood 1960
Marginal environment	Due to increasing need for efficiency, people living in marginal environments would have been forced to intensively manage their plants and animals, resulting in their domestication in some instances.	Binford 1968; Flannery 1969
Food crisis	Due to the loss of Pleistocene species hunted by people, they were forced to manage their remaining animals more efficiently, leading to their domestication in some instances.	Cohen 1977
Wet and stable	Dry and unstable environment prevented agriculture during the Pleistocene, but environment became wetter and stable afterward, leading to plant intensification, domestication, and agriculture.	Richerson et al. 2001
	Population Expansion	
Population growth	As populations began to expand and demand for food increased, the exploitation of certain species intensified and led to their domestication.	Cohen 1977
	Changes in Organization and Management	
Efficient hunter/gatherers	A particularly efficient group of hunter/gatherers increased the yield of a particular resource, perhaps to the point of increasing dependence and ultimate domestication of that resource.	Winterhalder and Goland 1997
Scheduling changes	Changes in scheduling in the exploitation of wild resources created an overreliance on some resources, eventually leading to domestication.	Flannery 1972

Environmental Change

It is clear that there was a major environmental change at the end of the Pleistocene, when the climate became hotter and drier. Many species went extinct, and many others moved to other regions. No one questions that these changes impacted human populations. While the details of human adaptation to these changes are not fully understood, all known agriculture began at about the same time in many regions of the world. This coincidence is highly suggestive of climatic change having a major role in the development of agriculture.

The Oasis Theory

V. Gordon Childe (1936, 1942) put forward the Oasis Theory, which argues that as the environment generally dried up at the end of the Pleistocene, people, plants, and particularly animals were forced into close association in areas of remaining permanent water, or "oases." He believed this geographic association led to a close symbiotic relationship between people and certain species and eventually to the domestication of some of those species. Childe held that farmers were superior to hunter-gatherers, that an evolution to farming was natural and desired, and that people would adopt farming at their first opportunity (see Watson 1995:24).

Striking support for this theory has recently emerged from the Middle East. The earliest known agriculture occurs in the Pre-Pottery Neolithic A period in and around the huge earthquake-riven zone that includes the Jordan River Valley. A major climate change, known as the Younger Dryas, made the region suddenly much drier around eleven thousand years ago. Human populations shrank and concentrated around permanent water sources. Here they apparently began to sow the crops they had previously harvested wild but now had to produce (Leslie Quintero and Philip J. Wilke, personal communication, 2000, based on their ongoing research). In general, agricultural development takes thousands of years to reach fully productive levels (Sauer 1952; McNeish 1992), but the Jordan Valley case indicates that Childe's concept of spatial manipulation and oasis survival may have been critically important.

The Hilly Flanks Theory

In 1948, Robert Braidwood, motivated by the Oasis Theory, began a concerted effort to identify the location and date of the earliest farming. He found the earliest evidence (to that time) of agriculture along the hilly flanks bordering the Tigris-Euphrates river valley. Braidwood (1960) argued that Childe's catastrophic climatic change had not happened and that the grasses that would become wheat and barley thrived in the prime grass habitat that surrounded the valley. Living among the dense stands of grasses, people eventually developed the technology to domesticate those species (see Watson 1995:24–26), and the practice then spread elsewhere. To date, no early agriculture has been found in the hilly flank zones, but sheep and goats may have been domesticated there.

Margin Theories

Lewis Binford (1968) and Kent Flannery (1969; also see Redding 1988) later argued that domestication should not occur in areas rich with resources but located in marginal environments, where people had to work harder to make a living. These researchers proposed that in regions where resources were less abundant, people would pay closer attention to the species, manage them better, maintain a more intimate relationship, and eventually domesticate them.

A Food Crisis

Mark Cohen (1977; also see Boserup 1965) suggested that the mass extinction of giant animal forms due to climatic change at the end of the Pleistocene caused a "food crisis" that led to agriculture. In essence, when the large animals that hunters relied on died off, people had little choice but to farm. There are many problems with this theory. First, it seems clear now that reliance on large animals figured less than was once believed. Second, if the idea were correct, one would expect evidence of malnutrition among the hunter-gatherers of the time, but there is no such evidence. Malnutrition became common only after agriculture was widespread—when reliance on one or two staple grain foods led to unbalanced diets and susceptibility to famines when crops failed (Cohen and Armelagos 1984).

Most recently, it has been suggested (Richerson et al. 2001) that the climate of the Pleistocene, arid and highly variable, made agriculture impossible. Further, as the climate changed to conditions more favorable to agriculture (wetter and more stable), plant intensification began and inexorably led to domestication and agriculture, although sooner in some places than in others. This interesting idea has yet to be fully explored.

Population Expansion

Cohen (1977) also argued that by the end of the Pleistocene, the human groups had occupied all terrestrial habitats and that population growth had reached a point that "required" a larger and more stable food supply. Agriculture was the only way to support the growing populations. The knowledge that many hunter-gatherer populations grew large without agriculture weakens this theory.

Some sort of population size and pressure may have been a precondition or an effect, rather than a cause, of the development of agriculture (see Price and Gebauer 1995b:7).

Changes in Organization and Management

Particular groups are organized in particular ways, presumably in some general accommodation with their environment. However, if conditions changed, various aspects of adaptive organization would also have changed, and some of these changes could have initiated the process of domestication. The conditions that might have changed could be environment, population, technology, or religion, to name a few.

Efficient Hunter-Gatherers

Hunter-gatherers differ in their efficiency, at both the individual and the group level. If a particular group was unusually efficient or developed a new and productive technology, that group may have been able to increase the yield of a particular resource, perhaps to the point of increasing its dependence and ultimate domestication. It has also been suggested that individual groups made key decisions about resource use and risk management that led to dependence and ultimately domestication (Winterhalder and Goland 1997).

Scheduling

Another explanation (Flannery 1972) proposes that simple scheduling changes could send a group down the road to an overdependence on a specific resource and ultimately to the domestication of that species. For example, if a group on its seasonal round stayed too long in a particular locality (e.g., a resource patch), they might arrive at the next patch too late to efficiently use the resources there. Under those circumstances, the group may have been forced to stay in the first patch and to increase the intensity of the use of the species there. If this were a recurring circumstance, the resource in that first patch may have eventually become domesticated and the group dependent on the resource and devoted to its cultivation.

Trade

Another theory (e.g., MacNeish 1992) took a different tack. Richard MacNeish's theory is complex, but it includes a recognition that agriculture arose independently in several areas that are ecologically very complex, with many different zones of altitude and moisture. They are also in the central parts of large regions where trade was active. Under these conditions, agriculture grew out of trade or exchange. People exchanged products in their local zone with other people in other zones, nearby or far away. One can imagine that these traders wanted to maintain a secure supply of their trade goods. This theory too has been criticized, largely on evidential grounds (Watson 1995). The debate is not over (see also Rindos 1984; McCorriston and Hole 1991; Blumler 1992; McCorriston 1992).

TYPES OF AGRICULTURE

For general purposes, agriculture can be defined as the raising and use of domesticated plants and animals. Agriculture has three major forms: horticulture, pastoralism, and intensive agriculture. The differences between the three are largely a matter of scale (table 6.2).

Within cultures emphasizing plant cultivation, **monoculture** (single-species planting) and **polyculture** (multiple-species planting) are the primary systems used. Monoculture results in a greater yield but is highly susceptible to pests, disease, and soil exhaustion, any of which can result in catastrophic crop failure, such as the Irish potato famine. In contrast, polyculture seems to be more stable (because of greater diversity) but has less yield. **Silviculture** is an agricultural system using domesticated trees. All of these systems have relatively low biodiversity (e.g., cornfields) and usually replace systems with much greater biodiversity (e.g., rain forests).

Horticulture

Horticulture (discussed in detail in chapter 7) is low-intensity agriculture involving relatively small-scale fields, plots, and gardens. Hunter-gatherers can employ horticulture on a part-time basis without abandoning hunting and gathering. While horticulture does involve tilling, planting, and harvesting, it does not necessarily require an irrevocable commitment to domesticated species.

Both plants and animals can be raised in horticultural systems. Such a system generally does not involve the use of equipment (e.g., plows), draft animals, or mass labor. The food is raised primarily for personal consumption rather than

Table 6.2. Types of Agriculture

Type	*Emphasis*	*Characteristics*
Horticulture	plants, but some animals, hunting and gathering often remaining important	small-scale, individual production; only human labor
Pastoralism	animals, but some plants and some hunting and gathering	small- to large-scale, generally mobile
Intensive agriculture	plants, but animals sometimes raised intensively	large-scale, use of labor supplements (animals or machines)

being traded or given to a central authority. Being small in scale, horticultural systems generally support smaller populations and generally less complex political systems than intensive agricultural systems support.

Pastoralism

Pastoralism (discussed in detail in chapter 8), sometimes called **animal husbandry,** involves the herding, breeding, consumption, and use of domesticated (or managed) animals such as sheep, cattle, reindeer, camels, horses, llamas, and yaks. Plant cultivation generally is not part of this adaptation, but plant goods and food may be obtained from other groups. Pastoralists, sometimes called **nomads**, may either be sedentary or practice a seasonal round based largely on the availability of forage for the animals.

Intensive Agriculture

Intensive agriculture (discussed in detail in chapter 9) involves a full commitment to domesticated species, although wild resources remain in the economy. Intensive agriculture is the large-scale cultivation of plants, often with the use of animal labor, equipment such as plows, and irrigation or other water diversion techniques. Some pastoral activities may be a part of intensive agricultural systems. It may well be that horticultural techniques are employed by intensive agriculturalists, as in the small personal gardens that are common in Western societies.

THE IMPACT OF AGRICULTURE

All organisms interact with and affect their environment. Humans, with their culture, impact it on a much greater scale than other organisms do (see treatments of human impacts on the environment in Goudie 1994; Meyer 1996; Redman 1999; Chew 2001; also see Fagan 1999). The practice of agriculture involves the genetic modification of plants and animals, and so impacts on those natural populations are substantial. Further, agriculture involves the manipulation and management of landscapes on a scale typically much greater than practiced by hunter-gatherers. This almost certainly includes changes in technology and the intensity of use of specific areas. These impacts increase as agriculture intensifies, with the greatest effects coming with intensive agriculture and industrialization (see table 6.3).

Impacts on the Natural Environment

Perhaps the most significant impact of agriculture on the natural environment is the transformation of landscapes, which results in an overall decrease

Table 6.3. Summary of the Environmental Impact of Agriculture

Category	*General Impacts*
	The Natural Environment
General	loss of biodiversity and habitat, pollution by chemicals
Plants	loss of habitat for most species, extinction of some species, domestication and proliferation of a few species
Animals	loss of habitat for most species, extinction of some species, domestication and proliferation of a few species
Water	loss of freshwater for natural habitats, pollution of most water sources
Soils	erosion of topsoil in agricultural lands, exhaustion of soils
Landforms	alteration of landforms, resulting in habitat loss
Atmosphere	pollution
Climate	long-term global warming through loss of forests and development of industrialized cultures
	People and Cultures
Population	huge long-term increase in population, subsequent sedentation and urbanization
Resource dependence	ongoing trend for reliance on a fewer number of resources, populations more frequently subjected to famine due to crop failure
Workload	increase in workload for most, decrease for some (elite, rich)
Disease	dramatic increase in crowd diseases affecting billions
Health	increase in general health for many people, little change for others, generally longer life expectancy
Warfare	increase in scale of warfare, with increasing effects on population and environment
Knowledge	loss of traditional knowledge, increase in specialized agricultural knowledge, explosion in overall knowledge

in biodiversity (Vitousek et al. 1997:495). This impact has been especially large in the last hundred years (see Matson et al. 1997). As agriculture expands into new areas, their relatively diverse and generalized natural ecosystems are destroyed and replaced with much less diverse and specialized agroecosystems. This process often results in the loss of entire natural ecosystems and the destruction of the habitat of many species. This loss of habitat constrains species, eventually leading to their extinction. It has been variously estimated that current extinction rates are between a hundred and a thousand times faster than in the recent past (Vitousek et al. 1997:498). Leslie Sponsel and colleagues (1996a:3) report:

> By 1989 the annual rate of [tropical] deforestation had reached 142,200 square kilometers, which represents 1.8% of the 8 million square kilometers of remaining forest, and the rate of deforestation is even accelerating. . . . Current rates of defor-

> estation exceed 0.4 hectares [about one acre] per second. . . . As a result of habitat destruction as many as 10,000 species may become extinct each year, a level unprecedented in all of geological history.

Overall biological diversity is decreasing, and any loss of diversity has long-term negative effects. One of the benefits of anthropological work with a culture is the recordation of biodiversity and its preservation through contemporary conservation programs (e.g., Western and Wright 1994; Orlove and Brush 1996). Interestingly, domesticated species are also imperiled by the increasing specialization and homogenization of agriculture. Of the many domesticated livestock breeds that once existed, many are now extinct, and the diversity among the animals used for food is decreasing.

A second major impact, a serious problem only since World War II, is pollution, from both agricultural and industrial sources. Agricultural pollution includes dust and pesticides in the air, chemical fertilizers and pesticides entering water systems through runoff from fields, and silt washed into rivers and streams from eroding fields. The quantity of silt pollution entering rivers and streams is so great as to clog waterways and destroy habitat.

Impacts on Plants and Animals

The biggest problem for most plants and animals is the loss of habitat that comes when natural lands are converted to agriculture. Land is cleared, and individuals either are killed (plants) or have to move (many animals). Eventually, there is nowhere left to move to, and extinction is possible. About 80 percent of the deforestation in the world is done to make land for agriculture.

Impacts on Waters

Today, more than one-half of the accessible freshwater on the planet is used for human purposes, primarily agriculture (Pimentel et al. 1994:204; Vitousek et al. 1997:494; Gleick 1998). Much of this water is diverted from rivers and streams into reservoirs and irrigation systems. The construction of reservoirs results in the loss of terrestrial habitat but creates freshwater aquatic habitat. River flow is often decreased below the irrigation systems, with a resultant loss of riparian and wetland habitats.

In addition, in the United States, groundwater is being pumped 25 percent faster than it is being replenished (Pimentel et al. 1994:205), with the result that water tables are being lowered at an alarming rate. Lowering water tables can cause the ground surface to subside and natural water sources, such as springs, to dry up, which kills species dependent on groundwater.

Impacts on Soils

Agriculture has a number of impacts on soils. First, removing vegetation and plowing break up soils and allow them to erode more easily. Water erosion can be serious, but wind erosion can also be a major problem, as demonstrated by the dust bowl in the central United States in the 1930s. An inch of topsoil takes about five hundred years to form; in the United States, it is being lost at rates sixteen to forty times faster (Pimentel et al. 1994:203).

Second, agriculture will eventually result in soil "exhaustion," the condition in which the loss of organic matter and nutrients in the soil precludes further crop production. To avert this, fields have to be allowed to lie fallow or crops have to be rotated. However, the pressure for short-term crop production sometimes rules out these options and results in rapid exhaustion.

Lastly, the absence of soil deposition can be a problem, as exemplified by the situation in Egypt. For thousands of years, the agricultural system in Egypt relied on the natural yearly flooding of the Nile River to provide irrigation water and a fresh layer of topsoil, a pattern that assured that soils were not exhausted. Since the construction of the Aswan High Dam in the 1960s, the Nile no longer floods, and no new soil is deposited. Today, Egyptian farmers are forced to use chemicals to maintain the fertility of the soil, an additional expense and a source of pollution affecting the fish populations that serve as food.

Impacts on Landforms

Humans have altered between about 40 and 50 percent of the land surface of the Earth, and this transformation is the "primary driving force in the loss of biological diversity worldwide" (Vitousek et al. 1997:495). Land is cleared, burned, leveled, terraced, and built upon. Rivers are rerouted, valleys flooded behind dams, and wetlands drained. Today, between 10 and 15 percent of the land surface is occupied by crop agriculture or urban centers and another 6 to 8 percent by pasture. All of these landforms have been drastically altered.

Impacts on People and Cultures

The impact of agriculture directly on humans has been considerable. With the increased carrying capacity that resulted from food production, population began to grow and has now reached a rate of growth of 2 percent per year. As the population of agriculturalists grew, people began to settle together, to build cities, and to clear more and more land for farming. Agriculturalists expanded their territory into lands suitable for farming, displacing the hunter-gatherers who lived there. The hunter-gatherers were forced to move, be assimilated, or be killed, and many groups just disappeared as distinct cultures.

Farming is a difficult profession, and farmers generally had to work longer hours to get the same number of calories as hunter-gatherers. Competition for resources, particularly land and water, increased, and warfare to control those resources also increased. As a result, a great quantity of resources (human and otherwise) has been invested in military matters in the name of preserving lifestyle.

Crowd diseases began to kill millions of people packed into dirty urban centers; at least twenty million died from the Black Death in Europe between 1346 and 1350 (Cantor 2001:7), and over twenty million more from the flu in 1918 (Crosby 1989:207). Even for the surviving, the misery was immense.

Agriculture had the further effect of narrowing the human food resource base from hundreds or even thousands of species in some societies to only a few dozen in agricultural societies. Today, wheat, rice, corn, potatoes, sugar, soybeans, and cotton account for most of the world's agricultural production, and a handful of other grains and legumes account for most of the rest. Animal raising depends primarily on cattle, pigs, chickens, and sheep. This is a dangerously narrow base, and it continues to narrow. Supermarkets may have a few exotic species in their produce racks, but the human diet is becoming increasingly dependent on fewer plants. Some of these unused species became extinct, and the knowledge required to use others was lost. Being reliant on a few species makes crop failure an ever-present and very serious problem, and famine can result.

One could argue that agriculture provided the food necessary to allow humans to develop the complex cultures seen today, to pursue science and art, and to improve the quality of life for all humanity. While it is true that the quality of life has improved, that benefit is enjoyed by a minority; most people still live in poverty and in conditions worse than the hunter-gatherers they replaced.

CHAPTER SUMMARY

When people began to manage certain species so intensively that they began to manipulate and control their genetics, those species became domesticated, and food production began. This process originated in a number of places at about the same time, and those people and societies involved in the process began to shift toward farming lifestyles and new organizations.

The reasons behind the shift to food production are not fully clear. Ideas include environmental changes that created opportunities or forced people to alter their subsistence practices; growth of population, which required new food supplies; and changes in organization and management, which enabled the shift.

Several types of agriculture can be defined: horticulture, pastoralism, and intensive agriculture. Horticulture is low-intensity agriculture involving relatively small-scale fields, plots, and gardens cultivated with human labor. Pastoralism involves the herding, breeding, consumption, and use of domesticated (or managed) animals. Intensive agriculture is the large-scale cultivation of plants, often with the use of supplemental labor and irrigation.

The practice of agriculture has significant impacts on both the natural and the cultural environment, including loss of biodiversity and habitat, extinction of species, erosion, and pollution. Impacts on people and cultures include growing populations and resource pressures, increasing disease, rising workloads, lengthening lifespan, more war and associated suffering, and a loss of some knowledge (but with a gain in other knowledge).

KEY TERMS

animal husbandry
horticulture
intensive agriculture
monoculture
nomad
pastoralism
polyculture
silviculture

7

Horticulture

Horticulture is low-intensity, small-scale agriculture involving the use of relatively small fields, plots, or gardens. Groups practicing horticulture will often have populations in the many thousands, live in one place all year, and frequently have a tribe-level political organization. Some hunter-gatherers will employ aspects of horticultural practices as a minor aspect of their subsistence system.

Horticulture generally involves the use of individual human labor and small hand tools such as digging sticks rather than of mass labor, draft animals, or equipment such as plows or tractors. Crops are grown for mostly personal consumption, although some of them might be traded or given to a central authority, such as the chief.

HORTICULTURAL TECHNIQUES

Both domestic plants and animals are raised by horticulturalists, but the emphasis is on the production of plant crops. Crops are mainly grown in small fields, called gardens, and such gardens may be solitary or part of a larger system of gardens. In addition, various small domesticated animals, such as pigs, chickens, and dogs, are often raised by horticulturalists. Wild resources may also form an important component of a horticultural economy.

Gardens

Gardens are small fields cultivated by individuals or small groups of people. The specific size of a garden is limited by the available labor. Some gardens are just small plots of land cleared and planted with little investment of time and labor. However, most gardens are used repeatedly and tend to be located in the same place over time, and techniques such as crop rotation, fallowing, and the use of fertilizer are used to maintain fertility. Thus, most such gardens require a large investment of labor but a relatively small investment in land.

Well-managed gardens can be highly productive and can sustain high population densities. For example, some root-crop gardens in highland New Guinea can support as many as 160 to 200 people per square kilometer (Brookfield and Brown 1963:119–122). Three basic types of sustainable gardens, chinampas, terraced gardens, and slash-and-burn, are discussed below. Any or all of these types can also be components of intensive agricultural systems. For example, many Americans today have small personal gardens in their backyards.

Chinampas

Chinampas are small raised fields or gardens used in a number of regions and extensively in Mesoamerica. They were built in water, as described below, and many types of plants could be cultivated in them. A single chinampa may be viewed as a small-scale garden. However, chinampas were generally constructed in combination and integrated into a much larger, intensive agricultural system, such as the one used by the ancient Maya (see chapter 9).

The classic form of chinampa (fig. 7.1) was constructed within a lake. First, long stakes were driven into the lake bottom to create a "form." Soil from the lake bottom was then dredged up and piled within the stakes. Layers of different types of soil were placed upon one another until the field had been raised up above the level of the lake. Once the chinampa was formed, willow trees were planted along its edges to help control erosion. Crops could then be planted. New soil was constantly dredged up and added to the chinampa, maintaining its fertility and productivity. As it was surrounded by water, the chinampa was self-irrigating, with the varying layers of soil serving the purpose of drawing water to the plant roots. Additional chinampas would be built and arranged in a pattern that created a system of canals between them (see fig. 7.1). The construction and maintenance of such a field require a great deal of labor.

The Mexica employed this form of chinampa in Lake Texcoco as a major component of their intensive agriculture. The Mexica chinampas around their capital city of Tenochtitlán have been erroneously referred to as "floating" gardens, and a remnant of that system, the Floating Gardens of Xochimilco, is a tourist attraction in Mexico City today (see Werner 1992).

A similar form of chinampa was constructed in swampy areas, such as the lowland Yucatan. These chinampas were created by digging ditches around a small patch of land and placing the soil in a pile in the middle, raising the patch up a few feet out of the water and creating ditches around it. A series of such

FIGURE 7.1
Mexica workers building a classic chinampa, with a system of them in the background (from a sixteenth-century drawing). (Courtesy of Corbis)

fields would be constructed in a waffle-like pattern, with the ditches linked to form a system of canals, probably connected to a major body of water, such as a river or a lake (fig. 7.2). As the ditches or canals slowly silted in, they were cleaned and dredged, with the resulting silt and organic debris being added to the fields. Thus, soil renewal was continuous, and the fields were self-irrigating and very productive. As with the other chinampas, the construction and maintenance of these fields required a great deal of labor. We are now beginning to realize that the ancient Maya made extensive use of these chinampa-like systems, as did past peoples in other regions, such as South America (see Bray 2000).

The canals present in both types of chinampas were purposefully created and served three major functions. First, they were transportation routes, with small boats being used in them to move people and products. The canals had to be maintained at the proper width and depth for the boats. Second, the canals served as passive irrigation systems: their water saturated the soil and so watered the crops. Lastly, the canals created a habitat for a variety of other resources. Fish,

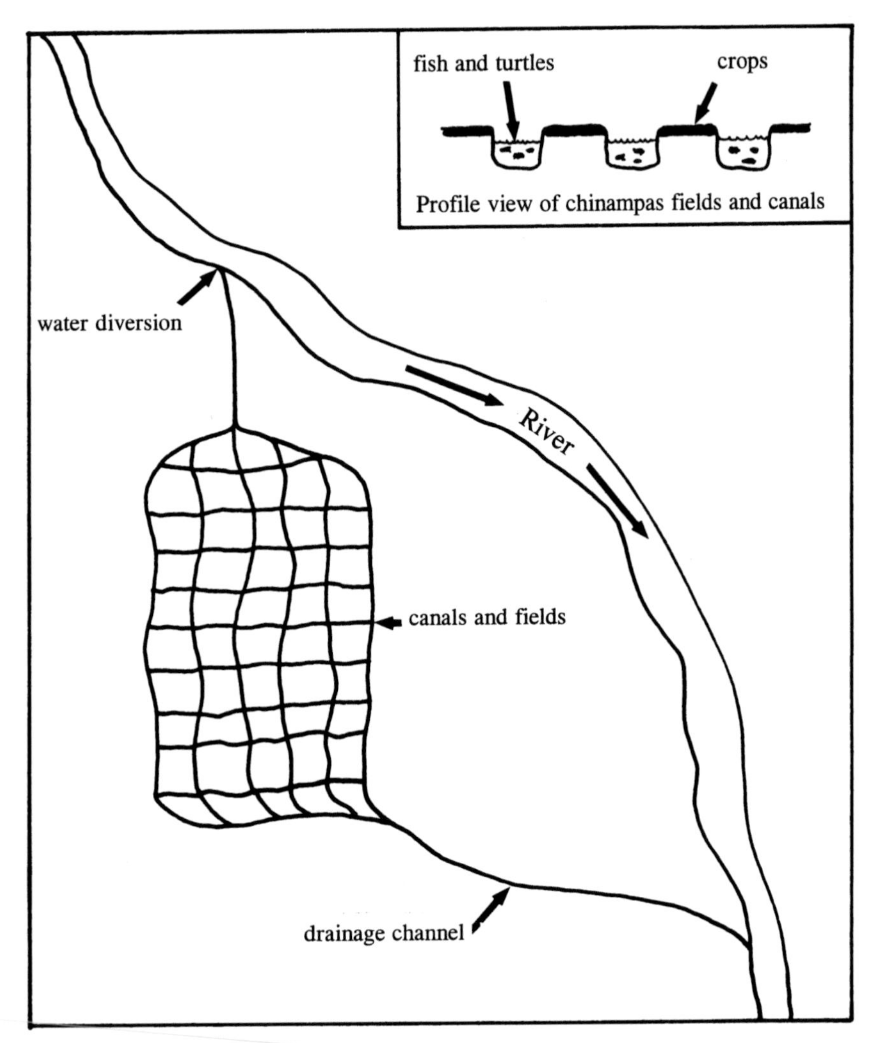

FIGURE 7.2
A plan-view schematic of a system of raised chinampas, fed by a diverted water source with a profile view of the raised fields and interconnecting canals (no scale); the crops are grown in the raised portion with fish, turtles, and aquatic plants being harvested from the canals.

shellfish, turtles, frogs, waterfowl, and other aquatic animals inhabited the canals and fed on pests and agricultural debris from the fields. In addition, reeds grew along the edges of the canals and were harvested and used for a variety of purposes, including matting and thatch. These various wild resources formed an important addition to the economy and were encouraged to inhabit the canals. Although not under direct genetic control, such resources could be considered "domesticated" in the broad sense of the term (see discussion in chapter 4), as their habitat and populations would be created and controlled.

It has been argued (Erickson 1986) that in highland Bolivia, an ancient system of chinampas was used to regulate the air temperature of the fields. The water in the system would have stored heat during the day, and the gradual release of that heat during the night would have kept the crops warm enough to prevent freezing. In this way the use of chinampas, instead of traditional fields, would have extended the already short growing season and permitted people to raise crop species not otherwise possible.

Terraced Gardens

Terraced gardens are small fields usually constructed on sloping terrain. Small retaining walls of rock would be built to impound both soils and water. Such fields are not to be confused with larger, much more extensive systems of terraced fields, such as those used in rice agriculture. Terraced gardens generally offered three advantages: (1) level planting surfaces, (2) erosion control, and (3) deep soils (Smith and Price 1994:175).

Terraced gardens were, and are, widely used by many cultures. The Chinese (see case study in chapter 9) employed small terraced gardens around their towns, and these gardens produced the majority of their vegetables. Rice was produced in much larger and more complex terraced fields. The use of terraced gardens was widespread throughout Mesoamerica. In addition to the small terraced gardens on slopes, Mesoamericans utilized small raised gardens, made by building small enclosures from low rock walls and filling the enclosure with soil (Smith and Price 1994). Terraced gardens could be very productive, and managed properly with the addition of new soil and organics, they could remain so for a long time.

A number of other gardening techniques were also employed. For example, aboriginal farmers in the American Southwest employed a variety of small gardens (see Forde 1931; Hack 1942; Kennard 1979; and Bradfield 1995 for descriptions of the Hopi system). Expedient gardens would be used anytime a sufficient amount of water could be captured from a thunderstorm. Small check dams would be constructed across washes to capture both sediment and water. Water would rush down the wash, be slowed or trapped behind the dam, and drop its load of wet soil, instantly forming a field of wet and fertile soil in which crops were planted. If a stream overflowed and created a patch of wet soil, crops would be planted. Sand dunes that had been rained on were planted with special types of long-rooted corn to take advantage of the mulching capabilities of the dune. Such single-use gardens were spread across the landscapes, and the crops in them would often fail. However, the investment in them was small, and their number and dispersion across the landscape served to reduce general crop failure. More permanent gardens were built and maintained around areas with more reliable water, such as permanent streams or springs.

Slash-and-Burn

Slash-and-burn, sometimes called shifting cultivation, is a technique used to create a garden or small field within a forested environment. Slash-and-burn is practiced in areas of poor soil, primarily in forests and woodlands, as many forested regions, particularly rain forests, have poor soil due to the constant rain, which washes away topsoil and nutrients and prevents the formation of rich soils.

As soils are often poor, slash-and-burn fields are usually productive for only a short time. People generally grow crops in a field for a couple of years or so, then abandon the field and move to a new one. Otherwise, a huge amount of time, labor, and organic fertilizer must be invested to maintain crop yields. It is just easier and more cost-effective to move on.

Slash-and-burn fields are generally small, sometimes several acres in size, and are created by one or a few people clearing a small patch of forest. The technology utilized in clearing is usually limited to axes, bush knives, and fire, with fire doing much of the work in reducing the vegetation. Nevertheless, a considerable amount of labor is required to make a field, and the successful use of slash-and-burn as a system (see the discussion of swidden, below) requires a great deal of land since fields are abandoned after a year or two.

At the beginning of the process of creating a field, the trees must be dealt with, as their canopies cause too much shade for crops. The trees can be killed (by girdling) and left standing, cut down and burned, or cut down and the wood used for other purposes, such as firewood and house construction. Next, the brush is cleared, or slashed, from the field (fig. 7.3). The resulting dead vegetation is allowed to dry and then burned. The burning is done at a time of year when the climate is dry enough to allow burning the slash but not dry enough to allow the fire to spread into healthy forest. Alternatively, firebreaks can be made. Not only is the field cleared; it is also fertilized by the ash, and logs too large to burn completely are left in the field to retard erosion. Crops are then planted with digging sticks.

Crop Choices. It is most common for cultures using slash-and-burn to practice polycultures, as yields and results are better and more reliable than with monoculture, which is highly vulnerable to flooding, pests, and diseases. However, sometimes just the staple crop is planted, particularly by the transient farmers of today.

Traditional farmers most commonly plant their slash-and-burn fields with many different plants, creating a forest-like environment, with grain crops, root

FIGURE 7.3
A slash-and-burn field within a forest in the Yucatan of southeastern Mexico; note that the palm trees were not killed, presumably because they do not cause enough shade to worry about. (Photo by E. N. Anderson)

crops, vegetables, herbs, fruit trees, cotton, and much else, perhaps as many as a hundred species or varieties in a single plot. Multiple crops complement each other: some thrive in shade, others in sun; some like bare soil while others do not; some fix nitrogen, others need more nitrogen; some grow on others (vines on small trees); and some harbor insects that eat the pests of the others (Conklin 1957; Geertz 1963). Root crops provide safety in case fire or hurricane devastates the field. This sort of mixed planting is an insurance policy. If one or more crops fail, the plot is not a total loss.

If the field itself is not a mixed system, there is inevitably a relatively large and diverse garden around the home. That garden would have the multilayered structure of the forest. It also has plants and animals occupying many different niches: sun or shade, thin soil or good soil, and wet or dry places.

Weed and Pest Control. The control of weeds is more difficult with single crops than with multiple crops because the crop occupies only a single niche, leaving all the others open for weeds. With many crops, more of the niches are filled and unavailable for weeds. In many cases of multiple-crop gardens, little or no weeding is needed.

Pests are also easier to control with multiple crops. Single pest species have fewer targets and can do only limited damage to any one crop. Moreover, some of the other crops may harbor other insects that prey on the destructive species, thus providing some built-in insect control. Some larger pests, such as deer, are controlled through hunting, a process that provides needed animal protein in the diet.

Moving Fields. The major problem in slash-and-burn fields is soil exhaustion. Soils are generally poor to begin with, and the choice of crops can influence the speed of exhaustion. For example, seed crops such as corn are more demanding on soil nutrients and wear out the soil faster than root crops. When a field reaches a point of declining production, it is abandoned, and a new field is created. It will take some time, depending on local conditions, for the abandoned plot of land to regain its fertility. If a sustainable system of field use, fallowing, and eventual reuse is developed, the system is called a swidden (discussed below).

Small plots are easily reclaimed by the forest if a sufficient amount of forest is left nearby to serve as a source of colonizing plant and animal species. In the large-scale forest clearing now occurring around the world, too little forest is left in place, land gets overgrown with weeds, and the forest takes a very long time to recover.

Current Problems with Slash-and-Burn. Although slash-and-burn agriculture has been practiced for thousands of years with a relatively minor impact on the forest, its greatly increased use by increasing numbers of people has made it a major problem. In all cases, slash-and-burn results in the elimination of the forest within the boundaries of the field. However, if the field is small relative to the remaining surrounding forest, the field is quickly reclaimed by vegetation, and the forest suffers no real permanent damage. However, if too much of the forest is subjected to slash-and-burn at any one time, the forest cannot recover and can be permanently degraded or even destroyed: the situation in many of the rain forests today. In addition, slash-and-burn agriculture can totally destroy valuable tree species, so the technique is hated by loggers.

Moreover, though it has been said that slash-and-burn cultivation is energy-efficient, this is true only of the human labor component. Fire does most of the work but consumes a great deal of energy in the form of biomass. Today, when energy sources are running short and greenhouse gases are causing concern, this consumption of wood represents another problem—although a minor one in comparison to burning forest and brush to create cattle pasture and plantations.

The Swidden System

Swidden is an integrated system of sustainable agriculture, or **permaculture**, that incorporates the slash-and-burn technique as its major component (e.g., Conklin 1957; Spencer 1966). Alone, slash-and-burn is a method that creates a one-time field, but the swidden is a sustainable system. **Swidden** is often referred to as *slash-and-burn cultivation* and is known by other names as well, including *shifting cultivation* and various local names. Using swidden, slash-and-burn fields are left fallow long enough to recover and are then reused in a regular and sustainable cycle. Small gardens may also be an important component of a swidden system.

A swidden system requires a great deal of land, most of which is fallow at any given time. For example, if a farmer uses five-acre fields, each producing for one year, and each requiring ten years to regain its fertility, the farmer will have to have fifty acres fallow with five in production, plus some land for the house and other small gardens (fig. 7.4).

The management of a swidden system—such as deciding when to burn, where to put the next field, when to plant, and what to plant—is a fantastically complex and difficult undertaking. To properly manage a swidden system, farmers must be highly skilled and have an incredible knowledge of local plants, animals, soils, weather, and cultivation methods. Studies in the Philippines (Conklin 1957), Mexico (Berlin et al. 1974; Alcorn 1984; Breedlove and Laughlin 1992), and Brazil (Balée 1994) show that people using the swidden system often know literally thousands of plant and animal species—not only how to identify them but how to use them and when to find them at their best. This finding has led to a great deal of research on swidden systems (e.g., Anderson 1990; Dove 1993a, 1993b; Balée 1994), and most studies show that the use of swidden is very efficient and highly adapted to local forest environments. It is a reasonable way to use forest—probably the best for some. However, these virtues apply only so long as population density remains low and proper plot rotation persists.

Knowledge of forest succession and of when a former field can be reused is an integral part of swidden management, and each individual system must be managed under its own unique circumstances of water, soil, and species. Within the system at any one time, most of the fields are in varying degrees of recovery, and each has a slightly different biotic community. In effect, then, the presence of a whole series of fallowed fields in different succession stages greatly increases the biotic diversity within the field system and provides increased opportunity for resource exploitation.

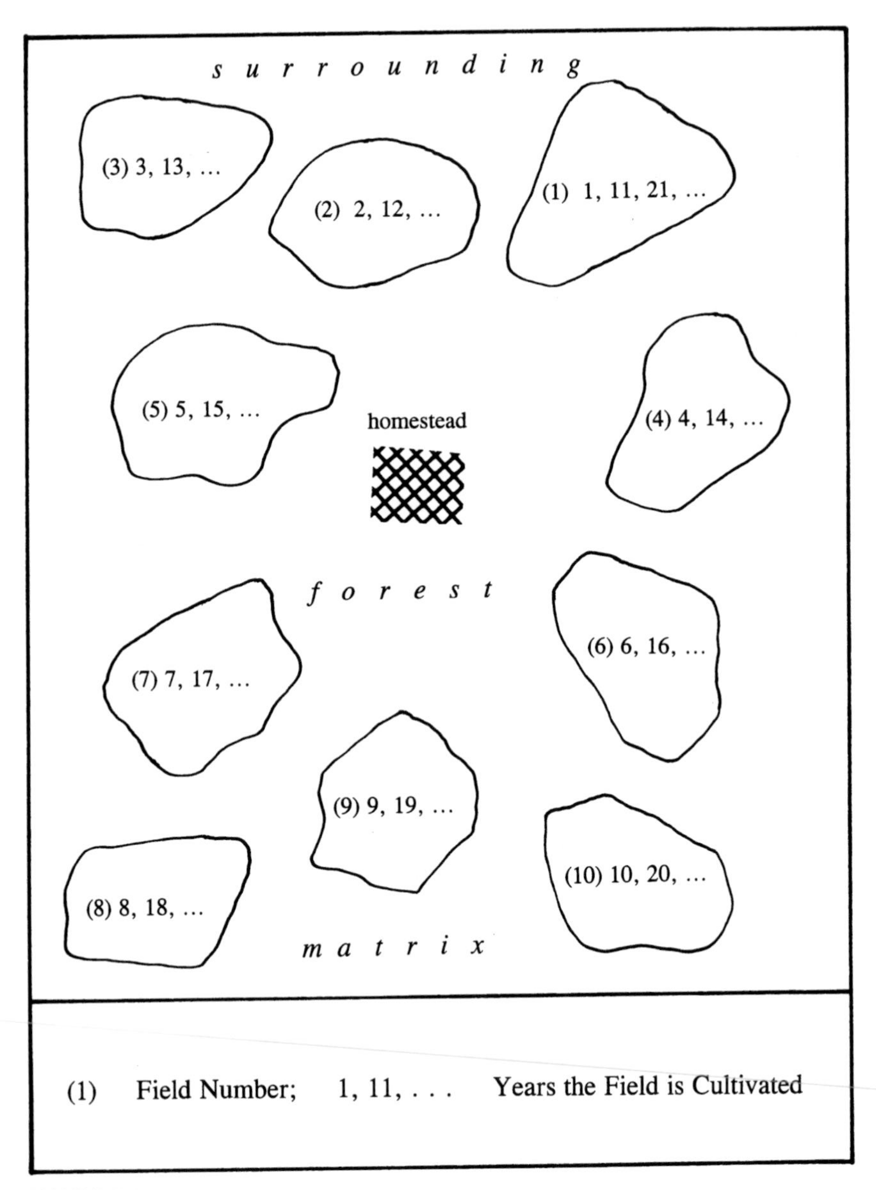

FIGURE 7.4
An example of the land requirements of a swidden system with a ten-year cycle. Field 1 is cultivated in year 1 and then left fallow until year 11; field 2 is cultivated in year 2, left fallow, and cultivated again in year 12; and so on. Thus, the system requires continual control and management of one active and nine fallow fields, with only 10 percent of the land under cultivation during any particular year.

If managed properly, swidden is a sustainable system. However, when population density is high, people may be forced to recut their fields before full fertility has been regained. Weeds, pests, and soil exhaustion eventually combine to ruin the yields, as in Thailand and the Philippines. Rampant logging on top of slash-and-burn cultivation greatly decreased the once valuable forests of these countries. Their destruction led, in turn, to massive flooding of more productive farming areas downstream.

In some areas—the temperate forests of the eastern United States, for one—the use of swidden gave way to more intensive farming methods. In tropical areas of poor soil and hilly relief, however, no other method works except establishing an artificial forest of commercial and subsistence tree crops. This sort of tree cropping is locally practiced, but it is not a perfect solution. For example, coffee is the most successful tree (or bush) crop in many cool, tropical mountains, and so it was promoted as a great way to intensify; therefore, after about 1950 almost every nation with any tropical mountain lands began to grow it. The result was overproduction and a worldwide crash in coffee prices.

Field Rotation

One of the keys in a successful swidden system is properly managed field rotation. When the crop yield from the existing field falls below a certain point (always dependent on the local conditions), the farmer abandons the field and opens a new one. To do this, the farmer has to be able to judge both the fertility of the existing field and the potential fertility of the new field under consideration. The condition of each of the fields in the system has to be monitored and considered in the overall field rotation plan. As long as field rotation is properly managed and population density remains fairly constant, a swidden system can be used over the long term, with the farmers living in small permanent villages and utilizing the surrounding forest on a rotating basis.

Another approach is to use a much larger-scale rotation system, with large tracts of land being rapidly used and exhausted. In a short time, the entire village would be moved to an entirely new location, leaving the original area to recover and to be reused generations later (see Carneiro 1960). In some cases, however, the movement of settlements may be related more to social factors or availability of animal protein or firewood than to land exhaustion.

However, if a group moves but keeps expanding into virgin territory rather than reoccupying past field complexes, a predatory society may develop, with warfare being practiced to drive other groups out of potential farmland, as the Iban did in Borneo.

USE OF WILD RESOURCES

Most horticulturalists utilize a wide array of wild resources. Hunting and gathering are done in areas away from the fields and gardens to obtain supplemental foods and materials for manufacture and medicine. Fallow fields are also exploited because different species of wild plants and animals will colonize them during the various stages of forest regrowth.

Hunting is often better in abandoned fields than in mature forests because new growth is attractive to browsing animals, such as deer. In Mesoamerica, a great many deer are hunted in such fields, and these animals form an important source of meat for the farmers. Thus, even "abandoned" fields are still used for some purposes.

Intensive horticulture is typical of Oceania. Grain crops do poorly or are susceptible to typhoons, so root crops such as taro and yams are preferred. Tree crops such as breadfruit, pandanus, and coconuts make up most of the rest of the vegetable food base. The horticultural systems on some islands became extremely intensive over time, with terracing of whole mountainsides, control of streams, and excavation of low wet areas for marsh-loving crops. Fishing is extremely intensive and usually managed in a very careful fashion; sustainable management is often compelled by taboos (*tabu* is a Polynesian word) and other religious mechanisms (Johannes 1981; Ruddle and Akimichi 1984). A fishing expert of a Micronesian island may know details about the habits and ecology of hundreds of species of marine life.

An interesting pattern emerges from the archaeology of Polynesia. On island after island, initial settlement led to wasteful and destructive use of resources, followed by a population crash; then, use of resources gradually became wiser and more constructive, allowing the population to rebound and become even greater than before (Kirch 1994, 1997). This is a clear case of something North Americans know from the history of European settlement: a pioneer fringe tends to be characterized by carelessness with nature's bounty, but ultimately reality catches up, and people have to economize and cultivate intensively.

ENVIRONMENTAL MANIPULATION AND RESOURCE MANAGEMENT

Horticulturalists actively manipulate the environment, such as by the construction of gardens and rotation of fields. However, the scale and impact of such alterations are less degrading to the environment than clear-cutting forests for cattle ranches. In some cases, specific localities can be significantly and substantially altered by horticulture. Examples would include terraced fields and uncontrolled and unmanaged use of slash-and-burn fields.

Passive environmental manipulation is probably more important to horticulturalists than it is to many hunter-gatherers. The control of weather, water, and sun (abiotic elements) is critical to agricultural success, and some efforts are commonly made in ritual to manage those forces.

Horticulturalists actively and intensively manage the specific domesticated species that they rely on. Some management of wild species may also occur, such as hunting deer in fallow fields.

RELATIONS WITH OTHER GROUPS

In general terms, it is possible for horticulturalists to share habitat with hunter-gatherers and pastoralists if scheduling of land use and details of resource competition can be worked out (see the case study on the Mbuti in chapter 5). However, it seems that such accommodations are relatively uncommon. Horticulturalists cannot coexist with intensive agriculturalists, because the two occupy too close a niche, with fertile land and water being critical resources.

CHAPTER SUMMARY

Horticulture is small-scale, low-intensity agriculture conducted using only human labor. Horticulturalists generally grow food, primarily plants but a few animals, for their own consumption. Many horticultural groups have relatively large populations and fairly complex sociopolitical organizations.

The primary type of field used by horticulturalists is the garden, a small field cultivated by one or a small group of individuals. Types of gardens include various kinds of raised fields called chinampas, terraced fields built on sloping terrain, and slash-and-burn fields in forested environments. Construction and maintenance of gardens require a great deal of effort, and in the case of slash-and-burn fields, gardens have to be frequently abandoned and new ones created. Decisions on what kinds of crops to plant, crop rotation, or field abandonment are critical elements of horticulture. Groups will often utilize more than one garden type, both to achieve diversity and to utilize different landforms.

In some cases, people will utilize gardens in an integrated system, such as a large group of chinampas connected by canals. A number of groups employ the swidden system, a coordinated use of slash-and-burn designed to rotate through a series of fields, which are used, allowed to recover, then used again in a sustainable system. Proper management of a swidden system requires good crop and field rotation and a fallow period long enough for fields to recover. Without good management, a swidden system will fail.

Horticulturalists also utilize wild resources to some extent, often as a major component of the economy. Environmental manipulation and resource management are common but are usually done on a small scale.

KEY TERMS

chinampa
garden
permaculture
slash-and-burn
swidden

CASE STUDY

CASE STUDY: THE GRAND VALLEY DANI OF HIGHLAND NEW GUINEA

This case study presents an interpretation of Grand Valley Dani ecology that emphasizes warfare and pig consumption. This interpretation goes beyond a simple model of adaptation to interconnect economic, sociopolitical, and ritual behaviors into a single ecological system.

The Dani comprise a series of related groups of horticultural people living in the central highlands of Irian Jaya, the western portion of the island of New Guinea controlled by Indonesia (fig. 7.5). One of these Dani groups, the Grand Valley Dani (hereafter called Dani), became famous in ethnographic and ecological literature through the films and books that resulted from the work of the Peabody Museum of Harvard University Expedition between 1961 and 1963. A number of studies of the Dani have been conducted (Gardner and Heider 1969; Heider 1970, 1979), and several movies about the Dani have been produced, including the well-known 1963 film *Dead Birds*. The pursuits of the Dani were summarized by Heider (1970:10) as being "war, farming, and pigs."

The people of highland New Guinea were not discovered by the outside world until 1933, when a group of Australian gold miners entered the interior of the island. They encountered about a million native people living in a large number of small horticultural societies. The gold miners carried a movie camera with them and filmed

CASE STUDY

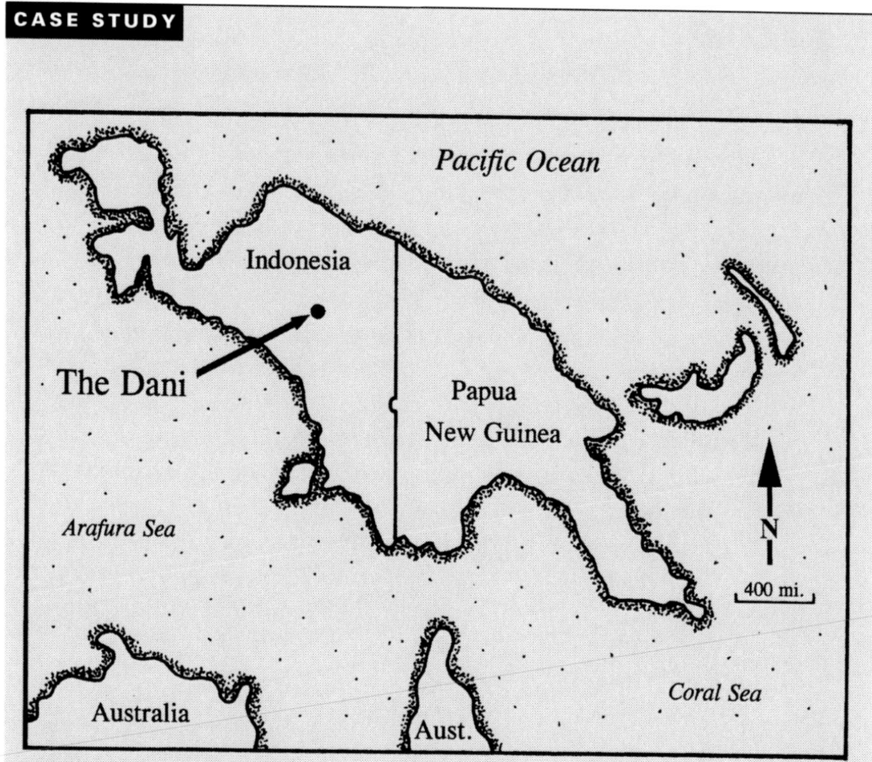

FIGURE 7.5
Location of the Grand Valley Dani in New Guinea.

the astonished natives. That film footage was made into a film, *First Contact*, and a book by the same name was written (Connolly and Anderson 1987).

THE NATURAL ENVIRONMENT

The heart of Dani territory is the wide and fertile Grand Valley, lying more than five thousand feet above sea level and surrounded by the great mountains that form the spine of New Guinea. The Balim River flows through the valley, only occasionally overflowing. Located just south of the equator, the valley receives about eighty inches of rain per year and has a moderate temperature. The Dani do not recognize any distinct seasons.

The land was at one time densely forested, but thousands of years of horticulture have converted much of it into fields, grasslands, and

CASE STUDY

second-growth scrub. Few wild animals remain in the valley itself, having been hunted out long ago, but a relatively pristine forest surrounds the valley and is occupied by a large array of wildlife.

SOCIOPOLITICAL ORGANIZATION

About fifty thousand people live in the Grand Valley, all speaking the same language and having a similar culture. They are organized into about fifty separate groups, or confederations, and there is no united political organization in the valley. Each of the confederations has well-defined territories with guarded borders. Some confederations join together to form temporary alliances, and each confederation/alliance was always at war with at least one other.

The people live in small settlements or compounds of a number of families, which are protected by fences and gates. Among the compounds are "Big Men," males of influence but of little real power. The political and social organization is centered on the compound, with little overall complexity.

ECONOMICS

The primary focus of Dani horticulture is the growing of yellow sweet potatoes, what Americans call "yams," and about 90 percent of the Dani diet is derived from this tuber. Yams are rich in vitamin A and have good-quality protein and a fair amount of B vitamins, although they are short in minerals and vitamin C. However, the leaves of the yam are eaten and supply adequate amounts of vitamin C as well as some other vitamins and minerals. Even so, the Dani had to depend on other sources for much of the nutrient value in their diet, and they were chronically short of protein, calcium, and iron, even though they ate pork. They were short in stature, probably (in part) as a result of their dietary deficiencies.

The primary type of garden is located in the valley bottom and consists of raised fields providing drainage from excess water (fig. 7.6). Yams grow best in these well-drained gardens, and much of the Grand Valley has been converted into a vast system of these fields, carefully engineered to be the perfect environment for the yam (see Gardner and Heider 1969; Heider 1970:37). The raised fields are sur-

CASE STUDY

rounded by ditches that not only drain the water but protect the fields from pigs. The fields were constructed by cooperative work parties, using digging sticks and other simple tools. Under such cultivation, yams can produce many tons of food per acre, and given the climate of the valley, yams can be grown all year. About one-half of the valley fields are fallow at any one time.

Yams and other crops are grown in the two other types of gardens used by the Dani. Small gardens are located behind most houses. Some men will use slash-and-burn fields on nearby hill slopes, abandoning them after a few plantings. The valley is located just at the upper range of bananas and pandanus, both of which are also used as food.

Much of the Dani language is devoted to agriculture. They recognize dozens of varieties of yam and even more variety names and have countless words evaluating them. Heider recorded more than seventy variety names in one small community. Another vocabulary details the processes and methods of yam cultivation.

FIGURE 7.6
Aerial view of a village and garden complex, highland New Guinea, about 1938. (Photo courtesy of Bill Richardson)

CASE STUDY

Interestingly, the yam is native to South America, apparently having reached New Guinea only a couple of centuries ago, give or take a century. Before that, the Dani and the hundreds of other New Guinea mountain groups grew other root and tuber crops, notably taro (*Colocasia esculenta*) and perhaps kudzu (*Pueraria montana var. lobata*). More important in the lowlands, but also common in the highlands, were the native large whitish yam of the lily family, very different from the late-arriving yam species. These roots have been grown for more than six thousand years, very possibly for ten thousand years, in the New Guinea highlands, making New Guinea one of the original and ancient homelands of domestication. Taro and white yams are still grown in a small way by the Dani.

Some other highland groups grow many species of vegetables, but the Dani seem to be content with growing only a few species of vegetables in addition to the versatile, tractable yam. Some tobacco and gourds are also grown, and some of the gourds are used as penis sheaths, much valued as male apparel.

In addition to their primary horticultural practices, the Dani raise domestic pigs (an example of sedentary animal husbandry; see chapter 8). Pigs supply the vast majority of the animal protein in the diet. Pigs are tended close to the village by children and are fed the garbage and other materials that people generally do not eat. Pigs are butchered and eaten almost exclusively on festive occasions, but these were frequent in traditional times, and so pork consumption was common.

The Dani once practiced some hunting of animals in the forest, including wild pigs, and a number of wild plants were gathered. However, most of the valley has been stripped of forest to make gardens, and so wild foods are fairly rare. Trips to the forested mountains are uncommon.

WARFARE

When not cultivating, the Dani were apt to be fighting. Highly ritualized combat, the sort portrayed in the film *Dead Birds*, took place frequently but caused few casualties. Less common, but more dangerous, were ambushes of isolated individuals who were cultivating

CASE STUDY

or hunting far from home. Much more rare and far more serious was a nighttime raid in which enemies would infiltrate a village and try to kill everyone.

To defend against the enemy, the border was constantly guarded, with watchtowers built at intervals. Constant vigilance was part of daily life. Warfare was not conducted for land, food, or other resources but to avenge earlier deaths. The Dani believe that death is not natural and is always caused by the action of the enemy, either directly in battle or other violent confrontation, or through witchcraft. The souls of the dead require that they be avenged; otherwise, the village will be haunted by ghosts. The other side also believes this, and so there is never any balance; one or the other side is always seeking revenge for the last death.

This pattern of warfare is more or less widespread in the New Guinea highlands (see Meggitt 1977), and one of the primary reasons for the intensive, labor-demanding cultivation of the valley floor was the extreme danger of going more than a mile from home to cultivate. This danger presumably explains the presence of intense horticulture in an area of relatively simple technology and low population density.

CEREMONIES

Throughout the year, the Dani put on a series of large ceremonies or festivals, in which rituals are held and food is consumed, particularly pigs. These events are sponsored by important men for a number of reasons, including celebrating the death of an enemy or mourning a death in one's own group. They also can function to make peace between conflicting groups or to marshal support for one particular group. The ceremonies also provide a reason for intensive food production. For one reason or another, such ceremonies are held frequently, at least one every few weeks.

AN ECOLOGICAL INTERPRETATION OF DANI WARFARE

Roy Rappaport (1984; also see Dwyer 1990) argued that among the Tsembaga Maring, a group very similar to the Dani, ceremonies were timed to get rid of pigs when they got numerous

CASE STUDY

enough to damage the gardens. Alternatively, the Tsembaga may actually let pig populations build up when they want a festival. The Dani have more intensive pig rearing; all are carefully raised and tended, whereas the Tsembaga pigs are semiwild. Thus, the Dani control the process better and seem not to worry about pig population cycles.

It has also been argued (Rappaport 1984) that the pattern of warfare can be linked to the system of ceremonies and pig consumption. Pork is the primary protein source but cannot be eaten daily, as there are not enough pigs, so their consumption has to be somehow controlled yet provide the needed protein on a regular basis. Because the pigs are eaten only at ceremonies, the frequency of the ceremonies would serve to regulate pork, and so protein, consumption. The ceremonies are tied, at least in part, to the regular cycle of warfare, which would serve to regulate the frequency of ceremonies, and so pig consumption. At each death, each of the warring groups would hold a ceremony, one to celebrate and the other to mourn, and everyone would get some pork to eat. Given the nature of Dani warfare, ceremonies could be expected every few weeks. The argument is that protein consumption is regulated so that people get enough and so that the resource base (pigs) is not overexploited. Although this is an obviously materialist and functionalist interpretation, it does have merit.

When the Dani were forced by the Indonesian government to give up warfare in the mid-1960s, they apparently did so with little resistance. Heider (1979:112) thought that Dani life did not change much, men still sat around talking, and "frequent occasions were still found to eat pigs."

CASE STUDY

CASE STUDY: THE LOZI OF WESTERN ZAMBIA

The Lozi case study illustrates two important aspects of adaptation. First, the political organization of the Lozi divides the culture into halves yet ensures the equitable distribution of resources throughout the land. Second, the agricultural system is intensive and complex but still involves a seasonal movement of people and an adoption of seasonally specific agricultural tactics. The resulting system is unique.

The Lozi live along the Zambezi River in western Zambia (fig. 7.7) and have an economy that emphasizes horticulture, cattle, and fishing. They consist of some twenty-five separate tribes, four of which are the Lozi proper, with the others being groups assimilated into the Lozi culture. The Lozi are also known by a number of other names, including the Barotse and the Luyana. The Lozi population in 2000 was about half a million, about seventy thousand of which are the Lozi proper. This case study is based on the group as it was in the early 1900s.

The Lozi moved into the region sometime during the seventeenth century and imposed their rule over the local inhabitants. By the early 1840s, they were conquered by the invading Kololo, but they regained their independence in 1864. In 1896, the British took control of the region and formed the colony of Northern Rhodesia. In 1964, the country gained its independence and was named Zambia.

The Lozi occupy some 150,000 square miles of western Zambia (now called the Western Province) and live primarily on the floodplain of the Zambezi River. Each year their villages are flooded, and the people are forced to move into the nearby forest until the water recedes. They then move back to the villages and reestablish their gardens for the next year's crops. Thus, they have a seasonal round that includes two separate systems of settlement and subsistence. More basic information on the Lozi can be found in Max Gluckman (1941, 1951), Victor Turner (1952), and Gerald Caplan (1970).

THE NATURAL ENVIRONMENT

The broad, sandy floodplain of the Zambezi River in southwestern Zambia is called the Barotse Plain. The plain lies below thirty-five

CASE STUDY

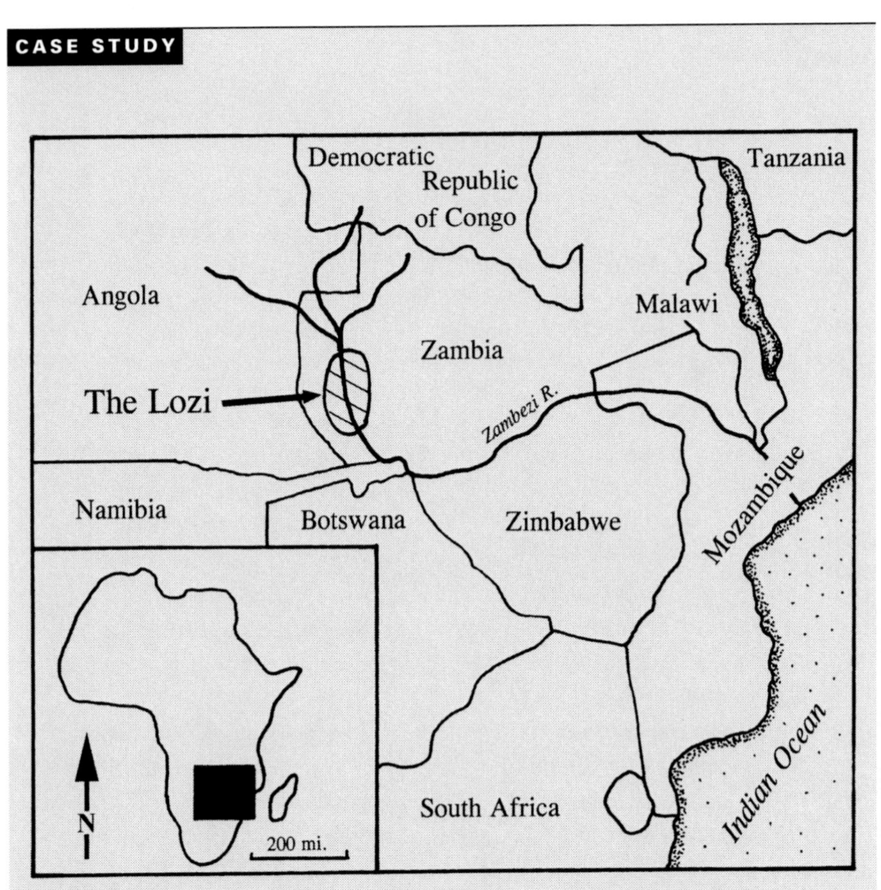

FIGURE 7.7
Location of the Lozi in southern Africa.

hundred feet, but the surrounding forested regions rise to a maximum of seven thousand feet. The area has relatively warm temperatures all year, with highs between 60 and 100 degrees Fahrenheit. Rainfall is variable, but the headwaters of the Zambezi River receive about forty inches per year. The plain is without trees except for some planted for fruit or shade. The grasses on the plain can grow as high as six feet.

The Barotse Plain is flanked by forest. The soils in the forest are poor in quality, consisting mostly of sand. The forest is traversed by a number of small tributaries of the Zambezi, and their small valleys are also inhabited by Lozi. The forest consists of teak and other trees plus many shrubs that form a relatively dense undergrowth.

CASE STUDY

At the southern extreme of this area, the Barotse Plain narrows significantly, and the Zambezi River changes direction to flow eastward to the Indian Ocean. One of the great natural wonders of the world, Victoria Falls (known locally as *Musi-o-tunya*, the smoke that thunders), is located where the river changes direction.

The region has three major seasons, dry, rainy, and hot, with the rainy season beginning about October/November and lasting until March/April. During the rainy season, the Zambezi River floods most, and sometimes all, of the Barotse Plain. In wet years, a lake some 20 to 30 miles wide, 100 to 120 miles long, and 15 feet deep is formed.

Lozi life revolves around the flood. The flood "covers and uncovers gardens, fertilizing and watering them, fixes the pasturing of the cattle, and conditions the methods of fishing. All life in the Plain moves with the flood: people, fish, cattle, game, wildfowl, snakes, rodents, and insects" (Gluckman 1951:11).

SOCIOPOLITICAL ORGANIZATION

The Lozi have a social system with classes, including royalty, the wealthy, the poor, and, prior to British rule, slaves. Most people live in small villages of between fifty and seventy-five inhabitants, and each village has a headman appointed by the ruler. Many of the Lozi villages are located on the Barotse Plain, but a number of Lozi live permanently away from the plain. The villages on the Barotse Plain are built on high ground, either on mounds especially built for that purpose or on old termite or ant mounds. Houses are constructed by plastering mud and cow dung over a wooden framework, a technique called wattle and daub.

The Lozi had a hereditary monarchy divided into two halves, North and South, although the actual territories of the North and South were geographically intermingled. Originally, each half was ruled by a king, assisted by a council of local leaders to advise him. The king lived in a capitol, a large village also on the Barotse Plain. Each king ruled one half but also had considerable influence in the other half. After 1864, the ruler of the South was a female, always a daughter of the king of the North. Each monarch had a council, and the governments of both the North and the South functioned as a single body to make important state decisions.

CASE STUDY

Both monarchs controlled land and resources throughout Lozi territory and collected taxes (in resources rather than cash) from all over their lands. These resources would be redistributed as needed. This dual system served to ensure that resources were evenly distributed throughout Lozi territory and that no one had a monopoly.

When the floods came, the monarchs would move their royal courts to the forest in a large public pageant called *Kuomboka*. On the date of Kuomboka, each small village on the Barotse Plain moved to its temporary village at the forest edge. The two monarchs coordinated their move to the forest so that all of the Lozi could move at the same time.

ECONOMICS

The economic system consists of three primary components: horticulture, cattle raising, and fishing, although some hunting and gathering is also important. Most of the gardens used by the Lozi are located on the Barotse Plain, as is most, and the best, cattle pasture. The main crop is corn (maize), but they also grow cassava, millet, sorghum, squash, tobacco, beans, and sweet potatoes, as well as bananas and mangoes. Cattle provide meat, milk, leather, and fertilizer, and the people also keep several small domesticated animals, such as chickens and goats. Dogs were kept and survived as scavengers. Fishing is a major activity, and some other hunting and gathering is also done. In addition, the Lozi maintain a substantial trade with other groups living in the forest. A detailed description of the Lozi horticultural systems was provided by David Peters (1960).

The Seasonal Round

The Lozi live in small villages located on high ground on the Barotse Plain for most of the year. However, when the Zambezi River floods, the Barotse Plain becomes a large lake, and the people have to move to higher ground in the forested areas on either side of the plain. Thus, the Lozi employ a simple and regular seasonal round, one that has them in their primary villages on the plain for about nine months and in their secondary villages on the edge of the forest for about three months each year. Each of these villages is usually occupied year after year.

CASE STUDY

Once the waters begin to rise, the cattle are moved to higher ground, with the herders living in small camps. About a month later, when the water gets high enough to flood most of the area and the insects become unbearable, the royal families move to the forest, a signal for the rest of the people to also move. The villagers pack what they need into their canoes and move to their forest villages to reestablish their households and begin work on their gardens in the forest. Some materials and food are left stored in the plains villages, and the people return by boat to obtain those materials as necessary. The forest villages consist of the same basic number of houses constructed in a manner similar to, though less substantial than, the plains houses.

The cattle are moved back onto the plains as soon as the water recedes, to permit them to take advantage of the new grass and to fertilize the agricultural fields. The land is dry enough for cattle but still too wet for people and planting. The people return to their plains villages about a month after the cattle. Upon their return, they must repair the flood damage done to their mud houses and prepare the fields for planting.

Gardens

The Lozi employ a number of different types and variants of gardens for the diverse conditions on the Barotse Plain and the forest. Most of the gardens are less than an acre in size, and the important conditions for the location and type of gardens are soil type and available water. Corn is the primary crop for the gardens on the plain while cassava is the most important crop grown in the forest gardens.

The Barotse Plain Garden System

The soils on the Barotse Plain are generally good and fertile, being renewed almost annually by new alluvial sediments dropped by the floodwaters. Most of the gardens are not purposefully irrigated; in fact, most have to be drained to be used. However, a few plots are supplied with irrigation water from a small system of canals, which are also used as transportation routes for canoes. Decisions on what and where to plant on the Barotse Plain are made depending on the level of flooding, the amount of new sediment, the soil moisture, and other factors.

CASE STUDY

The most desirable gardens, called lizulu, are located on small mounds or ridges, often old ant hills. However, these are the same places where villages are built, and there is competition between the two uses. Lizulu are cultivated all year if they remain above the floodwater, and during the flood season they are tended by people commuting from the forest in boats.

The most common garden types on the plains are similar to chinampas. The first type, called mikomena, are constructed in relatively poor soils by digging trenches around a small plot of land and piling the soil up in the middle to form a raised field. Some mikomena are "dry," but groundwater seeps into the trenches surrounding others. The mikomena are usually planted with sweet potatoes, then broken back down and planted with corn, then left fallow for at least six years.

The second form of garden, called li-shanjo (fig. 7.8), is similar to a chinampa, built in the same general manner as the mikomena, but in swampy areas, where peat is a common soil constituent. Relatively large tracts of land might be covered with these gardens, and the various channels dug around them would be connected to some natural watercourse to allow the water to drain. The gardens are prepared by burning the weeds and debris left from the last planting. Corn is the most common crop in these gardens, but sweet potatoes will be planted first in new or renovated gardens to increase the fertility of the soil for corn.

Litongo gardens are those placed in the relatively steeply sloped, sandy soils on the edge of the plain or in the forest and are essentially located in the ecotone between the plain and the forest. They come in several forms, dry, moist, and plains, depending on their location and water content. Most contain poor soils and are not highly productive. However, the moist litongo gardens are located on lower slopes of the plain, where water constantly seeps in, and are more productive. The Lozi mostly grow root crops in these gardens.

The small last major garden type is the ndamino, the kitchen garden, which is often placed on a moist litongo located near the village. The ndamino receives a "good deal of [household] refuse and the droppings of small domesticated animals" (Peters 1960:16). These gardens are quite fertile and are used for corn and sorghum.

CASE STUDY

FIGURE 7.8
A woman tilling a Lozi li-shanjo garden, ca. 1966. (Photo courtesy of Philip Silverman)

The Forest Garden System

The Lozi also farm the forests that ring the Barotse Plain, mostly during the three months that they live in the forest. The soils in the forest are generally poor and leach badly. Most areas have a covering of trees and thick brush. The primary type of garden in the forest is called matema, or bush gardens; they are slash-and-burn fields. The vegetation on a matema is cut between April and June, left to dry, then burned in September. Because of the low soil fertility, cassava is the primary crop of the matema, although millet is also planted when the rains come, usually in October. The matema system seems to have been adopted relatively recently but has been used long enough that most of the forest surrounding the Barotse Plain is now secondary forest.

CASE STUDY

Livestock

Cattle are the primary domesticated animal, and the Lozi keep a fairly large number of them. For most of the year, cattle are grazed on the Barotse Plain. In April, when the waters begin to rise and about a month before the people move to escape the rising waters, the cattle are moved off the Barotse Plain to pastures on the edge of or within the forest. The cattle are tended by herdsmen who often live in separate camps for the season. The animals are moved back to the Barotse Plain in June, about a month before the people return. The manure is used to fertilize the various gardens before the farmers return to begin cultivation for the year.

Use of Wild Resources

A number of wild resources were used by the Lozi, several of them very important. Some of these resources were obtained by the Lozi themselves while others were obtained by trade with other groups in the forest. Wild plants formed a relatively minor aspect of the Lozi economy.

Fish from the Zambezi River were the most important animal and were a major protein source. Fish were taken all year round by a variety of methods, including traps, nets, and spears. Game animals were hunted both on the plain and in the forest. Antelope, elephants, hippopotami (still a favorite, as the meat is thought to have aphrodisiac properties), crocodiles, and a number of small mammals and birds were hunted using spears, harpoons, and pit traps. When the water began to rise on the Barotse Plain, animals trapped on high ground were driven into the water and killed by men in canoes using spears. Today, few game animals survive on the Barotse Plain.

ENVIRONMENTAL MANIPULATION AND RESOURCE MANAGEMENT

The Lozi manipulate their environment in a number of ways. Active manipulation includes the construction of some artificial mounds on the Barotse Plain on which to build villages, which then generally survive the annual floods. Flooded villages require a considerable amount of rebuilding. A number of other fields and facilities are

CASE STUDY

built on the plain, but they are commonly washed away by the floods and have to be reestablished. Thus, much of the Lozi activity on the plain has little long-term impact. The slash-and-burn system in the surrounding forest, however, has resulted in the conversion of the original forest into secondary growth.

The Lozi actively manage both abiotic and biotic resources, including soil, livestock, crops, and some wild animals. Of interest is the soil management. Soil fertility is of great concern, and efforts are made to maintain fertility by the addition of animal dung, the rotation of crops of sweet potatoes and corn, the addition of ash from the burning of detritus, and a system of fallowing. Water is also managed, mostly by draining flooded fields.

DISCUSSION

The Lozi have developed a complex system of adaptations to their changing seasonal conditions. In the dry season, they occupy one ecozone (the valley) and practice irrigation agriculture on the valley floor. In the wet season, they move to a different ecozone and adopt a different subsistence system that revolves around forest gardens. This adaptation involves a regular movement of villages and requires a dual land base. Thus, each village uses both valley and forest ecozones plus the ecotone between them each year.

The sociopolitical system is adapted to this seasonally fluctuating economic system. The overlapping power of the dual monarchs ensures the distribution of resources throughout Lozi territory. The pomp and ceremony attached to the movement of the villages out of the plain and into the forest reinforce the social nature and ecological necessity of the move.

8

Pastoralism

Pastoralism is that form of agriculture in which the practitioners specialize in, and obtain their primary subsistence from, the husbandry of one or a few domesticated animal species. These species are invariably herbivores: cattle, horses, sheep, llamas, alpacas, goats, camels, reindeer, and similar animals. Plant cultivation often forms one component of pastoralism but is not generally dominant. In some cases, however, such as reindeer, the species of focus is not domesticated.

A precise definition of pastoralism is elusive, with the primary disagreement centering around the proportion of horticulture to agriculture in the economy and the degrees of mobility (see Krader 1959:499; Khazanov 1984:7, 15–17; and Cribb 1991:15–17, 20; also see Spooner 1973; Weissleder 1978; and Goldschmidt 1979). The term "nomad" has generally been used by anthropologists to refer to mobile pastoralists and should not be applied to hunter-gatherers (see Krader 1959:499). A brief history of the study of pastoralists has been presented by Neville Dyson-Hudson (1972:2–7; also see Waller and Sobania 1994).

Pastoralists and their animals have developed a long-term mutually beneficial relationship (see Krader 1959:501). Animals provide humans with products such as meat, milk, hide, dung, wool, and labor and with services such as companionship and the transportation of people and goods. Humans provide animals with protection from predators, a steady food supply, health care, an expanded habitat, and assured reproductive success.

Pastoralism requires a great deal of land as a pasture base. It is generally more productive (calories per acre) than most hunting and gathering but less productive than farming. However, pastoralism can be very efficient in areas unsuitable for farming. Pastoralists utilize their animals to convert unusable biomass from one trophic level into usable products at another trophic level:

Grasses that humans cannot digest are converted into milk and meat that they can eat. Even though using the animals involves an additional trophic step, it is highly efficient in such circumstances because people cannot use the grasses anyway. However, the use of supplemental feed, such as corn, that humans could directly consume is a very inefficient use of resources, and few pastoralists do it.

Domesticated animals also serve as efficient storage facilities, food resources "stored" on the hoof (making them mobile as well). In many cases, they could also be considered examples of social storage, wealth "stored" in animals owned by individuals.

GENERAL SOCIOPOLITICAL ORGANIZATION

There is no typical pastoral sociopolitical organization; pastoralists vary from loosely organized tribes to almost state-level societies. The Mongols of central Asia were loosely organized groups of families and lineages until Genghis Khan united them into an imperial state. Within a few generations after his time, they had reverted to their former organization. However, many pastoral groups maintain a loose tribe-level sociopolitical organization, one that preserves the independence of individual herding groups but ensures some overall control of the system of pastures.

The household, consisting of either a nuclear or an extended family, is the usual primary unit of social organization, and most are patrilineal. Social units have to be flexible in size and membership to be able to adjust to the shifting size and composition of herds. People tend to marry someone from a nearby family, both to maintain animal ownership and to cement alliances. Most groups maintain at least two major settlement types: villages, where most people live, and stock camps, where the herders tend the animals.

The population of pastoral groups can vary greatly, from a few thousand to hundreds of thousands, but pastoralists usually have a lower population density than plant cultivators because pastoralism is often less productive per acre than farming. However, in some areas such as in the central Andes and in Iran and central Asia, pastoralists, supplemented by some agriculture, have maintained fairly large and stable populations for long periods of time (see Yamamoto 1985).

In a pastoral society, most, if not all, of the population is engaged in animal husbandry. Most groups have a fairly strict sexual division of labor, with the men generally tending the animals, often at stock camps. The women are often left in the village to maintain the household and tend the lower-status animals.

TYPES OF PASTORALISM

Here we define three basic types of pastoralism (table 8.1) in which animals form the basis of the economy (roughly following Khazanov 1984:17–25). These types form a continuum from what many consider to be "real" pastoralists, those who are highly mobile and completely reliant on their animals, to "sedentary" or "village" pastoralists, those who are settled and for whom agriculture forms an im-

Table 8.1. The Basic Types of Pastoralism

Type	*Major Features*	*Mobility and Settlement Pattern*	*Examples*
	Primary Pastoral Systems		
Nomadic	almost all of the resources derived from animals and their products, some trade for other products	highly mobile, seasonal round with few permanent settlements	Saami
Seminomadic	bulk of resources used derived from animals and their products, supplementation by some horticulture, hunting and gathering, and trade	generally mobile, seasonal round but with some of the population remaining in permanent or semipermanent villages	Maasai
Semisedentary	animals and their products source of many of resources used, but horticulture, hunting and gathering, and trade very important	some mobility by specialized task groups, most of the population in settled villages	Navajo
	Pastoral Components of Larger Agricultural Systems		
Herdsman husbandry	animals important but farming the dominant activity	animals raised in pastures distant from the main agricultural centers, tended and moved seasonally by task groups	Basques, ranchers in the United States
Sedentary animal husbandry	animals important but farming the dominant activity	animals raised in a static location	Dani, dairies in the United States

portant aspect of the economy. Individual groups will alter their adaptations depending on conditions, and so the categories must remain flexible.

The first type is **nomadic pastoralism**, in which animals and their products form virtually the only resource. Some gathering of wild plant resources would be conducted, but no agriculture would be practiced. Such societies would consist of small, highly mobile groups that follow their livestock across the landscape. Nomadic pastoralists are uncommon, but the Saami (or Lapps), reindeer herders of far northern Scandinavia, are a good example (see Spencer 1978; Beach 1988).

Seminomadic pastoralism is the second major type of pastoralism. Seminomadic pastoralists have a seasonal round in which animals are moved from pasture to pasture, a pattern called **seasonal transhumance**. In some groups, the entire population, human and animal, will move to new pastures, meaning that the new pasture must be of sufficient size and quality to support all of the people and animals. In other groups, just the men will move the animals while the women stay in a permanent village tending gardens. In this system, animals are by far the most important resource, but some horticulture may be practiced to supplement the animal products. Trade of animal products to other groups would also be common. This is the most common form of pastoralism (see the case study of the Maasai below).

The third major type is **semisedentary pastoralism**, sometimes called agropastoralism. In this type, the animals are still the main resource, but horticulture forms a major aspect of the economy. Many of the people in such a group would live in a permanent village and grow crops. The animals would have to be moved around from pasture to pasture, but a relatively small number of specialized herdsmen would do that work. An example of this type of pastoralism is the Navajo of the American Southwest (see case study below).

Two other types of pastoralism have been defined (see Khazanov 1984:17–25), but they are really pastoral components of agricultural systems. In **herdsman husbandry**, animals are important, but agriculture is the predominant economic activity, and the majority of the people are sedentary farmers. The animal herds are tended by herdsmen in pastures distant from the main community. An example of this type of pastoralism is the Basque in France and Spain (see Ott 1993).

Sedentary animal husbandry is also a pastoral technique that forms a component of an agricultural system. This form is more accurately described as agriculture with some stock raising because it consists of full-time farmers who also raise some animals. A small-scale example is the Dani, described in chapter 7. The dairy industry in the United States is a good example of a large-scale practice.

THE GEOGRAPHY OF PASTORALISM

Until recently, pastoralists occupied a large portion of the Old World, extending from East Africa east to China and north to Siberia. Spooner (1973:6–7) defined five very broad ecological zones occupied by pastoralists in the Old World. Pastoralists relying primarily on a single species of animal occupy four broad regions of the Old World (fig. 8.1): (1) northern Eurasia, where reindeer are the primary stock; (2) the eastern Mediterranean region, where sheep are the major animal; (3) portions of North Africa and the Arabian peninsula, where camels are the primary stock; and (4) sub-Saharan Africa, where cattle are the main stock. A fifth general region, where pastoralists work multiple stock, including horses, occupies the broad arid region extending from North Africa east across southwest Asia and to Mongolia.

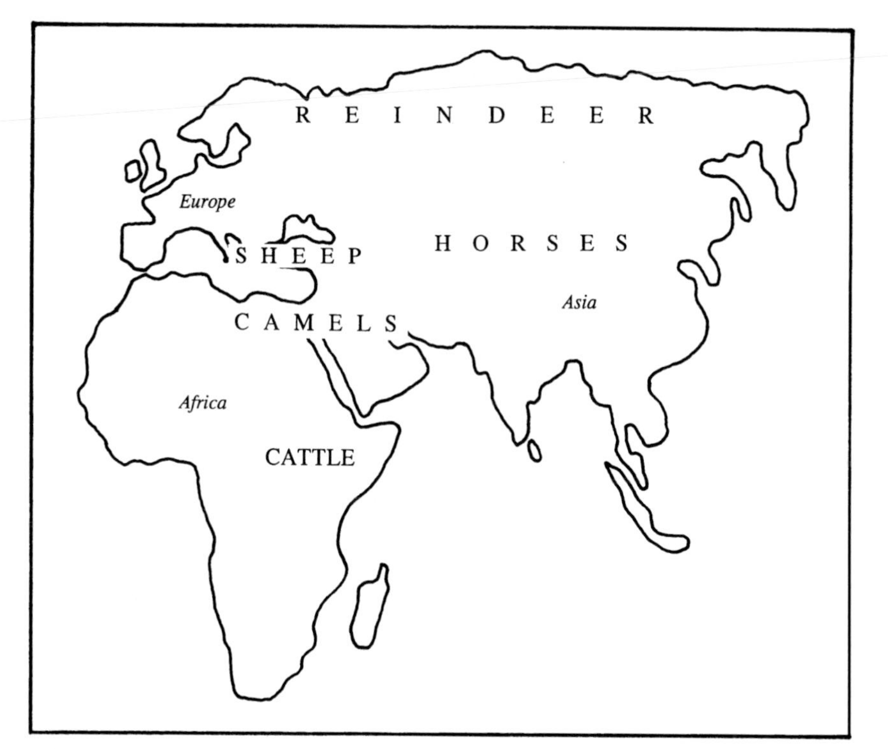

FIGURE 8.1
A general geographical distribution of pastoralists in the Old World by broad ecological zones.

There are only a few native pastoral groups in the New World. The Navajo herd sheep and some cattle in the American Southwest (see case study below). A number of groups herd llamas and alpacas in the mountains of South America. Other local pastoral economies in South America include the Gaucho economy of Argentina and neighboring countries and the Goajira of the Goajira Peninsula on the Venezuela–Colombia border. The Gaucho way of life developed as cattle ranchers moved onto the pampas, the vast, dry plains of Argentina. Gauchos were, originally, cowboys of mixed Spanish and Native American ancestry who developed a unique way of life following the herds of cattle across the unfenced ranges.

The Goajira are a Native American people who adopted herding soon after contact with Europeans. They raise cattle, horses, and small stock in their arid and isolated peninsula. Until recently, they maintained a strikingly distinctive way of life, but, as elsewhere, the old ways are now changing fast.

The areas occupied by pastoralists are shrinking all over the world due to constant and increasing pressure for them to adopt more sedentary lifestyles. This pressure is due partly to a loss of pasture to expanding agriculturalism and also to a desire by governments to exercise control over the pastoral populations within their borders. National governments now seek to regulate and "help" pastoralists by forcing them to settle in villages and adopt agriculture. (A discussion of pastoral sedentarization processes may be found in Salzman 1980.)

THE ORIGIN OF PASTORALISM

It is generally believed that pastoralism evolved from an agricultural base (see Lees and Bates 1974; Smith 1992). Perhaps some early farmers decided to specialize in animals, either by choice or because there were too many people for too small a grain harvest. Perhaps the first pastoralists were outcasts from agricultural communities, forced for some reason to live with the animals on the fringes of the town. Whatever the specific reason, it seems that early pastoralists were probably people who left farming and began moving their animals around the landscape to take advantage of areas not used by the farmers. It is probable that pastoralism emerged as a standard way of life soon after the domestication of sheep and goats some ten thousand years ago.

SOME PARAMETERS OF PASTORALISM

Pastoralism is a complex endeavor that requires a great deal of knowledge about both animals and the environment, including the availability of water and pasture and the presence of competing groups (Bates and Lees 1996:153). The important considerations include pastures, types of animals, herd composition and size, movement of herds, and products.

Pastures

The primary land resource for pastoralists is the pasture, a general area where animals can find the foods they need. Thus, like farmers, pastoralists are usually tied to specific plots of land. Many people equate a pasture with a grass-covered area where animals such as cattle or horses graze. However, a pasture is just a place where food for a particular animal is present. Pastoral animals occupy two general dietary niches: grazing and browsing. **Grazers** eat primarily grasses and low-growing plants while **browsers** eat primarily the foliage from bushes and trees. If the animals are browsers, a good pasture would consist of bushes and trees, not grasses. Nevertheless, the term "grazing" is commonly used to refer to the feeding behaviors of both grazing and browsing domesticated animals.

In all cases, pastures must be properly managed to prevent overgrazing. The primary issue is the carrying capacity of the pasture. The determination of carrying capacity is based on water availability, pasture type and quality, and the type(s) of animals to be herded, all of which can vary by season. One must be careful not to put too many animals or the wrong animals on a pasture or keep them there too long. As many pastoral groups move their entire livestock populations to new pastures, they must ensure that the pasture is capable of supporting all of the animals. In essence, then, the herders must monitor the nature and condition of pastures and note the succession stages of the various plant species in them to know when to have animals on them and how long the animals can be there. Another consideration is the slope of the pasture, as the herders do not want to excessively tire their animals on steep-sloped pastures.

If pastures are owned or controlled by individuals, the decision-making process would rest with that person. If pastures are communally owned, however, some sort of centralized control would be needed. Fredrik Barth (1964) showed that among the pastoralists of Persia (now Iran), while individual households owned and controlled their own herds, the tribal chiefs generally regulated the assignment of those herds to pastures. This practice served to prevent over-exploitation of the pastures by any one segment, as overgrazing could endanger the pasture system of the entire group. In addition to the regulation of grazing, pasture assignments functioned to control animal population, which in turn helped to control human population by limiting the food supply.

However, pastures are sometimes poorly managed. If individuals limit their herd sizes for the good of the community so as not to overexploit a pasture, they depend on the owners of other herds to do the same. If not everyone cooperates, there is little incentive for individuals to limit their herds, and short-term gain takes precedence over long-term stability (not a good situation). In such cases, the total population of livestock may crash in response to overgrazing, drought, disease, or other conditions that change the carrying capacity of the available pastures. The animal population will fall below the carrying capacity of the pastures, and the human population will follow suit. As the pasture recovers, both the animal and the human populations will grow until the next crash, and a boom-and-bust cycle will result. The recent drought in eastern Africa that so severely affected the Maasai is a good example of this phenomenon (see case study below).

Types of Animals

Ecological conditions influence the types of animals that can be raised in certain areas. The availability of proper pasture, water, and accessibility to land will limit and tether both animals and their herders. The choice of which animal(s) to herd is also greatly influenced by tradition. Some domesticated animals, such as pigs, chickens, and dogs, are not herded in a pastoral manner and are not considered as pastoral species.

Grazers

Cattle (*Bos* spp.) are primarily grazers and require fairly good pasture and a great deal of water to do well. Cattle eat the blades of the grass and leave the roots in the ground, allowing the grass to rapidly regrow and be eaten again. Cattle will also do some browsing of the leaves of trees and bushes. Domestic cattle require considerable human labor, particularly in areas with snow, because cows will not dig through even shallow snow to the grass below and have to be fed. Well-fed cattle will produce milk, a major product, in quantity. Yaks (*Poephagus grunniens*) are related to cattle (both are bovids) and behave in a similar manner, except that yaks live at very high altitudes in central Asia and will paw through snow.

Horses (*Equus caballus*) have the same basic habits and needs as cattle but generally require less human labor than cattle. Horses do less browsing than cattle and so have a greater impact on grasses. Cattle and horses can share pasture as long as the total number of animals does not exceed the overall carrying capacity of the pasture. Like cattle, horses can be used for meat, skin, and milk, but they are more highly valued for transportation.

However, horses and cattle cannot always be in the same pasture. In the Yucatan Peninsula, the Spanish had a very difficult time grazing their horses, as there was little suitable grass. However, they eventually found that horses would eat the leaves of the ramón plant (*Brosimum alicastrum*, ramón being Spanish for horse fodder). Cattle in the same region would eat the waxim plant, but horses could not, as waxim contains a chemical that makes the hair of the horses fall out. The cattle can eat it, as their four-chambered stomachs can detoxify the chemical.

Sheep (*Ovis aries*) are also grazers and will eat many types of vegetation. However, when sheep eat grass, they will often pull up and eat the roots, killing the plant and leaving the land bare. Thus, rather than quickly regrowing, plants must colonize the pasture anew. This process takes time, and so pasture recovery is more difficult. Sheep are thus more likely to overgraze and do not share pasture with other grazing animals well (having an overlapping niche and habitat). This is one of the problems that led to the enmity between cattle and sheep herders in the western United States in the past, and it remains a problem today.

Llamas (*Lama glama glama*) and alpacas (*L. g. pacos*) are the two species of domesticated camelids herded in the mountains of South America (see Flannery et al. 1989). Both animals are grazers, but the llama is more tolerant of lower elevations and is widely used as a pack animal and for meat. Alpacas need to remain above three thousand feet and are valued more for their fine wool than for their meat.

Browsers

Goats (*Capra* spp.) can both graze and browse, making the species very versatile and adaptable to most pasture types and able to take advantage of ecotone pastures. In that sense, goats are excellent secondary animals that can complement any herd. Goats produce milk that can be made into cheese. Goats, eating almost all vegetation, tend to seriously overgraze pastures and have to be carefully managed.

Camels (*Camelus* spp.) are browsers. The one-humped camel (*C. dromedarius*) is native to the Middle East and Africa while the two-humped camel (*C. bactrianus*) is native to central Asia. Camels can get along well on poor pasture and with relatively little water. Both water and fat are stored in the hump(s), and camels can survive in severe conditions for long periods of time. Camels are used for meat, skins, and milk, but the milk has too little fat to be churned into butter. Like horses, camels are most valued for transportation.

Reindeer (*Rangifer tarandus*) browse on the short vegetation of the tundra, primarily lichens. A number of groups in the far north of the Old World herd reindeer. While the herders own the animals and view them as being domesticated, at least in the broad sense of the term (e.g., Ingold 1980), most reindeer are not controlled genetically but are essentially tamed wild animals that are still more or less hunted by people. Small herds of tame animals are kept for milk and meat and for use as decoys when their "owners" hunt other reindeer. Pasture for reindeer is available all year, and the animals move around the landscape on their own to take advantage of better areas. Some groups of reindeer herders, such as the Chukchi of eastern Siberia, basically track and intercept wild reindeer as they migrate to new pastures rather than purposefully moving them to new pastures.

The distinction between hunter-gatherers and pastoralists is very fuzzy with reindeer herders. Reindeer are actually wild animals and are really hunted. However, the herders focus so specifically on reindeer as to function much as pastoralists do.

Herd Composition and Size

A herd is "not simply an aggregation of available animals" (Spooner 1973:9), and there are a number of things that must be considered in deciding on its composition and size. The types of animals herded depend on several factors. Initially, the natural environment has to be considered, because not all animals can successfully inhabit all ecozones. Even if a species is suited to the ecozone, "the economic expediency and effectiveness of herding them in specific ecological conditions" (Khazanov 1984:27) have to be considered. Once the choice is made, tradition and cultural preference play a major role in continuing the use of the animals. However, as conditions change, groups must be able to adapt by changing herd composition.

The size of the herd is also dependent on a number of factors, with the carrying capacity of the pasture being a primary issue. As different species of animals vary in their ease of control, susceptibility to disease or predators, or attractiveness to raiders or rustlers, the size of the herd may be dependent on the available labor to control and/or defend them. These factors are important considerations because disease or theft can instantly decrease the number of animals one has, reducing "a rich household to poverty overnight" (Bates and Lees 1996:154). In addition, each pasture will have a natural limiting factor that will restrict herd size, and the number of animals cannot exceed that limit (following Liebig's Law of the Minimum).

It may also be desirable to maintain certain sex and age ratios in the herd. In some cases, herds are culled, with most young males being killed for their meat and hides and only a few being kept for breeding purposes. Most of the females are kept to reproduce and to provide milk. In addition, animals will form attachments and relationships with each other. Due to species differences in life span, each species requires a different herd size to allow for individual animals to live together long enough to form these relationships (Spooner 1973:9).

Finally, there are social factors in determining the size of a herd. The requirements of the family owning the herd will greatly influence its minimum size. A herd might be developed to be as large as possible, even if overtaxing the pasture in the short term, so that it can be split into two herds. Such a split of herds may also be done to allow the human social unit to also split.

The number and types of animals are also status markers. A person might choose to maintain a herd of cattle for status reasons even if those animals were ill suited to the pasture type. For example, in Mongolia, wealthy families kept larger stock and tended to keep sheep while poorer families had smaller animals and tended to keep goats. All families tried to keep as many horses as possible, even if unneeded or uneconomic (see Khazanov 1984:25–26).

Movement of Herds

No natural pasture can support herd animals all year (Spooner 1973:21), and pastoralists must generally move their animals to different pastures depending on the season. The movement of animals from pasture to pasture may be decided on the basis of a number of factors. If a pasture is exhausted, the animals will have to be moved to a new one. In many cases, pastures in the mountains are used in the summer, and the animals are moved to pastures at lower elevations in the winter. Movement might also be based on rainfall, with animals being moved from dryer to wetter pastures. A seasonal migration might also be undertaken to maintain political autonomy, to avoid disease or pests, or to exploit other resources (Khazanov 1984:39).

Most pastoral groups are mobile and practice a seasonal round (see figs. 5.1 and 5.2 in chapter 5) within a well-defined territory, called **tethered nomadism.** However, all movements are flexible to some extent as conditions change, and people must be able to adapt to them. In some cases, only a segment of the population, such as the herders, moves with the animals in a seasonal transhumance system.

The movement of people and animals to different pastures requires careful planning that incorporates information on pasture condition (resource monitoring). Such trips range from as few as ten miles to as many as a thousand miles (Spooner 1973:21). Among the most important decisions made is where to move, when to move, and how long to stay: a decision string similar to that made in patch choice models. In some cases, such as in central Asia, there is nowhere to move the animals in winter. In those cases, fodder is gathered and stored to feed the animals during the winter.

Pastoralists must be able to judge the quality of water and pasture before the animals are moved because one does not want to invest the time and energy to move a herd to a place that cannot support it. Thus, like hunter-gatherers, pastoralists must monitor pasture areas so that sufficient information can be available when a decision about moving the animals is made.

Pastoral Products

The primary resources derived directly from animals are meat, blood, milk, hides, hair, wool, and dung. The major secondary uses of animals are transportation, of humans and goods, and labor, such as pulling plows.

Animals provide a number of foods used by people. Meat is a major animal product, but obtaining it results in the death of the animal; it is a "one-time" use. If meat is the major goal, it would be most efficient to butcher most of the males for their meat and hides, keeping a few for breeding purposes, and retain the females for reproduction and milk production. As the females die naturally, their meat and hide will then be utilized. This is the approach taken by the cattle industry in Western countries.

A more efficient use of the animals and their products is to utilize them without killing them, ultimately getting more food value from each animal. One example of this approach is the production of milk as the major product, called "**milch pastoralism**." Milk is obtained from females and is consumed or made into butter, cream, or cheese. If the product goal is just milk, females and only a few males are needed, as in the contemporary dairy industry.

Blood is another major product that can be obtained without killing the animal. About a quart of blood can be taken from an animal of either sex about once a month without damaging its health. If females are being milked, the blood extraction might be primarily from males. The blood can be consumed alone or can be mixed with milk for consumption. Some pastoral groups will not butcher healthy animals but will milk and bleed them, using those products, along with meat from aged animals, as their main foods. In East African pastoral groups, the primary component in the human diet is milk, with meat second and blood third (Khazanov 1984:64; also see Galvin et al. 1994).

The procurement of skins or hides also requires that the animal be killed. Hides can be used for a large number of products, including covers for shelter, clothing, shoes, and a variety of other leather products. Hair and wool can also be made into many products, but unlike hides, hair and wool can be removed from live animals and so are renewable resources.

Dung can be a major animal product. Dung can be used as fuel; in the construction of houses, walls, or other structures; or for a number of other purposes. Dung also serves as a fertilizer and helps to maintain the productivity of the pasture. In addition, dung can be traded to farmers for use in their fields. In some cases, farmers will arrange for pastoralists' animals to be grazed in their fallow fields so the fields can be fertilized.

USE OF NONPASTORAL PRODUCTS

No pastoralists can subsist exclusively on pastoral products; all depend to some extent on plant products in their diet. Such products may be obtained through hunting and gathering, horticulture, and trade. Interestingly, some pastoral groups tend to disparage the consumption of nonpastoral foods, seemingly as a mechanism of retaining their image of superiority over farming groups.

Most pastoralists practice some horticulture or agriculture, and all trade animal products for agricultural crops, usually some type of grain, and/or tools. Depending on the type of pastoralism practiced, horticulture might constitute a major activity. Gardens might be tended full-time by a portion of the population, or crops might be planted, left unattended, and harvested when the group passes through the area again in the seasonal round. In the Middle East, dates are important crops. Dates have to be pollinated and harvested at specific times but can be left unattended most of the year.

Some pastoralists, such as the reindeer herders in the far north, do not utilize any agricultural products but supplement their pastoralism solely by hunting and gathering wild resources for food and other purposes (see Ingold 1980).

Pastoralists rarely live in areas where fish are a significant resource. Some groups, such as the Navajo, taboo fish outright. A few coastal groups in the Middle East use fish, but, worldwide, the only significant pastoralist-fishers are the reindeer herders of northern Europe and Siberia; many of these live near rivers and rely heavily on fish, often dried for storage.

ENVIRONMENTAL MANIPULATION AND RESOURCE MANAGEMENT

Large-scale environmental manipulation is very common in pastoral groups, primarily through the actions of burning to create pasture and of grazing to maintain it. Without active grazing, much of their pasture would revert to woodlands or forests. Most pastoral groups, particularly the more mobile groups, generally invest their labor in their animals (Bates and Lees 1996:154) rather than capital improvements. However, water and pasture are also critical resources, and both are managed and manipulated to some degree. A frequent project is the construction and maintenance of water sources, some of which can be quite elaborate. A less common manipulation is pasture modification and improvement. However, the investment of time and material in pastures tends to tie groups to a piece of land and so decrease mobility. For more sedentary groups, such capital improvements are more common, with water sources, pastures, agricultural fields, and settlements receiving more investment.

Intense resource management, specifically of animals, is a hallmark of pastoralists. Most groups will utilize natural landscapes for pastures and so will invest relatively little time in environmental manipulation, preferring instead to invest their labor in increasing the number of animals and/or their products (Bates and Lees 1996:154). Controlling the animals and their movement, the kinds and amounts of food eaten, and the timing and circumstances of death are important management goals.

Exercising control of breeding is also a major management goal. One can prevent certain animals from reproducing by killing them before they can breed, castrating the males, or keeping potential mates isolated from one another. The purpose of controlled breeding is to maximize desired traits and minimize or eliminate undesired characteristics. Desired traits might include increased production of products, such as milk or hair, or an ability to flourish in poor-quality pasture.

RELATIONS WITH OTHER GROUPS

Pastoralists often occupy or utilize at least two ecozones: one where they live for most of the time and one where they pasture their animals. Usually, pastures are located at some distance from the main habitation area, and the regions between the pastures are occupied by other people, whether hunter-gatherers, other pastoralists, or agriculturalists, the latter being the most common.

To move their animals, pastoralists must often traverse the territory of these other groups (Khazanov 1984:33) and so must maintain good relationships with them. During those occasions, products will likely be traded between the two groups. If the second group is agriculturalist, the pastoralists may graze their animals in the fallow fields so that the dung can be used as a source of fertilizer.

Pastoralists may also share some ecozones with other groups, particularly farmers, rather than just traversing them. This sharing is sometimes done on a seasonal basis; each group occupies the area at a different time, so no conflict occurs. In some cases, the pastoralists will live intermingled with the farmers, grazing their animals in fallow fields, along roads, or wherever they can until it is time to move to new pastures. One could argue that the two groups occupy different niches within the same habitat (Khazanov 1984:34). However, Daniel Bates (1971) has argued that the patterns of shared land use should be viewed as a function of the balance of power rather than of ecology.

A NOTE ON THE IMPACT OF GRAZING

It is probable that grazing (here including browsing) has damaged the environment from the beginning of animal domestication. Devastated forests and woodlands in the early Middle East seem to have resulted from herding activities (Köhler-Rollefson 1988), although fire and agriculture have also contributed.

Throughout the world, sheep, goats, and cattle have eaten their way through the plant communities of local ecosystems. Unless carefully managed, those animals overcrop their favorite foods. Overcropping is an especially serious problem when the food in question is one of the dominant species in the local environment. Cattle eat willow and cottonwood shoots in the western United States, leading to elimination of these forests, which are the homes of many species of wildlife, and to massive erosion of streambeds once held in place by these trees.

As with the swidden system, grazing has been studied by many cultural ecologists, and the general conclusion is that light grazing with frequent moving of stock does little damage. Herding peoples in East Africa, for instance, have maintained a stable and balanced relationship with their environment for a long time (Dyson-Hudson and Dyson-Hudson 1980; Netting 1986; see the case study on the Maasai, below). When a given area is grazed to the point that cattle cannot find their preferred forage, the herders move on. The East African cattle complex (Herskovits 1926) is an ancient and complex system tied to local agriculture and beautifully adapted to the East African environment (for a history of its development, see Marshall 1990 and Mace 1993). Unfortunately, population growth and political restrictions on grazing and mobility have devastated this equilibrium (see McCabe et al. 1992; Campbell 1995).

Alpine herders in Switzerland (Netting 1981) maintain equilibrium in more populous surroundings by carefully moving the herds to summer pasture, then winter-feeding them on hay. Many other systems around the world practice intensive stock raising without environmental damage. By contrast, most of the Middle East has been turned into a desert in which nothing lives except shrubs too spiny for even goats to eat. Normally, this process involves the cutting of firewood, random burning, war, and so on, as well as herding, but herding was the final death stroke in the Middle East (see, e.g., Harlan 1992; MacNeish 1992; Williams 1996).

In the United States and much of Latin America, grazing was and is utterly uneconomic. It is, however, financed by government or international subsidies, and so ranchers are often politically powerful. Much of the western United States, as well as Latin America, has been deforested or turned from lush grassland into unproductive scrub by badly managed grazing (see Gillis 1991 for a brief, balanced treatment; Jacobs 1991 and Tucker 2000:285–341 provide anti-grazing positions; Kaus 1992 and Weeks 1992 provide prorancher treatments and show how government policies may help or hinder reasonable grazing practices).

In some cases, though, ranchers can be the major protectors of the local ecosystem rather than its local destroyers (e.g., Kaus 1992), depending on local conditions and economic incentives. Where cattle production is not subsidized, and where other uses of the land are recognized as valuable, ranching often does relatively little damage to the landscape. Where ranchers are paid by the taxpayers to overgraze at the expense of all else, the ranchers will do so.

Grazing has been subjected to all-out attack (e.g., Rifkin 1992) as a purely destructive use of the land. However, grazing is, after all, a natural thing. Hundreds of species of grazing animals exist, and grass has evolved a mutualistic relationship with them. Humans cannot use grass or woody vegetation for food. To get any food benefit from grasslands and brushlands, humans have to cycle some of its productivity through herbivorous animals. It seems that creating and maintaining stable, sustainable grazing systems are not beyond human ability. Indeed, some seemingly technologically simple societies manage it very well. Present serious problems with overgrazing are due not to any innate evil of grazing but to corrupted or badly conceived management.

CHAPTER SUMMARY

Pastoralism is the form of agriculture in which domestic animals are emphasized, sometimes to the exclusion of other resources. In pastoralism, humans and animals have formed a long-term mutualistic relationship in which animals are guaranteed reproduction and protection and humans get food and other products. Many pastoral groups maintain a loose tribal organization, but the household is generally the primary organization.

Three major types of pastoralism can be defined: (1) nomadic, in which groups are very mobile and depend almost entirely on their animals; (2) seminomadic, in which groups are less mobile and animal products are supplemented by horticulture; and (3) semisedentary, in which groups are not very mobile and horticulture forms a major component of the economy. Two other forms, herdsman husbandry and sedentary animal husbandry, are pastoral components of larger agricultural systems.

Pastoralists specializing in one species occupy four broad regions of the Old World, including northern Eurasia (reindeer), the eastern Mediterranean (sheep), the Arabian peninsula (camels), and sub-Saharan Africa (cattle). Pastoralists working multiple stock occupy the broad arid region extending from North Africa eastward across southwest Asia and to Mongolia. All of these groups are under pressure to abandon pastoralism and take up farming.

The primary components of any pastoral system include the use and maintenance of pastures, the types of animals (grazers or browsers) herded, the composition and size of herds, and the movement of herds. To be successful, pastoralists must make good long-term decisions in all of these areas. Pastoral products include meat, blood, milk, hides, hair, wool, and dung. In addition, most groups supplement these materials with other domesticated products, either grown or obtained by trade. Wild resources are also widely used.

Grazing, properly managed, does little damage to ecosystems, and some ecosystems, such as the Plains of North America, developed concurrently with grazing animals. However, poor planning, bad decisions, or other factors can result in overgrazing that can alter ecosystems until they are virtually destroyed. The modern practice of turning diverse forest ecosystems into pasture is causing considerable damage.

KEY TERMS

browsers
grazers
herdsman husbandry
milch pastoralism
nomadic pastoralism
seasonal transhumance
sedentary animal husbandry
seminomadic pastoralism
semisedentary pastoralism
tethered nomadism

CASE STUDY

CASE STUDY: THE MAASAI OF EAST AFRICA

This case study of the Maasai shows how pastoralists can operate, even when the populations of both humans and animals are large. The Maasai employ milch pastoralism, in which the products of live animals, rather than their meat, are the primary foods. In addition, the Maasai example illustrates how an environment can "adapt" to humans to form a "natural" and stable system.

The Maasai are seminomadic pastoralists living in southern Kenya and northern Tanzania (fig. 8.2), both former British colonies that gained their independence in the early 1960s. Today, the population of the Maasai is about 350,000 people. To the Maasai, cattle form the basis of life. In addition to their use as food and materials, cattle are used as currency and to legitimize marriages and solidify social relationships. The Maasai are one of the pastoral cultures defined by M. J. Herskovits (1926) as part of the "East-African Cattle Complex." (Summaries of Maasai culture are available in Saitoti and Beckwith 1980, Spencer 1988, and Spear and Waller 1993.)

THE NATURAL ENVIRONMENT

The Maasai occupy a region in eastern Africa that contains two major ecozones; a relatively arid plain and better-watered mountain areas. The large plain lies at an elevation of about four thousand feet and contains an extensive grassland. In northern Tanzania, this plain is called the Maasai Steppe, with the famous Serengeti Plain being just to the west. The plains generally receive less than twenty inches of rain per year. To the west of the plains, the elevation increases, there is more water, and forest intermingles with the grasslands. These highlands are mostly occupied by farmers.

SOCIOPOLITICAL ORGANIZATION

The social and political systems of the Maasai are flexible, with the elder men having most of the political authority. The Maasai have some fifteen primary territorial groups, whose members have priority rights to use the pastures under their control. The use of one group's pastures by any of the other groups must be negotiated with the owners.

CASE STUDY

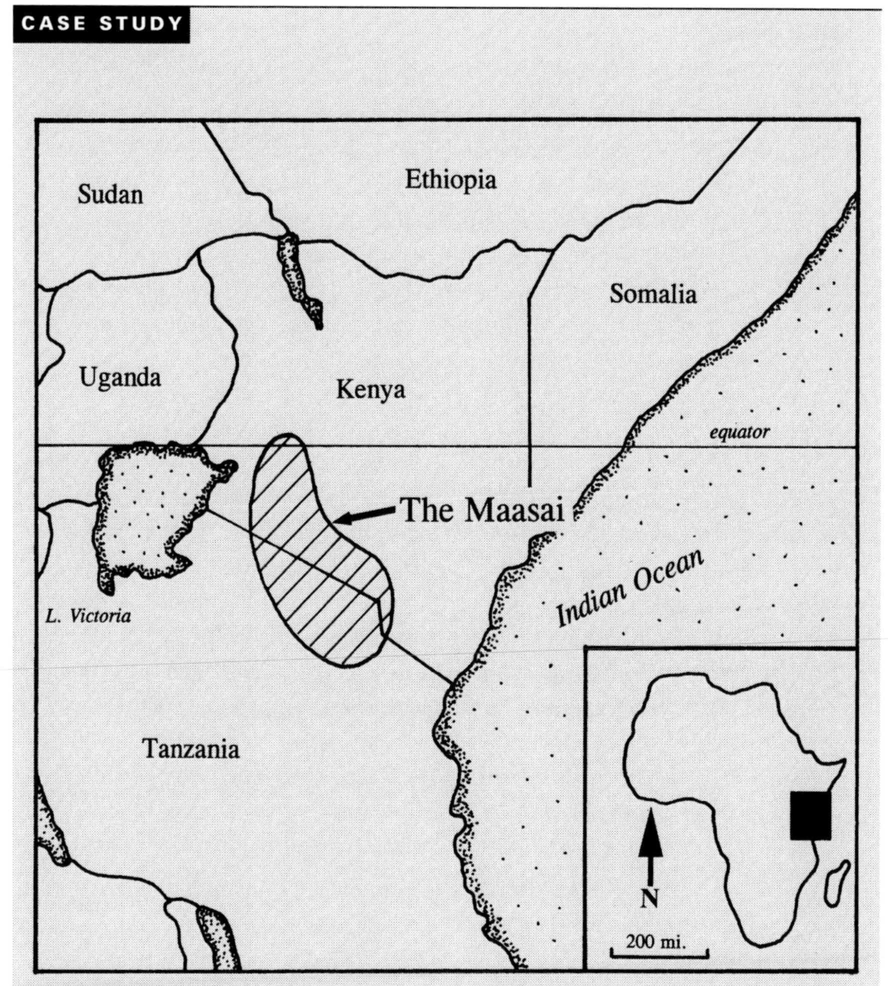

FIGURE 8.2
Location of the Maasai in East Africa.

The status and wealth of a man are determined by how many cattle he owns and what grazing rights he controls. Cattle are handed down from father to son. A man wants to marry as many women as he can afford, and cattle are given as gifts to the family of the bride. Children, especially male children, are also important as status markers, and a man who owns cattle and has male children is considered very wealthy.

Maasai society is organized around age-grades, a group of people passing from one status to another as they age. Males tend livestock

CASE STUDY

until they are about age twenty, when they become warriors and eventually get married. At this time, they generally stop tending herds but will continue to do so if they do not have younger brothers. Later, the men will become "elders" and possibly the political leaders of their family groups. Cattle are rarely killed, but a cow will be slaughtered to celebrate the passage of an individual from one age-grade to another. Traditionally, males had to kill a lion as a passage to manhood. As those animals became more scarce, the rite of passage has become more ritualized, with lions not being killed so often—although in some areas they are still killed when they menace herds.

A number of related families live in a small village, which is designed and constructed to protect the cattle. The village has a perimeter fence built of thorn bushes, and the small houses, built from bent poles covered with grass and plastered with cow dung, are constructed along the inside of the fence. The center of the village is reserved for the cattle. Smaller camps are made by the young men when they are herding cattle away from the village.

At one time, warfare was an important aspect of Maasai culture. Men were raised to be warriors, and once they had achieved that status, they generally stopped herding cattle. The primary function of a Maasai warrior was to protect the cattle from rustlers and to rustle cattle from non-Maasai people.

ECONOMICS

Although the primary economic pursuit of the Maasai is cattle herding, some horticulture is also practiced, supplemented by the gathering of wild plants. The cattle herds are tended by young males on foot. If the herds are located in pastures away from the village, small temporary camps are used by the herders. If the cattle are close, males take care of the animals during the daytime, and women are responsible for the herds at night. Women are also responsible for milking the cows and helping them give birth.

Pastures

Pasture is a critical resource, particularly in the relatively dry lowlands. Animals must be moved fairly often, and a complex system

CASE STUDY

of who can have how many animals in which pasture is controlled by the local leaders. Pasture condition has to be constantly monitored, and the Maasai closely assess their animals' general health and ability to produce milk in order to determine the quality of the pastures.

Types of Animals

Cattle are clearly the most important animals to the Maasai, and most of the effort made in stock raising is expended for cattle. In addition, however, several smaller species, such as sheep and goats, are included with the cattle herds. These other animals occupy the same habitat but have slightly different niches and so provide the herder with greater efficiency of pasture and water utilization.

Herd Composition and Size

Herds of cattle consist mainly of females and a few bulls. Most male cattle are castrated when young and kept for meat and labor or sold. The bulls in a herd are of different ages to prevent them from fighting. The size of a herd is dependent on the available pasture and water, but the constant goal is to have as large a herd as possible.

Movement of Herds

The movement of livestock is based on a number of factors and is carefully planned. A principal factor in the decision is the seasonal variation in rainfall, with animals generally being moved to the highlands during the dry season and to the lowlands during the wet season. Decisions on livestock movement must also involve other local conditions and negotiations with other groups for access to their lands. This system is designed to reduce the risk for any one group by ensuring that livestock are dispersed so as not to overgraze regions.

Many animals are lost each year to a variety of diseases, and disease control is a major factor in deciding when and where to move animals. Malignant catarrh fever (MCF) is a disease fatal to cattle (see Evangelou 1984:77). For centuries, the Maasai believed that MCF was carried by the wildebeest and spread when the animals had their calves. To reduce the exposure of their cattle to MCF, the

CASE STUDY

Maasai planned their seasonal movements to keep their cattle away from the areas used by the wildebeest while calving. Thus, through careful planning, the Maasai were able to utilize the same range as the wildebeest, but at different seasons. Recently it has been confirmed that MCF is indeed carried in wildebeest placentas.

Pastoral Products

The Maasai utilize all parts of their cattle. The primary products are milk, blood, and meat for food, urine for medicinal purposes, dung for fuel and for building houses, horns for containers, and hide for clothing, shoes, and rope. Milk and blood are most commonly eaten as they are the most efficient use of the animal. To obtain blood, the Maasai pierce the jugular vein of a cow or bull and remove about a quart of blood. Typically, the neck of the animal is lightly tied to make the vein bulge, and the vein is then shot with a tiny arrow. The blood runs into a cup. Bark is often added to prevent coagulation and spoilage—a useful technology for this high-cholesterol food because many barks have the ability to reduce cholesterol in the blood. In a healthy animal, blood can be removed about once a month. The blood is often mixed with milk for consumption. If an animal dies or is injured or sick, it will likely be butchered for meat and other products.

Horticulture

Plant foods are important to the Maasai, and they practice some horticulture in specific and confined localities. Small gardens are maintained around the villages, and women tend to eat more plants than men. In addition, the Maasai trade animal products for grain from their neighboring farmers.

Hunting and Gathering

The Maasai collect many wild plants for food, medicine, and other purposes and have a great deal of knowledge of regional botany. However, the consumption of wild game is frowned upon, and for all practical purposes, the Maasai do not hunt game for food and look with contempt upon people who do. Nevertheless, the Maasai do retain substantial hunting skills and hunt a number of animals, such as lions, that prey on their herds. In bad times (e.g., severe

CASE STUDY

drought), some individuals or small groups of people might have been forced into the niche of hunting wild animals for food. More recently, those people who, for whatever reason, cannot make a living as pastoralists, have emigrated to the cities to find work.

ENVIRONMENTAL MANIPULATION

In general, the Maasai practice relatively little large-scale manipulation of their environment. An exception is the controlled burning of tracts of land to eliminate brush and to encourage grass for better cattle pasture. This practice has had the effect of also encouraging some wild species of large grazers, such as zebras, to increase their numbers in Maasai pastures and has had a negative domino effect on much of the smaller wildlife.

In some areas, the Maasai have been prevented from burning because of a belief that the burning was detrimental to the large game. When the area was not burned, the brush took over, and the big herbivores moved away, leaving the land less productive both for cattle and for tourism. Without Maasai management, all productive use of the area declined.

RESOURCE MANAGEMENT

The Maasai basically have three major resources to manage, their animals, their pastures, and water. The animals are very intensively managed, with selective breeding and castration being important techniques. Poor-quality animals are culled, and the proper mixture of sex and age is maintained in the herds. The animals are also closely monitored to prevent or limit diseases.

The pastures are the second critical resource. Their condition is constantly monitored for sufficient information to make decisions about where to go, when to move, and what to use. The Maasai strive to prevent overgrazing with a complex system of pasture assignment.

Water is the third important resource, as cattle require a considerable amount of it. Water holes are important places and are owned by individuals, who modify and maintain them. Access to the water is open to all for the asking. Drought is always a concern. The

CASE STUDY

worst recorded drought befell in 1960 and 1961, when the Maasai cattle population declined from 630,000 to about 200,000. The cattle population had recovered to its predrought numbers by 1968, a recovery accelerated by Maasai herd management practices (Evangelou 1984:20).

RELATIONS WITH OTHER GROUPS

The Maasai believe that pastoral life is the best, and they have little but contempt for neighboring nonpastoral groups, especially hunters and gatherers. The surrounding pastoral groups are also considered inferior, and in the past those groups were raided for cattle.

DISCUSSION

The Maasai illustrate a number of important issues regarding pastoralists. To be successful, the Maasai must maintain a carefully coordinated and efficient system of management of animals, pasture, and water. They generally operate within a single ecosystem, the plains, but can adapt to the highlands if necessary. Through careful management, the Maasai are able to herd animals with different niches (e.g., cattle and goats) within the same pasture habitat. The Maasai also illustrate the efficient use of animal resources, storing their animals live to consume their milk and blood, rather than their meat, as the primary product. In this way, a relatively smaller number of cattle can support a larger human population.

In recent times, the governments of Kenya and Tanzania have been putting pressure on the Maasai (and other pastoralists such as the Ariaal of northern Kenya; see Fratkin 1998) to become settled farmers. In 1973, the government of Tanzania, where most of the Maasai live, required all people to move into "developmental villages" to get them out of the landscape and force them into farming. Due to pressure from herders, the policy was changed in 1974 to include "livestock developmental villages," thus allowing pastoral groups to continue their economies. Nevertheless, the trend away from pastoralism to settled farming is increasing.

CASE STUDY

In the last several decades, the Maasai have lost some 75 percent of their traditional lands and have been largely relegated to "reservations" (Olol-Dapash 2002). Some of this land has been taken by farmers while large tracts have been set aside as wildlife preserves and game parks. The Maasai cattle have been excluded from such preserves so that the wild animals would return and boost the tourist trade. Interestingly, without the cattle and land management by the Maasai, the brush began to take over the land, and the game that had lived there began to leave—just the opposite of what the governments wanted. Today, the Maasai are slowly being allowed to return to many such lands, to reintroduce cattle, and to resume burning, in the hope that the wild game will return.

CASE STUDY

CASE STUDY: THE NAVAJO OF THE AMERICAN SOUTHWEST

This case study of the Navajo illustrates a pastoralist adaptation within a desert biome. While successful, the Navajo have their share of problems with overgrazing, drought, and neighbors. In particular, the very success of the Navajo has brought them into conflict with the Hopi, a sedentary agricultural group. Here we see some issues of incompatible subsistence strategies and a clear illustration of cultural conflict.

The Navajo are semisedentary pastoralists living in Arizona and New Mexico in the American Southwest (fig. 8.3). They have the second-largest population (the Cherokee are more numerous) and the largest reservation of any Indian group in North America. The Navajo generally call themselves Diné, meaning "people of the surface of the earth." They moved to the Southwest about five hundred years ago and adopted various Pueblo, Spanish, and other traits, forming the present Navajo culture. The Navajo reservation overlaps that of the Hopi Indians, who are horticulturalists, and this

CASE STUDY

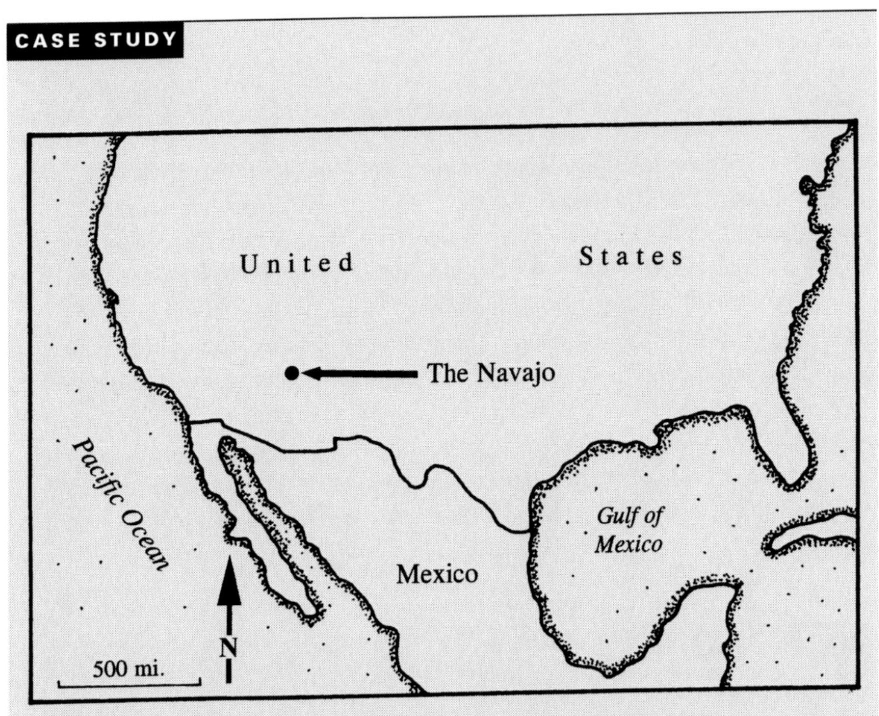

FIGURE 8.3
Location of the Navajo in southwestern North America.

relationship is quite interesting. (Recent descriptions of Navajo culture are available in Ortiz 1983, Iverson 1990, and Parezo 1996.) This case study describes a Navajo lifestyle that is slowly disappearing as more and more Navajo become employed in the larger American economy.

THE NATURAL ENVIRONMENT

The Navajo reservation is located in fairly rugged country characterized by mountains, deep canyons, many small mesas, and numerous valleys. Elevations range between three thousand and nine thousand feet. About half of the reservation is desert but contains grasses important for sheep grazing. The remainder of the reservation consists of mountains, mesas, and valleys that contain forests of pine and juniper, along with a great many other plants, including cottonwoods, willows, and many grasses.

CASE STUDY

SOCIOPOLITICAL ORGANIZATION

The Navajo are organized into a relatively large number of small, autonomous social and political units, sometimes called bands. Each band has a male headman to deal with outsiders, but the females are dominant in politics, with the family matriarch making most of the family decisions.

The primary social unit is the nuclear family, centered around the mother and her livestock and fields. Husbands generally move to land near the wife's mother and form a new family unit, or homestead. A number of these families related through a matriarch live in the same area, forming an extended family unit. These families are very mobile and move around to exploit changes in the availability of water, firewood, and grazing areas. The families will frequently move to new lands, and sheep camps are often established at some distance from the home during the summer.

Women do the household work and herd the sheep when they are close to the home, mostly during the winter. Men will herd sheep during the summer, when they are located at distant pastures. Men also tend to the horses and cattle, do most of the heavy agricultural labor, and conduct most of the ceremonies. Everyone assists in collecting water and firewood and in planting and harvesting crops.

Marriage is mandatory, as only married people are considered adults. The most important aspects of mate selection are economic and political considerations. Wealthy men might have two or three wives, who generally live in separate households near their mothers.

ECONOMICS

Prior to the 1860s, agriculture and raiding to obtain goods and livestock were major elements of the Navajo economy. After the Navajo were defeated by the Americans in 1864, the Navajo reservation was established (1868). At that time, the Navajo adopted herding, mostly sheep but some cattle, as their primary economic focus. The number of livestock increased rapidly, and by the 1930s, overgrazing by sheep had become such a problem that the government forced the Navajo to significantly reduce their sheep herds and encouraged them to increase the number of cattle.

CASE STUDY

All of the sheep owned by individuals within a homestead unit are kept in a common herd, and all of the homestead members share the responsibility for their care (Witherspoon 1983), with shearing and dipping being community efforts. The size and well-being of the sheep herd are the main factors in the status and power of the family or individuals within it.

Pastures

Pastures are generally located fairly close to the homestead, and the pastures, residences, and gardens of a family are called a "traditional use area" (Downs 1972:43). The pastures are controlled by families who share them with other members of the local residential unit. Each family unit is responsible for only its own livestock and the specific pasture used at any one time. If nearby pastures are not sufficient, as determined by the availability of water, families may move their animals to more distant pastures where there is more water.

Types of Animals

The main domestic animal is sheep, a grazer easily adapted to a variety of ecozones, which provides meat and wool for weaving. Wealth is measured by the number of sheep an individual owns (fig. 8.4). Other animals, including cattle, are also kept but are much less important than sheep. Horses are important for transporting people and materials and for social standing (everyone is expected to own at least one horse). Goats are often kept with the sheep herds and are used as a source of milk and cheese. Some chickens are also kept.

Herd Composition and Size

The primary animal is the sheep. It is generally desired that sheep herds be as large as possible, but this ambition has led to serious overgrazing and land degradation through loss of plant cover and erosion. Today the federal government closely monitors pasture condition and so regulates herd size. The cattle and horses do not have to be herded as sheep. Nevertheless, they require care, the job of men.

Movement of Herds

Four major factors govern the movement of sheep herds: (1) availability of water for the animals; (2) location of gardens; (3) availability

CASE STUDY

FIGURE 8.4
A Navajo woman on horseback herding her sheep in 2000. (Courtesy of Corbis)

of firewood; and (4) season (Downs 1972:44–46). If possible, sheep are kept near the homestead. During those times, the sheep are usually kept in pens for the night, driven to nearby pastures for the day, then driven back to the pens for the night. Dogs are used to help herd the sheep and to guard the herd from packs of wild dogs.

If water becomes a problem, both the homestead and the herds might be moved to locations with better water, although permission has to be obtained to use water on other pastures. A shortage of firewood might also prompt a homestead to move, with its sheep being moved, too. On the other hand, people may opt to stay near their gardens and move the sheep to other pastures without moving the homestead. In these cases, young men move and tend the sheep and may be away from the homestead for extended periods.

Pastoral Products

Sheep provide the major source of meat and all of the wool needed for clothing and weaving. Cattle provide only cash because

CASE STUDY

the Navajo do not butcher their cattle but sell them to others. Horses provide cash and transportation.

Horticulture

Domestic crops are much less important than the animals, but some horticulture is practiced. Each homestead has one or more agricultural fields, tended by one or more women. Corn is the primary crop, but beans and squash are also grown, along with wheat and oats. Fruit orchards are also common. In recent times, with overgrazing and livestock reduction, horticulture is becoming more important in the economy.

Hunting and Gathering

Hunting is not a very important source of food, although a few small animals, particularly rabbits and rodents, are hunted for meat. Although a minor part of the economy, hunting by young males is encouraged and serves to acquaint them with the landscape and the animals in it and to develop many of the skills needed for livestock herding. Some wild plants are gathered for food, but most plant gathering is for the purpose of obtaining herbs and materials for rituals, dyes, and medicines.

ENVIRONMENTAL MANIPULATION AND RESOURCE MANAGEMENT

The Navajo conduct very little environmental manipulation. However, the sheep cause a considerable amount of damage to pastures through overgrazing. The sheep strip the vegetation, producing considerable erosion and delaying pasture recovery. So much topsoil has been lost on the Navajo reservation that the federal government has been concerned about the silt clogging the Colorado River.

The Navajo manage their herds fairly intensively, with cattle and horses receiving less attention than sheep. Although sheep breed all year, the Navajo try to control the breeding so that lambs are born in the spring, when they have a better chance of survival. Otherwise, the sheep are managed to produce wool and meat.

The Navajo have tended to manage their pastures for maximum short-term sheep production. They allowed their livestock to seriously overgraze pastures, and then they moved on to new pastures. This

CASE STUDY

practice allowed them to have artificially large numbers of animals and to constantly expand their territory, which, one could argue, has been to their distinct advantage. However, since the 1930s, the federal government has regulated the number of animals the Navajo can have (see the next paragraph) and has limited the expansion of Navajo territory. Today, the government conducts considerable monitoring of pasture condition to gather information for making decisions about how many animals the Navajo can have.

The relationship between the Navajo and the U.S. government over grazing impacts has played a major role in Navajo ecology. When the Navajo reservation was formed in 1868, the Navajo brought with them a large number of grazing animals, mostly sheep. Overgrazing became an immediate problem. By the 1930s, overgrazing had become such a problem that the government forced the Navajo to significantly reduce their sheep herds and encouraged them to increase the number of cattle. Sheep were sold (at very low prices during the Depression) and sometimes even destroyed by the government without compensation to the Navajo. This stock reduction forced many Navajo out of ranching and into other work, transforming their entire economy. In addition, the wealth and status of women were lowered, since they owned most of the livestock (see Shepardson 1982).

The Navajo, who love their sheep, reacted extremely negatively to the federal reduction of sheep herds in the 1930s; subsequently, the Navajo have never trusted the government to regulate grazing. This distrust has led to accusations of bad faith on both sides, and range management has all too often been a casualty of the conflict.

RELATIONS WITH THE HOPI

The Hopi are horticulturalists living in the same general area as the Navajo. The two cultures occupy basically different niches within an overlapping habitat. The Hopi claim that the Navajo have taken up residence on lands that the Hopi consider their traditional territory. After its establishment in 1868, the Navajo reservation was expanded several times, and by 1934 it surrounded the Hopi reservation.

CASE STUDY

The Hopi have filed a number of claims and suits to force the Navajo to stop encroaching and to force the removal of those who moved onto the Hopi reservation since 1882, when it was established. The Navajo argued that the land is traditionally Navajo and that the Hopi were not using it anyway. The Hopi countered that the land has been Hopi for thousands of years, that the Navajo are trespassing, and that Navajo livestock are overgrazing and damaging Hopi land. The discovery of considerable coal and oil reserves in the region prompted the government to seek a settlement so that leases could be obtained (see Lacerenza 1988).

The Navajo–Hopi Land Settlement Act of 1974 partitioned the disputed lands between the Navajo and the Hopi, required a reduction in Navajo livestock (see Wood 1985), and mandated the relocation of Indians on the wrong side of the line, mostly Navajo who were on Hopi lands. The relocation of Navajo families began in early 1981, but some families remained very resistant, feeling that they have a strong tie to the land because many were born in the places they are being told to leave (see Schwarz 1997). Some people then argued that the Navajo have always been mobile, did not have a problem moving onto Hopi lands, and so should not have a problem moving off them (see Prucha 1984:1178).

Congress made a new effort to settle the issue and passed the Navajo–Hopi Land Dispute Settlement Act of 1996. This law allowed Navajo families still in the disputed areas to sign seventy-five-year leases with the Hopi and to live under Hopi jurisdiction. A majority of families did so, but a few did not. These families sued but lost, appealed, lost again, and appealed to the U.S. Supreme Court, which told everyone involved to settle out of court. The two tribes resumed negotiations, and no action was taken on the families. In April 2001 the Supreme Court dismissed the Navajo case, but the situation on the ground remains unchanged.

The Hopi and the Navajo have a number of issues regarding the land dispute, including access to a place of religious significance, royalties for coal and gas, and general frustration over losing lands to outsiders. Still, the degradation of Hopi lands by overgrazing Navajo livestock remains a key problem. (More information on

CASE STUDY

Navajo–Hopi land issues can be obtained from Kammer 1980, Feher-Elston 1988, Parlowe 1988, Clemmer 1991, Benedek 1992, Brugge 1994, and Schwarz 1997.)

DISCUSSION

The Navajo adopted pastoralism in fairly recent times, acquiring sheep and focusing their economy on that animal. Sheep treat pastures notoriously poorly, rapidly overgrazing them and making their recovery difficult. As long as the Navajo could expand into new territories and pastures, this short-term disadvantage was not a problem. Today, however, the Navajo do not have easy access to new lands and must adopt a long-term solution to finite and fragile pastures. Part of this solution is a reduction of the number of sheep and part is the replacement of some of the sheep with cattle. As these two animals occupy at least partly different niches, the pasture habitat can be better managed.

Also of interest is the competition between the pastoral Navajo and the horticultural Hopi. Although one could argue that the two groups occupy different niches within the same habitat, there is enough overlap (e.g., access to water and pasture for the lesser number of Hopi animals) that conflict has ensued. The solution to this issue will be political, but there is a great deal to learn about the interaction of the two lifeways.

9

Intensive Agriculture

Intensive agriculture is a large-scale and complex system of farming and animal husbandry often involving the use of animal labor, equipment, and water diversion techniques and the production of surplus food. Intensive agriculture represents a significant shift in the scale and scope of agriculture and reflects a fundamental change in the relationship between people and the environment. The use of animals and machines to supplement human labor is significant in intensive agriculture, although there are a few intensive systems that rely solely on human labor.

The changing scope and scale of intensive agriculture, coupled with increasing social and technical complexity, increase the range of options that humans have in adapting to the environment. Given a sufficient reservoir of potential action, some intensive agriculturalists have developed a worldview in which they are above nature and therefore less bound by nature than their less complex neighbors. This belief system may generate a feeling that nature can be, and so must be, controlled or conquered. This view is flawed, however, because all cultures are integrated with their environment and cannot escape the consequences of their actions (Flannery 1972) unless they relocate or exploit resources of distant places through trade or conquest.

Intensive agricultural systems tend to rely on a narrower range of domestic species than horticulturalists do, with increases in both productivity and risk. They will employ components of the other subsistence strategies—hunting and gathering, horticulture, and pastoralism—but these tend to be minor components of the system. A consequence of the increased productivity of intensive agriculture is an enormous increase in human carrying capacity (assuming food is the limiting factor, following Liebig's Law of the Minimum) and so huge increases in population. With population increase comes greater sociopolitical complexity, reduction of mobility, nucleation of settlements, and sometimes the evolution of state-level societies. While this cause-and-effect relationship among agriculture, population increase, and state development is overly simplistic, the trend is generally true.

CHANGES IN SCALE

The most dramatic difference between horticulture and intensive agriculture is that of scale. Intensive agriculturalists generally cultivate larger quantities of land, have larger populations, and impact the environment to a much greater degree. The use of animals to supplement the labor of humans significantly increased the scale and intensity of agriculture. The introduction of agricultural machines based on the internal combustion engine changed the scale again, this time in a massive way, ushering in contemporary industrialized agriculture.

Labor Supplements

In horticultural and pastoral societies, humans provide the vast majority of the labor needed to produce food, including clearing land, constructing fields, and planting and harvesting crops. People, by themselves, can do only so much. As animals and machines began to supplement human labor, the scale of agriculture increased dramatically, and ecozones previously too difficult for agriculture could be colonized by farmers, who often displaced other groups already in those ecosystems.

Animals were incorporated into some agricultural systems early on and quickly became an integral component of those systems. Human labor remained important, but animals are able to do the work of many people by carrying heavy loads and pulling plows through heavy soils. The use of animals as labor requires a support system of feed and care. In some instances, animals were fed food that people might ordinarily eat, but in many cases the animals ate substances that people could not, such as grass, making efficient use of those resources.

While domesticated animals had been around for a long time, large animals were required for agricultural labor, as animals such as dogs are too small to pull plows. However, large domestic animals were not available in all societies. The few present in the New World, such as llamas in the Andes, were used as pack animals but not for agricultural labor. Thus, the intensive agricultural systems in the New World, such as the Maya system (see case study below), were all based on human labor.

The use of machines is a recent development. Early machines were water or steam powered and did not have a dramatic impact on agricultural systems that still relied on animal and human labor. Those machines did, however, have dramatic effects on industrialization and transportation, which later affected agriculture. While some machines had been used in agriculture, the internal combustion engine, after about 1900, dramatically sparked the mechanization of farming, a process continuing today. Human labor is still important in mechanized agriculture but to a much smaller degree than before.

Unlike animals, machines do not compete directly for human food. However, like animals, machines require a support system. The machines and parts for them must be manufactured, repaired, and given fuel. All of this support requires a complex industrial infrastructure.

Technological Changes

Technological change has been an important aspect of the development of intensive agricultural systems. Technology has increased the efficiency by which farmers can grow crops and has opened new ecozones to agriculture. A number of simple technological changes had significant impacts on farming, including the use of metal axes, with which people could clear forests much faster than they had been able to with stone axes. To illustrate the point further: Europeans initially had little interest in colonizing the Great Plains/Prairies of North America, as the grassland sod was too thick to be penetrated by the plows of the time. In the early to mid-1800s, in fact, the Plains/Prairies were known as the Great American Desert, and white settlers passed through them to California and Oregon without major colonization. However, after the American Civil War, the development of heavy steel plows enabled farmers to penetrate the sod of the Prairies and to plant fields. These farmers became known as sodbusters, and they colonized large areas of the Prairies. The innovation of the heavy steel plow opened the Prairie ecozone to intensive agriculture and hastened the replacement of the Indians with white American settlers.

More dramatic change followed the development of machines powered by fossil fuels. Clearing a section of forest that would have taken many men a year with metal axes and horse-drawn wagons can now be done in a matter of days by a few men with gasoline-powered chainsaws and diesel-powered bulldozers. The machines changed the speed and efficiency of these activities and in doing so also changed the scale of environmental manipulation.

Finally, our technology is on the verge of allowing us to purposefully reshape an entire biosphere. Plans are being developed to send machines to Mars with the intent of creating an atmosphere breathable by humans, warming the planet, melting the polar ice, and creating oceans. We are fairly certain that this can be done, as we have inadvertently done the same basic thing to our own biosphere through the pumping of carbon dioxide into the atmosphere. Once this is accomplished on Mars, in several hundred years' time and at huge expense, Mars could be colonized by all manner of life from Earth.

Changes in Organization

As carrying capacity and population increased, social organizations became more complex. While there are large and complex groups of hunter-gatherers, horticulturalists, and pastoralists, even the most complex of these groups fall within the sociopolitical category of chiefdoms, and none developed a state-level organization. On the other hand, a number of intensive agriculturalists did evolve sociopolitical organizations classified as states.

A state (early states are sometimes called archaic states) is a society with elaborate social stratification (or at least, the classes of rulers and commoners) and a hierarchical and complex political system that was highly centralized and internally specialized (Marcus and Feinman 1998:4; also see Adams 2001). Many earlier researchers (e.g., Childe 1942; Steward 1955) used the term "civilization" and included a number of other criteria, including urban centers, writing, monumental architecture, craft specialization, bureaucracies, large populations living in urban centers, codified law, and a central authority with the ability to use force following the law. Each of the various criteria used to define a state is intended to demonstrate the sociopolitical complexity of the society and the ability of the rulers to call on the populace to do things. For example, if a group has cities, it follows that they must have had a complex infrastructure, a bureaucracy that can control the population, sufficient resources to support many people who are not farmers, and other complex organizations. The same logic is true of the other criteria.

All agricultural economies are based on a stable carbohydrate source, such as some sort of tuber or grain, and the agricultural systems of all state-level societies are based on some sort of grain crop. While grain crops provide a much greater caloric return per pound than root crops do (table 9.1), root crops generally yield much more per acre, offsetting the grain advantage. However, tubers are difficult to store in the quantities needed by states.

Theories on the Origin of the State

The reasons people evolved such complex state-level sociopolitical organizations have long been of interest to anthropologists (e.g., Steward 1955; Flannery

Table 9.1. Nutritive Value of Various Foods

	*Calories**	*Protein (g.)**
Grains		
Corn	361	9.4
Amaranth	358	12.9
Rice	357	7.2
Barley	348	9.7
Sorghum	342	8.8
Wheat	330	14.0
Roots and tubers		
Manioc (bitter)	148	0.8
Manioc (sweet)	132	1.0
Sweet potato	116	1.3
Yam	100	2.0
Taro	92	1.6
White potato	79	2.8

*per 100 grams (USDA 1963)

1972; Feinman and Marcus 1998; Sanderson 1999; Johnson and Earle 2000; Lamberg-Karlovsky 2000). All contemporary state organizations have their roots in past states, and so some comprehension of the origin and development of the state can add to an understanding of ourselves.

Managerial Models

Several models argue that state-level organizations developed out of the need to manage people, resources, or both. Agricultural surpluses, trade requirements, warfare, population expansion, and water have all been identified at various times as prime movers in the cultural evolution of states.

Karl Wittfogel (1957) proposed the Hydraulic Theory, in which water for irrigation was a critical resource. Wittfogel argued that in some instances, such as in large river valleys, competition for water for agriculture led to the development of irrigation management systems and even warfare between some groups. The model proposed that due to the need to manage and protect water, these societies evolved state-level political structures. Not all groups that used irrigation evolved state-level societies, and many researchers have rejected the hydraulic model (see Hunt 1988). However, David Price (1994) has argued that while some small-scale societies that used irrigation, called hydroagricultural by Wittfogel, did not evolve into states, others, called hydraulic by Wittfogel, did develop state-level organizations, suggesting that the general model may still be applicable.

Conflict Models

Conflict, in the form of competition or outright warfare, can be seen as a powerful incentive to organize. Warfare between political entities, particularly over critical resources, may have had a major influence on the development of complex organizations (Carneiro 1970; Cohen 1984). However, all state-level societies practice some warfare, and it is not clear whether warfare is a cause of, or a result of, states. Robert Carneiro (1970) also suggested that states might have arisen in very rich but very sharply circumscribed environments—usually fertile river valleys bordered by desert. Here, highly productive agriculture tempted conquerors and rulers, and the peasants had nowhere to run. State power would have been aided by the natural barriers of desert and mountain.

Multivariant Models

Researchers (e.g., Adams 1966; Butzer 1976) have suggested that causality in state formation was probably the result of a number of factors, with no single prime mover. Such a multivariate cause would involve a very complex set of interactions, with warfare, irrigation, and trade being major factors. The complex social and political organizations that characterize the state would have developed (in other words, would have been selected for, as in an evolutionary sense) from these interactions.

TECHNIQUES OF INTENSIVE AGRICULTURE

Intensive agriculture is cumulative in its use of techniques, employing all of the primary methods used by all other groups and being much larger in scale. The major new technique used by intensive agriculturalists is irrigation. Well-watered crops produce much greater yields than unirrigated crops. Higher crop yields can support more people, whose labor can help intensify agriculture to support even more people, and large numbers of people are a condition for the development of complex sociopolitical entities.

Irrigation

Irrigation generally involves the purposeful diversion of water from its natural source onto agricultural fields to provide water and any nutrients it may contain. Irrigation can be small in scale, such as the simple diversion of a small spring-fed creek onto a patch of grass, or very large in scale, such as the construction of hundreds of miles of canals. With irrigation comes the need to control water sources, a fact that inspired the Hydraulic Theory of state development (see discussion above). Some 40 percent of the food produced in the world comes from the 16 percent of the agricultural land that is irrigated (Matson et al. 1997:506).

A number of different types of irrigation systems are known. The first system is flood irrigation, also called natural or basic irrigation, and involves the use of floodwaters to cover and soak fields. While this method results in the irrigation of fields, it is not true irrigation, because no human constructions are used. Much of the irrigation of the Egyptian culture, from ancient times to the completion of the Aswan Dam in 1970, was flood irrigation. However, the Egyptians supplemented the natural flooding with pot irrigation and with the construction of basins to "store" floodwater.

Genuine irrigation involves the construction of facilities, such as dams, diversions, canals, or wells, to divert water to fields. Rivers or streams might be diverted, channeled into canals, and delivered to fields many miles distant. For example, between thirteen hundred and five hundred years ago, the Hohokam in the Sonoran Desert of the American Southwest employed a vast network of irrigation canals to support a large population and complex culture (see Crown 1990 for a review of the Hohokam). The ancient state-level societies of the Tigris-Euphrates and Indus river valleys also relied on extensive canal systems. The current Central Valley project of California involves moving water from northern California some four hundred miles to the south in a massive aqueduct.

In addition to surface water, people have made use of subsurface water. This water was extracted from the ground with the use of wells, pots, and, more recently, pumps. Such methods may constitute the primary irrigation system or may supplement others.

Dry Farming

Dry farming, somewhat of a misnomer, is the production of crops relying on water from rainfall; no constructed irrigation systems are used. Farming irrigated lands may have been the most important approach in early states, but today, about 84 percent of the world's agricultural land, including pasture, is dry farmed. In many areas, rainfall is quite sufficient as a water source, and in some areas, fields have to have systems to drain excess water. In other regions, such as the central United States, crops are utterly dependent on rain, and a dry year can result in crop failure.

The Use of Other Subsistence Techniques

Hunting and Gathering

All cultures do some hunting and gathering, but such activities are usually much less important in intensive agricultural economies than in horticultural or pastoral groups. Most intensive agriculturalists use very small amounts of wild resources, but some, such as ocean fish, are now extracted on a vast scale, using fishing fleets and industrial complexes. The economies of some countries, such as Japan, Iceland, and even the United States, are highly dependent on ocean fish.

In a number of instances, certain wild species, such as trout, have been domesticated and are now raised on farms.

Horticultural Techniques

Small-scale agricultural plots and gardens still form an aspect of intensive agriculture, as can be seen by the numerous backyard gardens in the United States. Some intensive systems are comprised entirely of an assorted combination of horticultural techniques, used intensively and in combination. An example is the Maya system described in the case study below.

Pastoral Techniques

Most farmers raise some animals for food, usually small species such as dogs, pigs, ducks, chickens, and guinea pigs. Intensive agriculturalists tend to raise much larger animals, such as cows, and while these are generally used for food, they are also used for a number of other purposes. These purposes include prestige (conveyed by some leather products), and labor, such as carrying loads (pack animals) and pulling plows (fig. 9.1).

Most intensive agriculturalists employ a component of pastoralism. Those using herdsman husbandry are sedentary farmers whose animal herds are tended by herdsmen in pastures distant from the main community. Those using sedentary animal husbandry are really full-time farmers who also raise some stock. This latter technique is widely practiced in the United States.

FIGURE 9.1
A farmer in Nepal plows a field with the help of two cows. (Courtesy of Corbis)

CONTEMPORARY INDUSTRIALIZED AGRICULTURE

The agricultural system used in the United States and some other industrialized nations is relatively new, having really come into its own after World War II. This system is highly specialized, focusing on a relatively few species, with corn being the most important of the crops. The vast majority of labor in this system is provided by machines (fig. 9.2) powered by fossil fuels, and machines to replace human workers continue to be introduced. Fields are very intensively used, and their fertility is usually maintained through the use of chemical fertilizers. Diseases and pests are controlled by chemical pesticides. Large-scale storage facilities and refrigeration allow crops to be stored for years and so mitigate fluctuations in productivity due to drought or other calamity. A complex transportation and trade system allows these crops to be moved around the world with little difficulty.

While this system is highly productive, it is also very polluting and inefficient. Chemical fertilizers, herbicides, and pesticides pollute water supplies and kill many other plants and animals. The system also results in the large-scale alteration of habitat and the loss of biodiversity. Further, the system is ultimately unsustainable. In the United States, in addition to the human labor, about three hundred calories of fossil fuel are expended to produce one hundred calories of

FIGURE 9.2
A large, air-conditioned combine harvesting a wheat field. (Courtesy of Corbis)

food, including the energy expended in packaging, transportation, and refrigeration (Pimentel et al. 1994:203). If the supply of fossil fuels became too costly or difficult to obtain (such as during a war in the Middle East), the entire system could collapse. Part of the solution may be to eat foods that are grown locally, increasing efficiency by decreasing transport costs (recall that acquisition cost is one of the elements in any optimal foraging model) (Schueller 2001).

Some recent trends seem hopeful. For example, there is a growing market for organically grown fruits and vegetables, those raised without the use of chemicals for fertilizer or pest control. One could view this trend as another form of species modification or intensification of resource management. Part of this trend is the small but growing use of biological pest controls, having "good" bugs eat the "bad" bugs rather than using chemicals to kill all the bugs.

Contemporary Use of Traditional Systems

Given the propensity of contemporary agriculture to destroy land and pollute water, many people are searching for alternative methods of food production. It is slowly being realized that some traditional techniques of agriculture can be highly productive and valuable, or even necessary, in many areas (Marten 1986; Wilken 1987), and ethnographers and agronomists are working together to document the vast storehouse of traditional agricultural knowledge (Atran 1993). For example, the system utilized by the ancient Maya (see case study below) could be adapted for use in rain forests and could support large numbers of people without the destruction of the forest.

Part of the problem is an arrogance on the part of Western farmers who believe that their practices must be superior. For example, on the island of Bali, in the southern Pacific Ocean, traditional water management practices were replaced with contemporary techniques. However, these proved so much less successful than the traditional system that the contemporary innovations had to be abandoned (Lansing 1991; Lansing and Kremer 1993). The traditional Chinese wet rice system (see case study below) is exceedingly efficient and productive, utilizing virtually all of the available resources within an area, but it is being replaced by industrialized agriculture. This technology will result in a short-term increase in productivity but will ultimately make the land less productive.

James Fairhead and Melissa Leach (1996) have shown that West African agriculture not only is far more sophisticated than had been assumed but has actually created many of the groves that outsiders were trying to "protect" from "underdeveloped" farming. Johan Pottier (1999) established a wider context for this high level of sophistication, describing the ways in which African agriculture is fine-tuned to a continent of poor soil and frequent drought and where "development" often means disaster. The geographers A. Grove and Oliver Rackham (2001) filled a large (and beautiful) book with ways in which Mediterranean peasants, often despised and regarded as land abusers by national elites, have actually maintained and efficiently cropped their rather hostile landscapes, often by intensive tree cropping (e.g., olives and almonds) and tree management (e.g., variously cutting branches, which can regrow, rather than sacrificing whole trees). Similar examples might be provided worldwide.

ENVIRONMENTAL MANIPULATION AND RESOURCE MANAGEMENT

The rate and scale of environmental manipulation are much greater in intensive agriculture than in other systems, and one of the hallmarks of intensive agriculture is the active large-scale alteration of landscapes. Entire ecozones have been so modified that they have little resemblance to their natural state. Whole valleys were flooded behind dams; entire forests were removed; rivers were rerouted; lakes were drained and plowed under; cities fill floodplains; plant and animal species continue to be driven to extinction; entire cultures have been absorbed: The list of impacts is long.

In addition, systems of passive manipulation of the abiotic environment increased considerably, with maintenance of the cosmos and control of the weather being major goals. Many of these systems required the commitment of considerable resources, such as the construction of temples, the support of priests, and, in some cases, the sacrifice of humans to the gods. In those belief systems, failure to adequately appease the supernatural powers could have resulted in an environmental catastrophe.

Intensive agriculturalists generally practice more intense resource management than horticulturalists. Agriculturalists' dependence on domesticated plants and animals is greater, but they focus on fewer species. Undomesticated species are usually unimportant.

RELATIONS WITH OTHER GROUPS

Intensive agriculture is generally incompatible with other systems of subsistence and rapidly replaces them. This occurs for a number of reasons. One is that being more intensive and possessive, intensive agriculturalists have a very different relationship with the land than pastoralists, horticulturalists, or hunter-gatherers do. Most intensive agriculturalists tend to simplify their ecosystems (see Flannery

1972:399); for example, through monoculture. Such practices so alter landscapes that other groups find the area unusable. Finally, intensive agriculturalists can successfully expand into the territories of other groups because they generally have larger populations and sociopolitical organizations and so can support a full-time military, often the vehicle of expansion.

CHAPTER SUMMARY

Intensive agriculture is large-scale farming, often involving the use of animal labor, equipment, water diversion techniques, and the production of surplus food. Intensive agriculture represents a fundamental shift upward in intensity of land use, size of population, complexity of sociopolitical systems, and human impact on the environment.

The change in agricultural intensity resulted in growth in population, increasing complexity in social and political organization, and eventually some groups' developing state-level societies. The processes by which this occurred are unclear, but several models have been proposed, including those based on management, conflict, and a combination of factors.

Intensive agriculture employs all of the techniques of horticulture and pastoralism, plus supplemental labor and irrigation, all combined into a complex and interrelated system. Such systems are quite flexible and resilient and can support billions of people.

After World War II, Western agriculture became increasingly industrialized and mechanized, with machines powered by fossil fuels providing much of the labor and manufactured chemicals much of the fertilizer. This type of agriculture is highly productive and has rapidly expanded to replace many traditional systems and to take over ecozones previously unoccupied by farmers.

Intensive agriculture has had a significant impact on the environment, much more than horticulture or pastoralism alone. Ecosystems become simplified, landforms significantly altered, and food economies more narrow. In the long term, issues of pollution, dependency on fossil fuels, and population growth will make contemporary industrialized agriculture unsustainable.

KEY TERMS

dry farming
irrigation

CASE STUDY

CASE STUDY: MOUNTAINS AND WATER—THE TRADITIONAL AGRICULTURAL SYSTEM ALONG SOUTH COASTAL CHINA

The traditional system of Chinese rice agriculture is highly productive and fine-tuned to the environment. This case study illustrates how the Chinese utilize every possible food source within a sustainable system. The encroachment of "modern agriculture" threatens to disrupt the traditional system and to replace it with one that is more productive in the short term but is not sustainable in the long term.

The Chinese word for landscape painting, *shan-shui hua*, means, literally, "painting of mountains and water." Indeed, much of China's landscape is divided into mountainous land and well-watered river and stream valleys. This is especially true in the southern part of the country. North China's most populous region is the flat North China Plain, but the plain does not extend south of the Wei, Huai, and Yangtze rivers that traverse the center of the country. South of these rivers, the ancient continental shield is highly dissected, and from the air, this portion of China appears like a choppy sea. Rainfall is intense, and almost all of the rain falls in summer. The climate is monsoonal: Hot, wet winds blow during the warmer months from the South China Sea; cold winds blow all winter from the dry interior of Asia.

In south coastal China (fig. 9.3) thousands of years of experiment and innovation by local peasants have produced an agricultural system well adapted to local realities. This system supports about two-thirds of China's vast population. It produces about one-third of the world's rice and perhaps most of the world's pigs, as well as enormous quantities of corn, sweet potatoes, peanuts, and many fruits and vegetables. (Information regarding the traditional Chinese agricultural system is available in King 1911; Buck 1937; Anderson and Anderson 1973; Bray 1984; Chao 1986; Wittwer et al. 1987; Anderson 1988; Ruddle and Zhong 1988; and Dazhong and Pimentel 1984, 1986a, 1986b. For archaeological information, see Ho 1975, 1988; Chang 1986; and Pearson and Underhill 1987.)

CASE STUDY

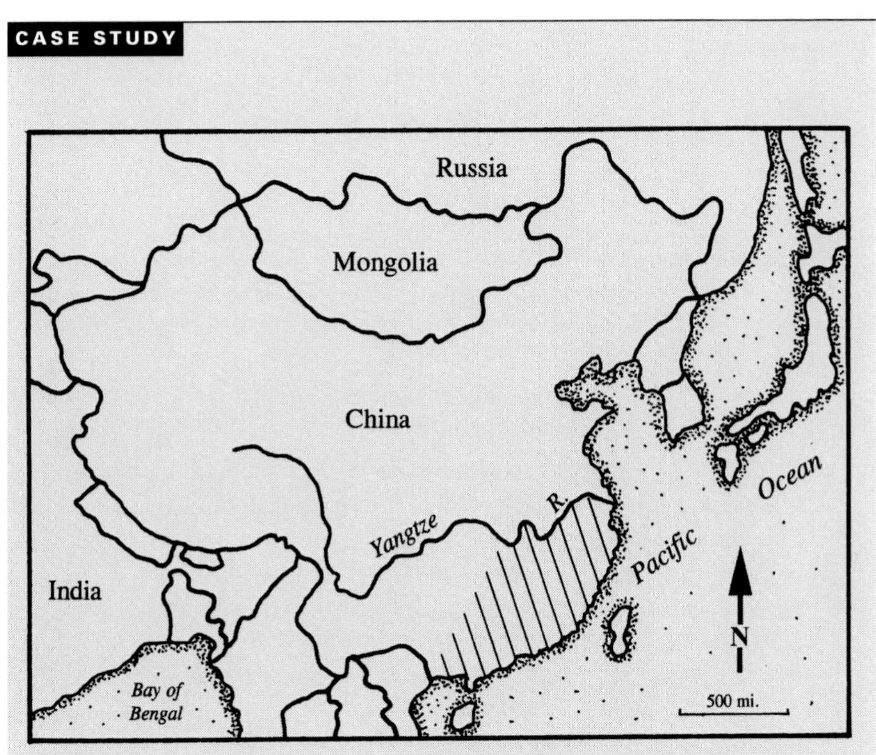

FIGURE 9.3
The region (lined) of traditional Chinese wet rice agriculture.

THE TRADITIONAL SYSTEM

We present here a somewhat generalized description of the way south China's rice production system functions on the ground (see fig. 9.4). Let us consider a typical valley on the south China coast just before the Western world brought agricultural chemicals and mechanization to the region.

This system is based on exquisitely fine-tuned choices of what to grow. The plant and animal production processes fit into each other neatly. They have come, through the millennia, to an accommodation. At every point, recycling and composting maintain or increase soil fertility and quality. Everything is composted: dung, old rope, agricultural wastes, ashes, and household debris. Even pottery, mud bricks, and tiles were ground up in the old days and added to the pile. Proper composting kills parasites, breaking their cycles of transmission and reducing risk of infection for people.

CASE STUDY

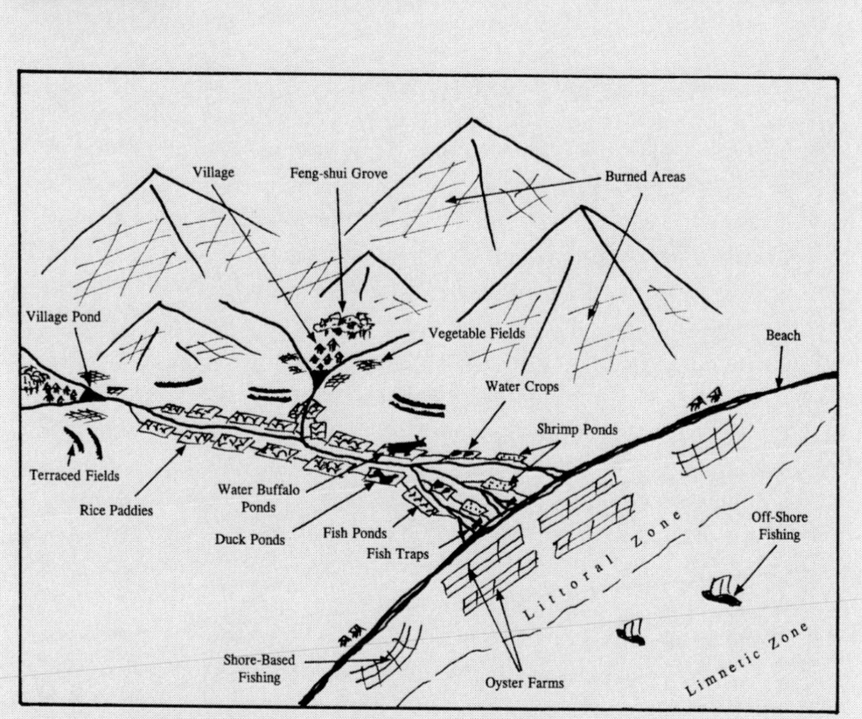

FIGURE 9.4
An idealized model of traditional Chinese wet rice agriculture. (Drawing by Blendon Walker)

Geography

Each valley centers on a large stream or river. Ninety percent of the watershed of the stream is mountain land, too steep to cultivate on any significant scale. Now and then someone creates a dry field or terraces the few patches of soil in this zone, but it is more often left in forest, brush, and grass. These areas are important, as they are where people hunt, gather herbs and other plants, and obtain firewood.

Farther down the valley, the hills become less steep, permitting the terracing of some of the slopes. Eventually, the stream begins to develop a floodplain, forming some land suitable for fields. As one travels farther down the valley, the terrain flattens out, and freshwater lakes and swamps appear. As the stream reaches the ocean, estuaries form. At the end, a beach and littoral zone and eventually open ocean are encountered.

CASE STUDY

Feng Shui: Manipulation and Management

The Chinese employ the approach of feng shui (wind and water) in dealing with their environment. Feng shui is a broad system in which humans fit into their environment, interact with it, and remain in some general balance with it (see Burger 1993). Among the many feng shui practices are land and crop management tactics intended to maximize the benefits of wind and water while minimizing the damage to the land; these practices have many important applications in the system of traditional agriculture. Feng shui regulates the placement of most things, including roads, towns, fields, and forests, to ensure harmony. Feng shui is also used within homes to properly place furniture and other items.

Feng shui has a strong magical or religious component. It is believed that good or evil fortune follows from one's choices of habitation. The siting and orientation of a house affect the health and wealth of the inhabitants. The placement of a grave affects the fate of the descendants of its occupant. Cutting down the trees (the feng shui grove) associated with a town would bring disaster to the town, and the older and more well grown a tree is, the more good luck it radiates.

Whatever mystical beliefs it may have accumulated over the years, feng shui is based on solid, pragmatic observation. Its magical overtones make it more persuasive—persuasive enough to convince millions of peasants to sacrifice short-term personal gain for long-term community benefit. Without the magic, the ecosystems in south China will degrade or collapse. Rational self-interest does not adequately motivate people to consider the long-range consequences of their behavior (Anderson 1996).

Hills and Forests

The hills are often burned, either to clear land or destroy dangerous animals or by accident. This frequent burning releases ash, which washes down the streams, helping to restore the fertility of the farmlands downstream. Thus, though this burning is unfortunate—it would be much better if the land were left in forest—it does have its advantages. It releases, in the ash, mineral nutrients pulled up by tree roots from deep rock and soil layers. Also, many hill

CASE STUDY

plants have, in their roots, large concentrations of bacteria that can take up nitrogen from the air and fix it as organic nitrates—something no higher plant or animal can do.

Of all of the nutrients that plants derive from the soil, nitrates are the ones needed in the most quantity, and they tend to be the great limiting resource for agriculture and plant growth (recall Liebig's Law of the Minimum). The plant cover of an area thus depends on its nitrogen-fixing bacteria. As plants burn or decay, most of the nitrogen goes back into the air, but a good portion gets washed down to the fields below. In southwestern China, though not in the southeast, some communities deliberately plant alder trees on fallow land. When the trees are tall, they are cut for lumber, and the land can then be farmed, taking advantage of the enormous quantities of nitrogen that are released as the leaves and roots decay.

The greatest problem with traditional south Chinese environmental management is its destructive effect on the tree cover. Much of the forested areas of southern China have been cut down, but because of feng shui, large groves of trees are left around and above towns and constitute most of what little forest cover survives. Nevertheless, a chronic shortage of wood persists.

Without the deeply religious influence of feng shui, the temptation to deforest is too strong. Where modernization has meant the loss of belief, feng shui groves have disappeared. All the predicted disasters have duly come to pass: erosion, water shortages and droughts, fertility loss, and the other well-known consequences of massive deforestation. Thus, feng shui, slighted by many contemporary Chinese as mere mixin (superstition), proves its value—whatever the skeptic may think of "good influences" and the mythical dragons that are often said to send them. It provides a classic case of the use of religion to sanction good ecological management, supporting the theory of Roy Rappaport (1984) that ecological information and management practices could be encoded into religion to ensure that they were followed.

Towns

Most people live in towns and houses located in the upper part of the valley, on land that cannot be used for fields. They are sited on

CASE STUDY

a comfortable, sunny slope above the streams, preferably near the junction of two streams, and just below the tops of the ridges that shield them from typhoons. The roads leading to the towns are winding to discourage evil spirits and other evils such as soldiers, tax collectors, and robbers. Buildings are located near reliable water sources but above areas that flood. Graves are located on high, steep slopes, where they do not compete for land used for food production. Feng shui groves are left around and above the towns and provide shade, firewood, erosion control, fruit, some construction timber, and similar benefits.

The use of feng shui succeeds in keeping towns away from floodplains. This has the double value of protecting the towns and preventing the urbanization of critically needed rice-producing land (the Western world has yet to learn this lesson). In the great floods of June 1966, one of us (ENA) observed that all the traditionally sited buildings in the western New Territories of Hong Kong were above water, and almost all the new buildings were below it. Construction sites in which the builders had cut too deeply into the hill slopes were also observed. With the intense rain, the slopes failed and destroyed the sites and many buildings. The peasants said the cuts had "cut the dragon's pulse." Whether one believes in the dragon or in the equally mysterious geological concept of angle of repose, the effects are the same: Undercutting a slope brings disaster.

Gardens

Available land around the town is terraced for vegetable gardens. The vegetables are grown nearest the towns for several reasons. First, they need the good drainage—they do not grow well in the bottomland, where the rice grows. Second, they are heavy, and they need a lot of fertilizer, and carrying everything to and from the vegetable fields is difficult. Third, they demand intensive care—constant weeding, pest control, watering, and harvesting—and people need to be close.

Rice Paddies

Below, in the valleys, most of the land has traditionally been dedicated to producing the great staple food: rice. Peanuts and sweet potatoes are grown on land unsuitable for rice. In south Chinese dialects, as in most Southeast Asian languages, the word for "cooked

CASE STUDY

rice" is often used to refer to food in general. Rice has to be irrigated to yield well, which demands good control over water. The better managed the water levels are, the higher the yield. A difference of a couple of inches in water level can affect production. Moreover, rice has to start growing when water is shallow, finish growing in deeper water, and then dry off before harvest.

Every tract of land with adequate drainage and water supply is terraced and diked to catch and control water (fig. 9.5). Fields are plowed in winter—the dry season—but often again in summer. With the spring rains, the fields are flooded to very shallow depths. Rice is grown in seedling nurseries and transplanted into the fields. (Rice grows better and produces more grain if it is transplanted.) As the rice grows, more water is added to the fields. When the grain starts to mature, the fields are drained so that the rice will not rot. Harvest takes place, and the cycle starts again. Most fields produce two crops of rice a year; some produce three. Mulberry trees are grown on dikes between rice fields to stabilize the dikes and prevent erosion. The trees do double duty, as their leaves are used to feed the silkworms grown to make silk.

FIGURE 9.5
Rice paddies in China, 1978. (Photo by E. N. Anderson)

CASE STUDY

Lakes and Marshes

Lower in the valley, there comes a point at which rice cultivation is no longer economical because the water is too difficult to control; it ponds up too deeply or stays too long. Crops that need more water, such as lotus, water spinach, and other greens, are grown here. In addition, water buffalo and ducks are raised in these areas.

Still lower in the valley, where permanent freshwater in the form of lakes or marshes exists, wild and domesticated fish are intensively farmed. If quality fish are desired, ponds produce about a ton of fish per hectare (2.47 acres), but if buyers are satisfied with small, rather lean fish, yields can go up to seven or eight tons. The fish raised are species of carp, along with some catfish and mullet. Most of them eat vegetation and small animal life. The pond stocking system includes a number of different fish so that they can utilize every niche in the pond: floating vegetation, bottom fauna, microplankton, and even grass and weeds cut from the bank. The ponds are fertilized with night soil (human waste), soybean wastes, and other byproducts of the land economy. A detailed account of this amazing system has been provided by Kenneth Ruddle and Gongfu Zhong (1988; also see Anderson 1988).

At the mouth of the stream or river, there is usually an estuary and brackish swamp. Soil erosion from the deforested mountains expands the alluvial fan, extending the land outward into the water; much of south China's rice land is of recent origin, having been created from new alluvium over the last thousand years. Silt or sand is deposited; mangroves and reeds grow, preventing further erosion. Eventually the inner fringe of this new land can be drained and cultivated. Thus are born the "sand fields" (*sha-tian*) commemorated in many place-names along the south coast. In the meantime, the swamp is used for catching fish and shrimp. Brackish swamps are the most productive of all landscapes, enriched as they are by input from both land and sea.

The Sea

Out to sea, cultivated oyster beds continue the range of aquiculture as far as shallow water extends. Oystermen own long strips, aligned between landmarks on the coast. They create oyster habitat

CASE STUDY

by dropping tiles, bricks, or stones for the spat (floating oyster larvae) to attach themselves to. Each oysterman knows his attachments and his landmarks, so theft is difficult. Oysters are phenomenally productive but subject to predators, so management must be careful and continual.

Finally, the ocean is fished intensively by professionals. Living in their boats, they are a different class from the shore people. They rarely fish from shore, and the land people rarely fish from boats. The fishers have their own songs, traditions, customs, and classificatory systems. However, they rely on the land people for a variety of products, such as rice and vegetables. Likewise, the land people rely on the fishers for fish and other ocean products. In a very real sense, the two groups (land people and fishers) have formed a mutualistic relationship in a cultural and natural ecotone (the shore) much like the Mbuti and Bantu of the Ituri Forest (see case study in chapter 5).

About Rice

Rice is the foundation of the traditional Chinese agricultural system. The first known domesticated rice comes from the Yangtze delta, where both long-grain and short-grain varieties were grown about seven thousand years ago (see Higham 1995). Soon rice dominated the food production system of the region. Today, rice is the staple food of perhaps two billion people, the vast majority of them in monsoon Asia.

Nowhere is rice more dominant than in south coastal China. Here it provided, until recent times, 90 percent of the calories for the ordinary person. Fields produced two or even three crops per year. The lowland landscape was a solid sheet of brilliant gold-green—the color of plants photosynthesizing at maximum efficiency. Even in cloudy monsoon light, they are trapping the energy of the sun and using it to drive the chemical reactions that turn air, water, and mud into the world's most productive grain crop.

Under traditional conditions, paddy (i.e., wet-grown) rice yields 1,000 to 3,000 pounds per acre. Thus, a triple-cropped field could theoretically produce almost 10,000 pounds per acre, though it is doubtful whether such yields have ever been achieved. The best

CASE STUDY

figures are from Taiwan, where average yields of paddy rice increased from about 1,300 pounds per acre in 1900 to twice that by 1961 (Chinese-American Joint Commission on Rural Reconstruction 1966:26). The figure of 2,600 pounds per acre indicates about the most that was possible for a region-wide average with traditional techniques; since 1961, improved varieties and agricultural chemicals became widespread, and yields soared. Rice now yields up to 10,000 or even 12,000 pounds per acre per crop.

Rice responds very well to fertilizer and to control of water amount and quality. Lower figures (1,000 pounds per acre) are typical where fertilizer is not used, where the water or soil is salty (as on new "sand fields"), or where water levels are not well regulated. But at least it produces under such conditions. Few grains will produce at all in brackish water, but specialized rice varieties flourish there.

No other grain produces so well under traditional conditions. In China, however, corn, wheat, and millet approach it. All were grown on the drier uplands, where rice does not yield well. Wheat was often grown during the winter in areas where water could not be made adequate for rice or in areas too cold for a winter rice crop.

At these yield levels, it is possible to feed six or more people per hectare—around two thousand per square mile (about the density of an American suburb). Densities of this level are common in river deltas and other especially favorable localities. Even higher population densities have been recorded in areas where people could produce high-value handicraft items such as silk (Huang 1990).

Other Resources

Rice is mostly starch and alone does not provide adequate nutrition for people. Supplementary protein, vitamins, and minerals must come from other foods. In southern China, these foods include soybeans, peanuts, and other beans, whose protein complements that of rice. They also include vegetables, especially Chinese cabbages, which are among the richest in vitamins of all greens. Chiles, tomatoes, eggplants, squash, sweet and white potatoes, and dozens of lesser vegetable crops abound, providing further supplementation. All these are high-yielding under south Chinese conditions. The land is too scarce for wasteful uses. Chinese cabbage, for in-

CASE STUDY

stance, produced about eleven thousand pounds per acre under traditional conditions in Taiwan (Chinese-American Joint Commission on Rural Reconstruction 1966:100). Other vegetables were similarly productive.

The shortage of calories is not as limiting as deficiencies of protein, calcium, and iron. The Chinese recognize a condition ("cold ch'i," what we call anemia) and have developed a specific and effective treatment: feeding the sufferer pig livers (the addition of "warm ch'i"). Although the Chinese do not know that it is iron deficiency they are treating or why pig livers are an effective treatment, it works.

The domestic animals grown by the Chinese are those that can most efficiently turn farm and food wastes into meat: dogs, pigs, chickens, and ducks. Pigs and chickens were domesticated in China about seven thousand years ago (Underhill 1997:121), and ducks perhaps as early. These animals not only eat garbage, waste, and scraps; they eat weeds and pests. Chickens and ducks are often the insect and weed control in the fields and paddies. Some duck owners rent their ducks to other farmers for the ducks to eat the pests; the renters also get to keep the duck manure generated while the ducks are on their fields. Thus, dangerous pests are turned into good meat. By contrast, cattle are rare and are mostly plow stock. Sheep and other less-efficient meat producers are very rare. Unlike Western livestock systems, no animal is raised if it requires food humans could eat. Wild animals are encouraged but are cropped for food.

Many of the wild animals also have a pest-control function. When pesticides came to southern Asia, frogs were exterminated in many areas. This catastrophe not only deprived the peasants of an important food source—it also triggered an explosion of insect pests. The frogs had been more efficient insect killers than the chemicals were. The pesticides also killed fish and birds that were eaten by people. It was better to lose some of the rice to the fish and birds and get to eat them.

Wild resources include not only timber for firewood but also herbs for medicine and game for food. Even insects are eaten, and the giant waterbug is a delicacy in parts of south China and northern Thailand

CASE STUDY

(Pemberton 1988). Vines for tying, leaves for fertilizer, flowers for decoration, water plants for pig fodder, and countless other products are derived from wild and weedy plants.

Cultural traditions have evolved to fit well with the overall system. Cooking, for example, has come to be based on processes that require very little fuel: stir-frying and steaming instead of baking and long boiling. Food preferences run to vegetables and fish, not the "meat and potatoes" of Western diets. Choice vegetables cost more than meat in the markets of China.

Nutrients through the System

Consider the fate of a nitrogen atom in this system. It is trapped and incorporated into a nitrate molecule by bacteria in the root nodules of a hill plant. Eventually, as the plant burns or decays, this particular bit of nitrate happens to wash downstream. The water is channeled into a vegetable garden, and the nitrogen is picked up by a radish. Picked and carried into town, the radish is eaten, and the nitrogen atom is eventually excreted. The night soil is composted, mixed with other composted waste items, and returned to the land as fertilizer.

The nitrogen atom might simply cycle forever between vegetables and town, but eventually most nutrients escape downstream into the rice paddies. Here they are caught up in yet another cycle: rice to town, to compost, and back to the land. The rice is consumed by humans. The rice straw is eaten by cattle and water buffalo. Straw too tough to eat is made into ropes and sandals, and when these wear out, they, too, are composted.

In the rice paddies, more nitrogen is fixed. Blue-green algae, often growing on small floating ferns of the genus *Azolla*, are excellent nitrogen fixers. Chinese and Southeast Asian growers have long known that rice paddy pond scum makes excellent fertilizer. Vietnamese peasants even learned to transplant Azolla to new paddies.

Eventually escaping the paddy cycle, our atom washes on downstream and is cycled several times more through the fish ponds. Finally it leaves via the swamp (and more cycling) into the ocean. Here it becomes part of the outflow that fertilizes the marine waters and produces a rich fishery. Eventually it is incorporated into a fish. The

CASE STUDY

fish is caught, dried, and traded back up to the town at the top of the valley—and the entire story begins again.

Thus, a given bit of nitrate passes through many plants and many digestive tracts on its way to the sea. Few nutrients escape this system. Because new ones are constantly being incorporated into the system as they erode from the surrounding rock, it follows that the system keeps getting richer. Indeed, many of the valleys of south China have become more extensive and fertile in spite of the enormous nutrient drain caused by exporting food to cities and other far places.

DISCUSSION

This broad-spectrum use of the environment is ecologically healthy. The whole south Chinese landscape is cropped. Its productivity is maximized, and virtually everything is used. In contrast, Western-style agriculture normally eliminates most natural biota and sets up an artificial system producing very few crops. This system is productive for a short while but is unstable. Soil degradation, crop diseases, and changing economics that make a crop suddenly unprofitable can all be devastating in short order. The long-term result is desertified, abandoned land.

The diet produced by the Chinese system is also healthy—if one can avoid waterborne diseases and parasites. Living on rice, fish, and vegetables, Chinese males have one-sixteenth the heart disease rates of American males, and the cholesterol level of the average adult is 127 (it is over 180 in the United States). The people in southern China have less cancer than those in the north, who consume more wheat and meat. Life expectancy in China is almost as high as in the United States, although until recently China had only 1 percent as much wealth per capita (see Chen et al. 1990; or summary in Lang 1989).

This is not to say that China's system is, or was, perfect. Plants are tougher and more adaptable than they are productive—they have to be. Famine was always a possibility if rice crops failed. In recent times, improved pest control and the adoption of more productive rice strains have resulted in higher production. Animals also need

CASE STUDY

better disease control and better feed than they usually get in traditional China.

Moreover, Southeast Asia developed an even better system, based on extensive tree cropping as well as rice, vegetables, and fish. Forests are left on the hills but are often converted to economically useful species. Large orchards and tree plantations maximize use of dry, thin soil, not well used in south China. The Southeast Asian system is better adapted to the tropics, where tree growth is faster and more tree varieties occur, but it is locally found in south China. The real problems with tree cropping in China are the slow growth of the trees, the pressure to feed an increasing population immediately, and the danger of warfare. Trees, once cut in war, regrow very slowly.

Other superior systems have been devised and tested, but only on a small scale. So far in the world, only China and Southeast Asia have shown an ability to feed an extremely dense population for thousands of years without destroying the resource base. (But consider the Maya system, discussed below.)

The south Chinese system, and its close relatives or variants in Vietnam and Java, feed more people per unit of area than any other on Earth. Even the much-vaunted agricultural system of the United States does not do so well, because so much of the land is taken up with less-productive uses: extensive cattle ranching, wheat production, specialized but nutritionally worthless sugar growing, and the like. Also, Americans prefer more meat, which takes much more area and feed to produce.

But, in ecology as in other walks of life, "there ain't no free lunch." People must pay in labor. China's agriculture is fine-tuned, more like home gardening than American field cropping. Individuals lavish vast quantities of skill and knowledge, as well as sheer hard work, on the land. They must know when and where to grow everything. They must plant the next vegetable crop among already-growing crops—planting the seeds among the maturing leaves. They must know exactly what mix of varieties is ideal for their particular fields. Rice under traditional conditions requires fifty person-days of work per acre per crop in southeast China and more in some other areas (Buck 1937:302). In California today, rice is seeded by airplanes and

CASE STUDY

cultivated by tractors, and the labor requirement is a mere few hours per acre.

The sheer population growth in south China created problems (Huang 1990). In the absence of rapid growth in urban employment, peasants had no alternative but to stay on the land. Even emigration to America and southern Asia was not enough to relieve the pressure of population. Thus, the system had even greater labor assets and became ever more fine-tuned and skill intensive.

This population growth also resulted in a form of development that was overwhelmingly focused on crop varieties and cropping patterns rather than machines and chemicals. While the Western world was inventing threshers, harvesters, tractors, gang plows, artificial fertilizers, and thousands of other high-tech innovations, China was perfecting an organic or biological system. American agriculture, with all its machines, is intended to get the most production per worker because the United States is well supplied with land but not always with labor. China must get more per land unit by lavishing labor upon it (Hayami and Ruttan 1985). The system did not stand still; it continued to innovate and continues to do so today. When people face a limit, they breed a new variety, domesticate a new species, or develop a new and more intensive cropping pattern.

Agricultural development does not just happen; it is helped or hindered by government (Hayami and Ruttan 1985). The Chinese government, throughout history, has favored agriculture and tried to help the farmers. Land taxes were usually low, and occasionally they were waived altogether.

The Chinese imperial government, as early as two thousand years ago, performed experiments, issued extension manuals, recognized productive farmers, distributed superior planting stock, had international missions bring back useful plants, and in many other ways helped the agricultural process (Bray 1984; Anderson 1988). What Yujiro Hayami and Vernon Ruttan (1985) showed for the recent past was in fact true throughout history: The government favored biological technology as opposed to mechanical or chemical. Heavy taxes on industry, and lack of encouragement for technological innovation, often inhibited mechanical progress. Thus, China's agriculture was profoundly shaped by government policy.

CASE STUDY

Moreover, the Chinese government was perhaps the first in the world (before recent times) to organize charity and famine relief, both by distributing food and by working to develop agriculture (Will 1990; Will and Wong 1991). Private charitable societies also worked with the government on this effort (Handlin Smith 1987). This development effort helped agriculture, but it also allowed rampant population growth. Thus, in spite of the best efforts of the government, the fortune of the farmers did not improve. Their numbers increased; their income did not. It is a story all too familiar in the world today.

In China, as in all state societies, the "invisible hand of the market" is simply another imaginary supernatural being. In the real world, agricultural development takes place because of conscious choice by government officials and farmers. These people must choose in response to political, social, ecological, and personal factors as well as narrowly "economic" ones. The price of farm products must be considered along with charitable famine relief, population pressure, religion, ethnic traditions, and the fads that so often affect food buyers. The apparent irrationality of feng shui turns out to lead to rational site planning; the apparent rationality of Western monocrop agriculture leads to irrational destruction of millions of acres of land. Agricultural systems are more complex than they appear at first sight.

Many authorities think that China's pattern has led to technological stagnation or even decline (Elvin 1973; Chao 1986). With peasants growing poorer and harder-working all the time, there was no incentive to devise machines. There was also no incentive to mass-produce consumption goods—the peasants could not afford them. At best, the farm sector did not prove to be the driving force toward modernization that it became in the United States, England, and elsewhere.

Having said all that, agriculture in China is rapidly changing to a system modeled after the West, one that emphasizes chemical fertilizers, pesticides, and machines powered by fossil fuels. In the years since 1955, the use of fossil fuels in Chinese agriculture has increased a hundredfold (Wen and Pimentel 1986a, 1986b), and the use of chemicals is also rapidly increasing. The change is partly motivated by an attempt to obtain short-term increases in productivity

CASE STUDY

and by the overall effort by the Chinese to reduce population growth. By inducing families to limit the number of children, the government is shrinking the labor pool needed to maintain traditional agriculture, and the need to use other methods is increasing. Nevertheless, the shift away from traditional techniques, developed and perfected over thousands of years, will have a serious long-term impact on the environment.

CASE STUDY

CASE STUDY: THE MAYA AGRICULTURAL SYSTEM

This case study of the ancient Maya shows how a complex state-level society with a large population can adapt to a rain forest environment without totally destroying it. The lessons to be learned from the Maya can be employed by contemporary people living in rain forests and can slow, or even prevent, rain forest destruction.

The agricultural system developed by the ancient Maya was able to support a large population living in urban centers within a rain forest. The ancient Maya system incorporated a series of horticultural techniques that, used intensively and extensively, made the entire system one of intensive agriculture. While the ancient Maya are no longer extant, the contemporary Maya still employ most of the elements of the old system. An understanding of the ancient Maya system, which was actually quite variable from lowlands to highlands, could lead to the development of techniques and systems that could be utilized in contemporary rain forest settings, perhaps supporting large populations without the outright destruction of the rain forest. This is an important goal. (Additional general information on the Maya agricultural system can be found in Redfield and Rojas 1934; Kintz 1990; Wilk 1991; Fedick 1996; and Faust 1998.)

THE ORIGINAL VIEW OF MAYA AGRICULTURE

When the ancient Maya were discovered by archaeologists in the mid-1800s, they were thought to have had a relatively small and widely dispersed population that lived in small villages. The large

CASE STUDY

Maya sites with monumental architecture (pyramids) were believed to be ceremonial centers constructed by the people living within the jurisdiction of a center's priests. It was thought that only a small population of priests and other specialists lived in these centers.

Consistent with the belief that the Maya population was relatively small and dispersed, it was thought that a relatively simple swidden system, which is used by a dispersed population, was capable of supporting the Maya. Although little research was done on the agricultural system, it was hypothesized that some sort of a failure of the swidden system had led to the downfall of the ancient Maya.

A NEW UNDERSTANDING

Beginning in the 1970s, researchers began to realize that the forest contained an unusual distribution of certain plants, shrubs, and bushes not native to the area, certain tree species clumped together, such as the seed-bearing ramón (*Brosimum alicastrum*), the fruit-bearing sapodilla (*Manilkara sapota*), mamey sapote (*Pouteria sapota*), various species of citrus, and the mahogany (*Swietenia macrophylla*), desired for hardwood. People began to look more closely at the ethnobotany of the contemporary Maya for clues to the ancient Maya system.

A bit later, archaeologists began investigating areas outside the large sites with impressive architecture, and they began to discover small, low house foundations that had previously gone unnoticed in the thick jungle. Large numbers of houses were discovered, and the archaeologists soon realized that the sites, originally thought to be ceremonial centers, were actually cities with large urban populations.

It quickly became apparent that if these centers were actually cities and the Maya population was much larger than previously believed, a reevaluation of the Maya agricultural system was needed. It was clear that the swidden system originally thought to have been used by the Maya was not capable of supporting the three hundred to four hundred people per square mile who lived in the Maya region.

Researchers began looking much more closely at the agricultural system and discovered the archaeological remains of a number of

CASE STUDY

nonswidden field systems. Large numbers of small, stone-walled enclosures, the remains of small terraced gardens called *pet kot* (round wall of stone) by the Maya, were discovered on the sides of small hills. These features are very subtle and difficult to notice in heavy undergrowth. Systems of raised chinampas were found in a number of swampy places, the best studied being Pulltrouser Swamp in southeastern Yucatan and northern Belize (Turner and Harrison 1983; Sharer 1995; also see Gidwitz 2002). These were constructed by digging ditches in the swamp and piling the soil up in a waffle-like pattern (see chapter 7, fig. 7.2), raising the fields up out of the water and creating canals between them. The canals were dredged, and the soil from dredging was added to the field to maintain its fertility. The canals between the fields were colonized by turtles and fish, both of which were eaten.

While our understanding of the Maya agricultural system is still incomplete, it is now clear that it utilized at least four major approaches. The first was a complex system of orchards integrated into the forest (agroforestry). These trees were utilized for food, lumber, and beauty. Second, large numbers of small terraced gardens were constructed where possible. Third, the Maya constructed complex systems of chinampas in areas not normally thought of by Westerners as being suited to agriculture. Last, of course, the Maya extensively utilized a swidden system. The combination of these techniques into a unified system was highly productive and is classified as intensive agriculture. In addition, the Maya utilized a number of wild resources as part of their overall economy.

Environmental Manipulation

The Maya extensively manipulated the forest and in fact virtually remodeled it to suit their needs. Native trees were removed and replaced with other species, creating a mosaic that can still be seen today. Water was rerouted into fields and canals, altering the natural system. Finally, sections of forest were cleared for the construction of the many cities, towns, and other facilities used by the Maya. In spite of all of their extensive modification of specific areas, the Maya managed to maintain the overall diversity of the forest.

The Maya also conducted extensive passive manipulation of the cosmos to ensure continued good harvests. These practices included

CASE STUDY

a wide assortment of ceremonies, and recently it has been determined that some of these ceremonies incorporated human sacrifice.

Resource Management

The Maya practiced intensive management of their various domesticated crops. They also intensively managed their trees, planting them at various places in the landscape, even in areas outside their natural habitat.

Intensive management was also practiced on a number of wild species. The fish, turtles, and other species in the canals of the chinampas were controlled as to their numbers and locations, their habitat being expanded or reduced depending on conditions. However, as far as is known, these species were not genetically domesticated. There was also some management of other wild animals, such as deer. The animals were attracted to the vegetation of fallow swidden fields, where they were hunted. Proper management of those fields and of the hunting could ensure a consistent supply of deer.

Collapse

The culture of the ancient Maya did eventually collapse. It is currently believed that the cause of this collapse was population growth that finally overtaxed the agricultural system. Deforestation, shortening of swidden cycles, erosion, and silting in of chinampas are all candidates for causal factors (see Abrams and Rue 1988). However, there are other possible factors, including warfare and even depopulation due to European diseases. Whatever the cause, there is much to learn and apply from the Maya example.

THE CONTEMPORARY YUCATEC MAYA SYSTEM

As the ancient Maya culture declined, so did Maya population. The Maya people stayed in the same place but scaled down their agriculture, adopting more of a horticultural-scale system, with swidden being its major component. They retained much of the knowledge of the ancient system and continue to use its various components today. Most recently, Western-style intensive agriculture is being adopted, even if grudgingly, and the Maya region is being transformed much like the Amazon Basin and other rain forests around

CASE STUDY

the world (see "The Rain Forest Dilemma" in chapter 10). The Yucatec Maya (see Kintz 1990) are one of a number of contemporary Maya groups that occupy the same regions as their famous predecessors did.

The Natural Environment

The Yucatec Maya live in the Yucatan Peninsula, in the Mexican state of Quintana Roo (fig. 9.6). They live in a communally owned area known as an *ejido*, of some 14,330 hectares. The ejido, named Chunhuhub, lies in an ecotone between seasonally dry forest and permanently moist rain forest. Three major seasons are recognized. The wet season lasts from mid-May to November. From November to late February, it is cool and relatively dry, but with frequent cool rains. The dry season lasts from March to mid-May. Rainfall averages between fifty and eighty inches per year, depending on location. Temperatures range from 50 degrees Fahrenheit on cold winter nights to around 100 degrees in the height of the dry season in April and early May.

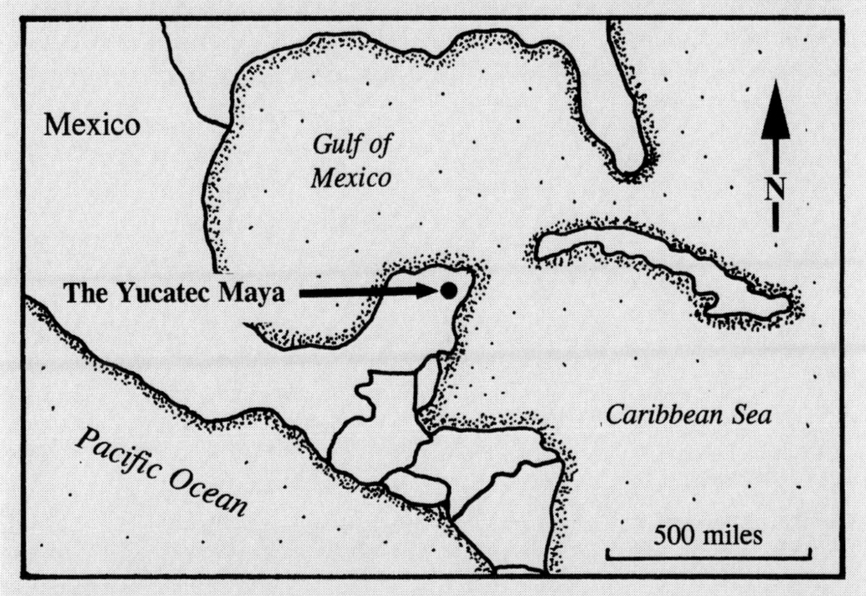

FIGURE 9.6
Location of the Yucatec Maya in the Yucatan Peninsula.

CASE STUDY

Sociopolitical Organization

Traditionally, the Maya were organized by patrilineages, with the leaders of each exercising political power. Today, the primary social unit is the bilateral extended family. The leaders of the various levels, towns, ejido, and municipio, are now democratically elected, although family leaders still have considerable say over things.

Economics

The Swidden System

The Yucatec Maya use a swidden system, generally called *sak'aab*, referring to the first stage of forest succession. The system involves the use of rotating fields called *milpas* (fig. 9.7), an Ameri-

FIGURE 9.7
A Maya milpa. (Photo by E. N. Anderson)

CASE STUDY

can misuse of the Nahuatl word for cornfield, and remains highly productive. Like any human activity, the swidden system impacts the forest, but it does not destroy diversity or cause extinctions. The Yucatec Maya continue to manage the forest and maintain a sustainable agricultural system. A concerted effort is now under way to record and preserve as much traditional knowledge as possible so that an intensive agricultural system, sustainable in a rain forest, can be developed.

Light, well-drained soils are preferred for milpas. The favored areas are low hills and valleys covered with dry forest. The wet, clayey, less-fertile soils of the tropical rain forest are not good for milpas—a fact that has protected most of the high forest from frequent cutting. To cultivate a milpa, the family first cuts the forest, then lets it dry. When dry, the vegetation is burned. Any large branches that do not burn completely are gathered up and reburned. The ash is then mixed with the soil.

In cutting the forest for the milpa, the family manages the trees. Certain fast-growing trees are cut and then used for firewood and building materials. Slow-growing food- and medicine-producing trees are not cut for wood, and some flowering trees are kept for aesthetic reasons. Other trees are retained due to their interdependence with other species. For example, some bees depend on certain trees for nectar, so those trees would be left and the honey collected at a later time. Cleared lands are then planted.

The men, sometimes aided by women, use digging sticks to poke small holes in the ash and soil. A few corn kernels, often with a bean or two, are placed in the hole and covered with soil. Chiles and so on are planted separately, in the more fertile, low-lying parts of the field. The simple sharpened stick is the only cultivating tool—it is ideal for the purpose; shovels and plows would only disturb the light soil and ash and allow them to blow away. Weeding is done a couple of times as the crop grows; machetes are used today, but in earlier times weeds had to be pulled by hand or mashed down with stone tools. The yields are consistently around a thousand pounds per acre.

Milpas are cultivated for two years and then abandoned. The Yucatec Maya recognize three succession stages of forest. A newly

CASE STUDY

abandoned field is called *sak'aab,* as mentioned above; a field two or three years out of cultivation is *hubche'* (brushwood); and then the forest grows up to *kaanal k'aax* (tall forest). The other stages are known (see table 9.2) but not used much by the Yucatec, and some Maya distinguish even more than six stages.

The success of the system depends on a thorough knowledge of the plants, animals, and soils. Older Maya have an encyclopedic knowledge of the local land. They recognize five or six major soil types, each with many subtypes, and they know what kinds of plants will grow best in each subtype. Some four hundred species of animals and close to a thousand wild plants are known.

Gardens

Dark, moist soils rich in organic matter are favored for orchards and gardens, and houses are built in those areas. The gardens are then placed around the houses, and dozens of species of plants are grown. The yields of fruit and vegetables in the gardens are far higher than in the milpas. The gardens produce up to a quarter of the food of a family and an even higher percentage of the protein and vitamins.

Crops

The Yucatec Maya grow many types of crops in their milpas and gardens. By far the most important crop in the milpas is corn, as it has been for the last five thousand years. Corn still provides three-quarters of the calories in the diet of the rural cultivators—a figure

Table 9.2. General Maya Forest Succession Stages

General Stage Fallow	*General Maya Name*	*Years of*
1	sak'aab	1 to 2
2	sak'aab'kool	2 to 5
3	kambalhubche'	5 to 10
4	kanalhubche'	10 to 30
5	kelenche'	30 to 50
6	kaanal'k'aax	More than 50 (mature or "tall" forest)

CASE STUDY

unchanged for millennia. In contrast to much of Latin America, beans are not very important; they do not grow well in Quintana Roo, apparently due to the poor soil. Squash and chiles, the other famous Mexican crops, are correspondingly even more important than elsewhere, and squash seeds traditionally provided much of the protein in the diet. Today, chickens and pigs have taken over that function, and squash seeds are largely a ceremonial food.

Milpas on light hill soils are usually used to grow corn alone. In better soil, especially in the valleys between the hills, beans, squash, chiles, papayas, tomatoes, and occasionally other foods are grown among the corn stalks. The beans climb the corn and fix nitrogen that fertilizes the soil. The squash covers the soil in between, shading it and killing weeds, and the chiles derive protection from the other plants. Thus, each of these plants depends on the others.

The gardens and houses are protected and shaded by some thirty species of tall fruit trees. More chiles and tomatoes are grown in the garden, along with chives, onions, cabbage, radishes, and medicinal herbs like rue, mint, and aloe vera. Some enterprising gardeners grow their own coffee and chocolate. Others produce cotton or experiment with plants like grapes, which do not normally fruit in Quintana Roo. Domesticated chickens and pigs, introduced by the Spaniards, plus native turkeys and Muscovy ducks live in and near the gardens.

Hunting and Gathering

The Yucatec Maya do some hunting to supplement the chickens and pigs raised around the houses. The primary game is deer and peccary (a small wild pig), which are attracted to active milpas and gardens and to recently abandoned milpas. In addition, some wild pigeons are hunted. The proliferation of shotguns has increased hunting success and put pressure on game populations. As the human population increases, this pressure intensifies, although game birds are adept at escape, and their numbers remain high.

A large number of wild plants are gathered and used for various purposes, including over 330 used for medicine, 122 for food, and 103 for construction wood. Firewood alone requires a book's worth of knowledge; each kind of wood is distinctive in specific heat, dry-

CASE STUDY

ing qualities, rate of burning, and usefulness in particular contexts. Another five species are used for basketry, two as rat poison, and one sticky-leaved plant for catching fleas.

Environmental Manipulation

The Yucatec Maya practice much less environmental manipulation than their ancestors. Today the population is lower and the major cities are no longer occupied. In addition, some of the agricultural techniques, such as terracing and field-and-canal systems, are used either less or not at all, reducing the need for manipulation. Still, the forest is managed for purposes of swidden and fruit tree production. The abandoned swidden fields are still managed for their use as hunting grounds for wild game.

Passive environmental manipulation is still important and is accomplished through ceremony. Until recently, these ceremonies were carried out to bring rain and to pray to—and later thank—the spirits of forest and field for the success of the crops. The *chaak* (rain gods) and *yum il k'aax* (Lords of the Forest) want to be remembered and appreciate honey mead, corn gruel, corn bread, and chicken stew offered with appropriate chanting and incense. A detailed round of ceremonies was part of daily life until very recently. Many farmers still hold some or all of these ceremonies, but modernization is taking its toll on the old ritual representation of agriculture.

Resource Management

The Yucatec Maya continue to manage their various resources very well. Some recent "improvements," such as vegetable farming and fruit growing, have not worked out except as changes to already existing practices. The swidden system has been effective at preserving forest and soils. Some areas cleared for more "modern" agriculture have not done well, losing soil and being overrun by weeds. An expansion of fruit orchards into these areas is one way to reclaim that land. Logging has impacted the forest, and mature mahogany trees have been all but eliminated. Reforestation is a continuing practice, but tree poachers continue to frustrate these efforts.

The Yucatec Maya also continue to employ passive resource management. They are still careful not to offend the spirits of the

CASE STUDY

forest by taking more resources than they need. They also employ prayer and offerings to the spirits to protect the fields and the crops as harvest approaches.

DISCUSSION

The Maya today practice a sustainable agriculture within a rain forest; however, it is well below the carrying capacity of the ancient system. The reasons are unclear; it has only recently been realized that this was the case. It seems that the ancient Maya exceeded the carrying capacity of the forest and that the system collapsed catastrophically (a bust cycle), with a drastic reduction in population. It may be that any recovery was interrupted by the Spanish conquest and disease. The Maya population may only now be recovering. Increasing populations may create more pressure to adopt ancient techniques.

10

Current Issues and Problems

As people, we all face a number of environmental issues and problems. The world environment is paying a high price for the system of intensive agriculture now widely adopted across the world. Modernization and development have resulted in increasing homogenization of both the natural and the cultural environments. Diverse and complex ecosystems are converted into simple ones through monoculture, with an associated loss of habitat and extinction of species. Long-term overgrazing and other mismanagement have turned entire regions, such as the Middle East, into deserts. Western land management policies tend to be shortsighted and destructive, committing us to an agricultural system that is ultimately unsustainable. Change in planning procedures is clearly in order.

Human activity has also led to changes in the climate of the Earth (see McIntosh et al. 2000). Deforestation, which dates from at least the Neolithic in Europe and Africa, has resulted in the alteration of long-term rainfall patterns across the region and is likely a major cause in the expansion of the Sahara Desert across most of northern Africa. While the Romans initiated industrial pollution (e.g., lead), it was a modest beginning compared to the pollution levels of modern industry. Contemporary pollution is in the process of causing global warming, although it is not yet clear what effect it will have. There is little hope, however, that it will be positive. We simply do not fully understand the impact of our actions; however, some causes are clear but ignored (e.g., the refusal of the United States and some others to reduce carbon dioxide emissions).

As traditional cultures are altered and ultimately vanish, everyone suffers the loss of their experience, knowledge, products, and solutions to problems. In the final analysis, the loss of any group weakens us all through the deterioration of our overall diversity and our adaptability. Few people recognize these issues as problems, but consider this analogy (from Ehrlich and Ehrlich 1981:xi-xii): An airplane is held together with thousands of rivets, each one attached to something else, all forming a complex system that keeps the plane from falling apart. What would you do if, just before the plane took off with you on it, you saw someone removing rivets from the wing? Would you just sit back and say "Oh well, a few rivets don't matter"? As you watched rivets being removed one after another, you would have the very real fear that the wing would fall off and the plane would crash. Think of other cultures and their adaptations as rivets.

Some things have improved. Today, there is a better understanding by the Western public of many environmental issues and a reduction in ethnocentrism and racism. Groups such as Cultural Survival have been formed to help indigenous people. Conservation organizations such as the Nature Conservancy are trying to save portions of ecosystems. Government efforts to improve the environment through air quality acts, preservation of endangered ecosystems, and the like seem to have increased. There is so much more to do, but there is also reason for hope.

Many of the problems faced by Western groups are also faced by traditional people. However, the circumstances are worse for traditional people, since it is *their* cultures that are threatened with extinction, *their* homes destroyed, *their* lands taken, and *their* children dying. They also have to deal with issues such as global warming (particularly bad for some Eskimo), but only after they get out of the way of the bulldozer. Here we discuss several of the major problems faced by all of us, but perhaps more directly by traditional peoples. Each of these issues is related in that together they form much of the root causes of many of the specific problems faced by traditional cultures.

THE TRAGEDY OF THE COMMONS

When short-term gain is sought at the expense of long-term return and an overexploitation of a resource results, a **Tragedy of the Commons** (Hardin 1968) occurs. A Tragedy of the Commons is a situation in which a resource is owned by a large group or not owned at all, resulting in its unregulated use. If the resource were valuable, there would be considerable pressure to exploit it as fast as possible so as to maximize short-term returns. If only one player took that approach, all of the others would be forced to join in and get theirs as fast as they could, before the resource was gone. Thus, a free-for-all develops.

Persons with weak political power—the poor (as in most traditional cultures), the young, and above all the unborn future generations—are at an enormous disadvantage. So are downstream users—the upstream people get to pollute a river, for example, unless the downstream users can sue or show force. The powerful, the rich, and the upstream thus tend to win out even if their use of the resource base is suboptimal (Murphy 1967).

No one can conserve the resource, no one has the authority to protect it, and each person has to take as much as he or she can. This situation leads to destruction of the resource base because profits accrue to the most predatory users. If some try to preserve the resource, they get forced out of the competition because the users get immediate profits and so prosper while the conservationists forgo immediate profits and so go out of business. If those (the rich, governments, or companies) with strictly short-term, cut-and-run interests have the power, they are apt to destroy the resource, and a much greater long-term payoff is sacrificed (Daly 1993; also see Repetto 1992; Dove 1993a; Moran 1993a, 1993b, 1996).

Many Tragedy-of-the-Commons examples can be cited, from the overlogging and overgrazing of public lands in the United States to overfishing the Grand Banks of New England to destroying the world's rain forests by mining, lumbering, ranching, and other activities (also see Tucker 2000). In these cases, governments and companies unite to exploit the resources, companies out of greed and governments out of corruption and/or a need for cash. The common people are pushed out of the way, sometimes by force.

One of the delusions very prevalent in environmental politics is the "jobs vs. trees" fallacy (Goodstein 1999). Clearly, cutting down all the forests and catching all the fish are bad for the economy in the long run. Even preserving aesthetic resources pays off, in tourism and increased property values.

A new field of social ecology (an aspect of political ecology) has been developed (Guha 1994), and considerable research on the management, or mismanagement, of resources is being conducted (Stonich 1993; Alcorn and Molnar 1996; Sponsel et al. 1996b; Stonich and DeWalt 1996). Andrew Vayda (1996), putting these political-ecological questions in a wider context, has discussed the complex effects of forestry management schemes on the ground in Indonesia.

The literature on preventing the Tragedy of the Commons is extensive and has recently burgeoned, especially in regard to fisheries. It has been found that traditional fishermen, from Anglo-Americans to Micronesians, almost always specify the Commons—that is, parcel them out on an ownership basis. "Sea tenure" has boomed as a field of study, just as land tenure did some years ago (see Johannes 1981; Ruddle and Akimichi 1984; Morrell 1985; McCay and Acheson 1987). E. N. Anderson and Marja Anderson (1978) have studied the opposite case, the breakdown and failure of a system.

AGRICULTURAL INVOLUTION

Clifford Geertz (1963; also see White 1983; Huang 1990), describing the agricultural system of central Java, coined the term "agricultural involution" to describe the pattern in which people work harder and harder to grow more and more on less and less land. In Java, he showed that colonialism was the cause; the Dutch forced peasants off the best land and created a labor demand that led to large families. The result was, of course, more people on less land. People were forced to work harder.

Involution is now generally defined as a rise in total production accompanied by a fall in per-capita income—because of a rise in population and/or extraction of "surplus" value by outsiders. The extraction of this "surplus" sometimes means that the survival needs of the farmers are not met, and a number of them starve to death every year.

Agricultural involution has occurred in traditional preindustrial societies suddenly shoved into the modern world (e.g., Indonesia and Mexico). Population increase and oppressive elites force people to work harder and harder, but in conditions of such extreme poverty that they cannot get ahead. Indeed, they tend to fall further and further behind, becoming even more poorly fed and more destitute. These pressures face traditional farmers all over the world.

AGRICULTURAL DEVELOPMENT AND INTENSIFICATION

Both the Tragedy of the Commons and agricultural involution result from development and intensification. One could view post–Ice Age human history as the history of agricultural intensification, and ecologists often quote Jonathan Swift to the effect that "whoever makes two blades of grass grow where one grew before is the greatest benefactor of man." Grass not only supports grazing animals but, of course, most of our staple foods are grasses—including wheat, corn, rice, and millets. The seeds of grasses are unique in having a particularly efficient balance of starch, oil, and protein for human nutrition and are easily storable and very easy to grow.

This intensification involves an enormous increase in capital and wealth but no real improvement in the average human condition. A variety of agricultural "systems" currently exist, each the result of particular circumstances of land, crops, labor, and available technology. A variety of models are proposed to explain some general trends in the development of such systems.

Yujiro Hayami and Vernon Ruttan (1985) have developed the best model yet, pointing out that people will always work on the part of the system that is of the highest cost. If land is scarce or expensive, people will work to increase the yield per acre. If land is easy to obtain but labor is scarce and thus expensive, people will invent labor-saving tools and machines. Hayami and Ruttan showed that East Asian agriculture developed with abundant labor and scarce land and thus has achieved the maximum (in terms of balanced diet, though they did not say it that way) out of an acre using a fantastic labor input, while Western agriculture has developed in a situation of abundant land (at least in the United States and other bigger countries) and scarce labor and so has concentrated on labor-saving machines. Some countries have both limited land and expensive labor, so they have both intensified use and adopted labor-saving technologies (e.g., Netherlands and Denmark). Conversely, areas with much land and cheap labor do not have to develop at all. This model neatly predicts the observed facts about hunter-gatherers and many traditional agricultural societies.

However, the model does not always work. It assumes a society is willing to work for change (rather than, say, robbing the neighbors). Consider Haiti: very scarce land, but not much improvement in agriculture in five hundred years! Instead of changing the system to support the increased population, the Haitian policy has been to allow the population to become worse and worse off, to the point of starvation. However, other things being equal, agricultural societies do clearly target their efforts toward reducing the worst problems (or highest costs). The theory that Hayami and Ruttan developed from this insight is, in their terms, the theory of induced development: society induces development in the specific area of highest costs.

The theory of induced development can subsume previous theories, such as that of Ester Boserup (1965) that population growth leads to agricultural intensification. Obviously, population growth does lead to agricultural intensification if society sees food as a limiting factor, has the capability to develop, and is willing to devote resources to development. Otherwise, population growth does not lead to intensification, as in the case of Haiti and other Third World countries. In contrast, a society with a static population will intensify agriculture if there are other reasons to do so, such as a desire to increase sales of farm products or an increase in ceremonial activity leading to greater need for ritual foods (Rappaport 1984).

This sort of activity maintains the production of food in New Guinea societies, where agriculture is highly intensified in spite of sparse population, no markets, and no elites—just rituals compelling the people to produce food by means of a sustainable system (Rappaport 1984).

Current Trends

Agricultural productivity worldwide has increased dramatically over the last two hundred years. There was not much difference between the agriculture of Europe in 1700 and that of the Middle East five thousand years ago—at least not by comparison with what came after, although the groundwork of the revolution was laid by 1700. Asia was far more advanced agriculturally in 1700, but by 1800 it was falling behind and by 1900 was "underdeveloped." Today, Asia is catching up, but Latin America and Africa are still less developed. The agricultural changes in the Western world were due to increased trade and commerce (and thus to demand), to new technological events stimulated by the same expansion of commerce, and to related factors.

The legacy is an agriculture that relies heavily on machines and land rather than on human effort. In other words, capital and land are consumed, even squandered, while labor is conserved. This situation is due to the relatively high price of labor although, relative to other occupations, agricultural labor is notoriously poorly paid, not just in the United States but everywhere. In Asia, agricultural innovation caused the price of labor to fall because the new technologies allowed more people to be supported per acre and per unit of capital input. Thus, it became economical to substitute labor for land and capital.

The result is that a two-track agriculture has developed in the world. "Developed" agriculture uses far more energy and other capital-intensive input, overusing and polluting the land and leading to a rapid decline of the resource base: It is an agricultural system that is unsustainable. "Underdeveloped" agriculture tends to be labor intensive, low in production per worker, but more fine-tuned to local ecosystems. The two systems lead to and reinforce each other, thus, "the development of underdevelopment." In the extreme forms of this relationship, the rich countries extract resources from the poor ones so that the rich can use resources lavishly and impoverish the poor ones by terms of trade, forcing them to rely on unskilled, poorly paid labor. This exploitation is supported by a world system of international trade—the terms much in favor of the rich—and security that permits the repression of the poorer countries.

This argument has socialist antecedents, but note that the richer socialist countries are not significantly better than the capitalist ones in their behavior!

Indeed, the countries of the former Soviet Union are the world's great environmental disaster story, a cautionary tale for other countries (Feshbach and Friendly 1992). Ecologically, the deforestation, soil erosion, "mining" of water resources, exhaustion of genetic stocks as crops get more homogeneous, and buildup of pests that are resistant to current pesticides are proceeding so rapidly that total ruin of the world's intensive agriculture and return of the world to an essentially medieval agriculture are now a real possibility.

Hope for the future lies in developing a way to use all the traditional techniques, as well as all the new techniques we can find, that are sustainable and that do not massively destroy the environment. Most traditional techniques are sustainable—or they would never have become "traditional"—although there are some very important exceptions. For example, overgrazing in the Middle East has been endemic and disastrous for some eight thousand years! We must learn and apply all we can, now, before it is too late (see, e.g., Wilken 1987).

THE RAIN FOREST DILEMMA

One of the most serious ecological problems of the world in recent decades has been the disappearance of tropical rain forests (see Anderson 1990; Croll and Parkin 1992; Moran 1993a, 1993b; Stonich 1993; Sponsel et al. 1996b; Tucker 2000). The destruction of rain forest has reached levels that can only be described as catastrophic. Rain forests cover about 16 percent of Earth but are enormously diverse and rich in life; they are home to perhaps 50 percent of all species. The loss of rain forest is not limited to species, however, but to entire ecosystems, which are very difficult, if not impossible, to reconstitute. Small-scale plots such as slash-and-burn fields are easily grown over by forest with little permanent damage, but large-scale deforestation results in the forest's being segmented such that the necessary colonizing species are too far apart to successfully interact and regrow the forest. In past centuries, destruction of tropical forests has occurred on a local scale, as, for instance, in some of the lands of classic societies like the Maya and Khmer. What is new is the rate of clearance.

Until recently, the forests were protected by barriers of disease, remoteness, and the sheer difficulty of clearing. Today, diseases that once frightened people away from the forests can be controlled, and the technology of clearing has changed from axes to chainsaws and bulldozers. This change has produced something of a pioneer attitude in most of the world. Like the early European settlers of the United States (and even the early people of much of Polynesia [see Meilleur 1996]), local and international interests alike are rushing to clear the tropics with little thought of the future. Before the mid-twenty-first century, all old-growth tropical forest not specifically reserved will be gone.

Issues

The destruction of tropical rain forests is a serious problem for several reasons. First, and most obvious, most of these forests could support long-term exploitation that would produce far more revenue than do the farms and ranches that replace them (a Tragedy of the Commons). Most groups living in forests have shaped the forest to some degree as a result of their long-term management (e.g., Anderson 1990; Balée 1994) and have often created a fine-tuned, well-managed landscape capable of sustained yield. Even with increased exploitation, forests could still be managed for sustainable yields (see Robinson and Bennett 2000; also see Cooke 1999).

Second, tropical forests produce a large number of products, including food (e.g., Pimentel et al. 1997). One important product is hardwoods, often worth hundreds of dollars per cubic meter. If these woods were selectively harvested rather than clear-cut, the forest could produce hardwoods for centuries to come. Countless nontimber forest products could be harvested sustainably (Plotkin and Famolare 1992). These products include such things as rattans (tough palm vines used for furniture and baskets), honey, medicinal plants, game meat for local inhabitants, poles for local construction, wild and semiwild fruit and nuts, and ornamental plants for the nursery trade.

Third, forests provide watershed and soil protection. Many tropical soils, stripped of their forest cover, dry out and become very hard and difficult to cultivate. Other soils wash away quickly or, due to chemical changes, become too acidic to grow much useful new cover. It is also possible that deforestation reduces rainfall downwind as the massive transpiration of the forests is eliminated. The desertification of the Sahara of Africa may be in part due to the clearing of the West African coastal forests.

Fourth, deforestation releases massive amounts of greenhouse gases as the trees are burned or allowed to decay. These gases have been a factor in global warming, although fossil fuels are a far more serious cause.

Fifth, tropical forests are very diverse ecosystems. They hold literally millions of species of animals, plants, fungi, molds, bacteria, and algae. Few of these species have been adequately studied, and many of them have enormous potential as medicines (Plotkin 2000). For example, our most powerful antibiotics—penicillin, chloromycin, and the like—are derived from molds, and many of the few rain forest molds that have been studied also show antibiotic properties. Literally tens of thousands of tropical plants have been reported to have medical effects. Thousands have been examined, and many have already produced valuable drugs, ranging from quinine and caffeine to the new cancer drugs (see, e.g., Plotkin 2000). Other tropical species have enormous potential for producing food, fiber, oil, fuel, and other commodities. But even the less obviously useful tropical species provide such an incredible wealth of beauty, interest, genetic potential, and evolutionary possibility that there are more than adequate reasons for saving them.

Sixth, and perhaps the most serious of all to anthropologists, the tropical forests are home to many hundreds of cultural and ethnic groups whose lives revolve around using the forests in some way. These groups range from bands of scattered hunter-gatherers to the huge, stable, ancient states of the Maya and Khmer. Many groups have been deprived of their forests in recent years. All too many of these groups have suffered from cultural and personal collapse. It is psychologically devastating to any group, be it a hunting band or the workers of an industrial city, to be suddenly deprived of its livelihood and its way of life. Small groups, impoverished by loss of their forest resources, have great difficulty in adjusting to a new world—especially if that world is openly hostile to them and their interests.

Loss of the forests would then be a tragedy for their inhabitants and for all of us. The effects on climate, food production, medicine development, and hydrology, taken together, would amount to a worldwide catastrophe far worse than any ecological crisis that humanity has endured so far.

The Dilemma

The forests have been cleared for a number of reasons, but primarily for the expansion of grazing and agriculture. Broadly speaking, there are three patterns. First, extensive, but not intensive, cattle ranching has been the major cause of deforestation in many areas (Downing et al. 1992). This practice often involves clearing the forest, planting grass, and running a few scrub cattle worth far less than the timber wasted in the clearing process. In other situations, with better soil and better facilities, viable ranches with quality stock can be developed, but even in these cases one doubts whether the cattle are worth as much as the forest was.

Second, contemporary commercial agriculture, usually oriented toward export to rich temperate-zone countries, has replaced forests in much of the world. The tropics supply the world's coffee, chocolate, rubber, palm and palm-kernel oil, coconut oil, and bananas, plus countless species of specialty fruit. They produce much of the tea, citrus fruit, and rice that enters international trade. On a darker note, the New World montane tropics produce the world's cocaine as opium flourishes in many tropical and subtropical highlands.

In some areas, wood is the most important of these export products. For centuries, forests have been selectively cut for valuable woods, but now whole forests are being clear-cut—often for pulpwood, fuel, or cheap, throw-away uses such as chopsticks or temporary construction siding. Mining and oil extraction are also local problems of consequence (Rudel and Horowitz 1993).

Third, and the most morally difficult to deal with, is the huge increase in the number of people using slash-and-burn cultivation in the forest to support themselves. Until recently, most of the real tropics were too sparsely populated for slash-and-burn cultivation to be a major problem. The relatively few people living in those areas utilized a sustainable swidden system and made no significant impact on the forest. However, recent population growth in developing countries and the migration of unemployed city workers to farms have led to shortening of, and then the elimination of, the swidden cycles. Now, so many people are using slash-and-burn that there is not enough land to lie fallow for eventual reuse. Instead, the field is abandoned, and the family moves on to the next patch of forest with no thought or intention of returning. Like a wave moving across a lake, the forest is cut, farmed for a year or two, and abandoned, and a new field is cut, leaving nothing but abandoned fields in its wake. Eventually, deforestation, weed invasion, and erosion lead to total ruin of the landscape. Eroded hills, covered with almost impenetrable tangles of worthless grass and brush, replace former forest. Some level lands with relatively fertile soils can be salvaged after years under grass (Dove 1981), but most lands cannot.

The problem is human. The poor farmers have little choice but to continue their farming, however destructive it is in the long term. They have to feed their families now. One cannot just tell them not to farm without providing some alternative way to make a living. Peasants who see their children starving are going to do anything to get land and food—now. This problem is real and serious, especially in densely populated areas like southeast Asia, but even there its role has been exaggerated (Dove 1993a). In more lightly populated areas, such as the Brazilian Amazon, it is less important than corporate and large-scale projects.

A Tragedy of the Commons

One wonders why people destroy million-dollar forests for ten-dollar cows. On the whole, there is an economic rationale behind this: the million-dollar return is very slow in coming, but the ten dollars is there now. Sustainable selective cutting of valuable woods produces a steady but modest stream of income, but destroying the forest for quick profit produces more in the short run. If people are in no position to reap the long-run benefits, they will opt for the short-term ones.

The biggest problem of all seems to be the short-term philosophies of many governments and multinational bodies. Multinational firms and international organizations are less than rational: they want results *now*, whether for their stockholders or for press releases to backers back home. The predatory instincts of multinational firms are well known but are often exacerbated by local tax systems and other institutions that virtually force the firms to destroy (Ascher 1999). Susan Stonich (1993; Stonich and DeWalt 1996) has meticulously documented the ways in which government policies and large landlord politics continue to force peasants in Honduras to cut and burn against their better judgment. Probably all anthropologists who have worked in the forested tropics have similar stories. Thus, the old question of whether "the rich" or "the poor peasants" are the worst offenders in clearing the forests is not very useful to ask. We have learned that the poor peasants often are driven to act because of forces unleashed by the powerful. These forces may be simple and straightforward: If the powerful have seized all the good land, the peasants have to do what they can. Or the problems may be much more complex: International coffee-pricing policies, and the response of the Mexican government to them, have affected the small-holder coffee economy of Mexico in negative ways (E. N. Anderson, personal research; Jan Rus, personal communication 1999).

Perhaps this state of affairs is understandable in the case of politicians, who may be looking ahead only to the next election or military coup. However, economic theory suggests that large landlords should be more concerned with long-term profits. They have enough wealth to allow them to look to the long term, and they should be interested in maximizing their overall wealth, not their immediate advantage.

Indeed, many of them are, but others are interested only in making a quick killing and then retiring to enjoy a genteel lifestyle. Still others may simply not understand what they could do with their forests. While indigenous people have an enormous knowledge of forest products, the immigrants often know nothing at all about the forest and thus see nothing better to do with it than clear it for cattle.

Moreover, much of the worst devastation of the tropics has been done by (theoretically) well-meaning international aid organizations, who clear land for pasture, build dams, and otherwise ravage the landscape without needing to concern themselves with the cost-benefit ratios of the projects in question. They have acted from mistaken charity or from a genuine commitment to "progress" or, in some cases, just to show that they are busy. Often the projects are economically irrational.

In Quintana Roo, Mexico, site of E. N. Anderson's recent research, the government has blown hot and cold. On the one hand, Anderson was shown a huge tract of land that had been cleared by the government to "open up land for cultivation." The government had deforested a very dry tract, scraped off the topsoil in the process, and then left it to the local villagers—and all this without providing any water! Twenty years later, the land was still a desert, and no one could figure out any use for it. On the other hand, and more recently, the state and federal governments have cooperated on some truly exemplary plans for sustainable forest management, local reserves, and wildlife protection. These plans are real models for the world, and one can only hope they succeed (Kiernan and Freese 1997; Flachsenberg and Galletti 1998; Galletti 1998).

Protecting local people from dispossession of their land by multinational mining or logging companies or by commercial farmers and ranchers has proved difficult. The multinationals, in particular, have vastly more money and political influence than do small local groups of forest hunters or farmers. In a few cases, perhaps especially tragic, local groups have been displaced to create national parks and reserves to "conserve" the forest. Fortunately, most conservation bodies are now aware that, if a group has been using and managing a forest for several thousand years, throwing it off the land is more apt to destroy the forest ecosystem than to preserve it (e.g., Western and Wright 1994).

Alternatives to deforestation have often been proposed (e.g., Anderson 1990; Committee on Sustainable Agriculture and the Environment in the Humid Tropics 1993; Dove 1993b). They include forest farming of various kinds, often with native tree crops. However, while these methods may preserve the structure of the forest and thus save the soil, planting too many trees of the same species could destroy almost as much biodiversity as clear-cutting does. The same can be said for the regrowth of forest in much of the Amazon (Moran 1993a, 1993b). It was feared that cutting the Amazon forests would lead to total desertification. This has not yet occurred; much land has gone out of production and has been rapidly reclaimed by forest. Unfortunately, this second-growth forest has only a tiny fraction of the biodiversity of the original high forest.

Sustainable logging for selected valuable woods remains the best option in many areas and is correspondingly the most widespread option currently. However, it, too, is damaging too many animals. Reforestation, successful in much of the temperate zone and dramatically so in Korea (e.g., Choe 1994), has been less successful in the tropics. The better-quality tropical trees grow slowly; the forest habitat recovers even more slowly. Soil degrades in the process and may take centuries to recover.

In practice, reserves—with traditional management firmly in place, where appropriate—are necessary and will be increasingly necessary in the future. It has been repeatedly pointed out that because the tropical forest is a world resource, the world community must pay to save it.

Tropical forest nations are often among the poorest in the world and cannot shoulder all the burden. The World Wildlife Fund, Nature Conservancy, Conservation International, the Rainforest Alliance, Cultural Survival, and other nongovernmental organizations (NGOs) interested in conservation have spearheaded the resulting multipronged attempt to unite the world community in saving what is probably its most valuable ecosystem. Cultural Survival has been particularly concerned about indigenous peoples and their problems and interests. Other NGOs have been sensitive in varying degrees to this concern, but most have rapidly improved their awareness—and consequently their behavior—in recent years.

The situation would be cautiously hopeful were it not for the runaway population growth in the tropics and the runaway growth of markets for tropical products in the rest of the world. As it is, the sheer pressure of needs and wants is increasing so fast that almost no one can plan comprehensively for the long term.

The above is not to be taken as implying that only the tropics face deforestation problems. Destruction of old-growth forests is progressing as fast in the United States and Canada as it is in most tropical countries (see Tucker 2000). This situation is doubly unpardonable. Rich nations do not need to destroy their resource futures. Moreover, most of the wood they produce is exported, cheaply, to countries that make expensive products from it and thereby reap almost all the profits. Tropical countries often note, pointedly, that at least their forests are being destroyed to feed starving people.

In essentially all conceivable cases of resource use, society must deliberately intervene to protect long-term, wide interests and thus optimize resource use (while protecting indigenous rights). This can be done legally, via suits and standing to sue, property rights, or conservation laws and enforcement; it can be done politically (via vote or equivalent); it can be done through religion, by cultivating an ethical system that is highly conservationist. In practice these approaches must all be invoked.

Moreover, education, information flow, research, and communication are all essential—in short, knowledge has to be shared. People have to know what their interest really is and be able to make key decisions on the basis of the information they have.

The contemporary world fails at two levels. First, we simply do not have an adequate institutional framework to deal with our rapidly increasing ability to wreck the environment. Second, we are failing more and more at treating the poor, the weak, and the downstream users fairly. We can profit from examining resource management cross-culturally. Traditional people know a wide range of tactics from which we can learn; and more generally, the comparison will show us what problems are universal and what solutions are most widespread. Research in this area, especially into the question of how knowledge is shared and passed on, is only beginning.

Considerable recent work turns on simple matters of what is done that effectively manages resources. Gerald Marten (1986) provided a major review of how traditional farmers in southeast Asia manage trees, soil, pests, and other things. Local studies that cover similar ground include Harold Conklin (1957), James Stuart (1978), Janis Alcorn (1984), and many others. These investigations address to some extent the question of what is done, how, and why. Special topics range widely, from crops that extend the range of cultivation (Harlan 1992) to fire and its role and the costs of fire suppression in natural brushlands (Minnich 1983). An excellent study of education for traditional food procurement has been conducted by Kenneth Ruddle and Ray Chesterfield (1977).

Literature on common property resource management has expanded greatly in recent years (Conference on Common Property Resource Management 1986; Alberta Society of Conservation Biologists 1988; Libecap 1989; Ostrom 1990; Hardin 1991; Oldfield and Alcorn 1987). It now appears that communities can be expected to regulate their resources unless pressures are too numerous or uncontrollable (Burton 1992; also see Ostrom 1990). Management must be carefully tuned, however, to allow maximum local management (Pinkerton 1989). Excessive top-down control, which alienates control from the people on the ground, is a virtual guarantee of failure and is probably the reason for the breakdown of world environment recently (Pinkerton 1989; Anderson and Burton 1992). On the other hand, there must be centralized goal setting. Also, some problems are now worldwide and must be handled on a worldwide basis, which makes the question of "community" a challenging one (Burton 1992).

Needed now are studies of traditional agriculture comparable to those of contemporary Third World agriculture (e.g., Ghai et al. 1979; Hopkins et al. 1979). Some anthropological research related to this approach covers marketing (e.g., Halperin and Dow 1977), nutritional effects of poorly planned modernization (e.g., Bryant et al. 1985), gender division of labor in agriculture (e.g., Burton and White 1984), the different responses to contemporary stress by traditional groups (e.g., Hayami and Kikuchi 1981), problems of agricultural intensification (Geertz 1984; White 1983), and much more.

In a world with plenty of food and many chances for expanding the supply (Avery 1985), there is no excuse for the worldwide poverty, hunger, and waste we actually observe (Brown 1981). The explanation for mismanagement is clearly social, not technological. Thus, studies of how people actually decide what to do have grown apace.

CHAPTER SUMMARY

We all face the challenges of a changing natural and cultural environment. The way individual groups confront and address issues ultimately impacts everyone to some degree. Problems such as the Tragedy of the Commons, agricultural involution, agricultural intensification, and deforestation should be the concern of everyone.

If there is any truth in the findings of cultural ecology, it is that religion and symbolism greatly inform human uses of the environment (Anderson 1996). People preserve their environment for aesthetic and moral reasons as well as for narrowly economic ones. Insofar as this is the case, we need to study traditional conservation and resource management. We have a great deal to learn—not only about how to manage Earth but about how to motivate other people to manage Earth better. At best, we will learn from each other. At worst, even the most cursory study of cultural ecology will open our minds to the many different ways to see the world.

KEY TERMS

agricultural involution
Tragedy of the Commons

Glossary

Abiotic: That part of the ecosystem that is not biological in origin, such as water, minerals, and land forms.

Active environmental manipulation: The purposeful, physical alteration of groups of species and of ecosystems on a large scale.

Active resource management: Physical management or control of resources to maintain and/or increase productivity.

Adaptation: Modification of the body, species, or culture in response to changing environmental conditions.

Agricultural involution: A pattern in which people work harder and harder to grow more and more on less and less land.

Allen's Rule: States that in warm climates, body shape will change to a higher surface-area-to-mass ratio for heat dispersion, that is, people will become taller and thinner in warm climates.

Anatomical adaptations: Long-term changes in genotype and phenotype through natural selection.

Animal husbandry: The herding, breeding, consumption, and use of domesticated animals.

Anthropogenic: Human-caused, e.g., changes in the environment.

Anthropological linguistics: The study of human languages, including the historical relationships between languages, syntax, meaning, cognition, and other aspects of communication.

Anthropology: The study of humans (biology, culture, language, past and present, etc.).

Archaeoastronomy: The study of the astronomy of past groups.

Archaeology: The scientific study of the human past.

Band: A small-scale society without formal leaders; the family is the primary sociopolitical and economic unit.

Bergmann's Rule: States that in cold climates, body shape will change to a lower surface-area-to-mass ratio for heat retention, that is, people will become short and stocky in cold climates.

Biodiversity: The number and dominance of species present in an ecosystem.

Biological anthropology: The study of human biology through time, focusing specifically on biological evolution and human variation.

Biomass: The quantity (mass) of living matter within a specified area.

Biome: A large, general, easily defined environmental zone (e.g., rain forest or grassland).

Biosphere: The global environment and interacting ecosystems.

Biotic: that part of the ecosystem that is biological in origin (plants, animals, etc.).

Boom and bust cycle: Increases in population to exceed the food supply, resulting in a rapid loss of population.

Browsers: Animals whose food consists mainly of the foliage of bushes and trees.

Calorie: Unit of energy (in food); one calorie is the energy needed to raise the temperature of one gram of water one degree Celcius.

Carbohydrates: Compounds in foods, including sugars, starches, and cellulose.

Carnivore: A species that is primarily a meat-eater.

Carrying capacity: A measure of the maximum number of individuals of a particular species that can be supported within a specific ecosystem for a specific time.

Central place foraging: An optimization model that assumed that a group stays in one central place, with specialized task teams traveling to the resources.

Chiefdom: A society with a relatively large population, permanent settlements, some central authority, and a stratified social structure, but no formal state institutions.

Chinampa: A small field constructed within or on a body of water.

Cognition: The way in which information from the senses is processed and interpreted; includes classification systems, decision making, and planning.

Collectors: A classification of hunter-gatherer groups oriented toward moving resources to the people; typically viewed as larger and more sedentary than foragers.

Cultural anthropology: The study of the many aspects of human culture, generally of extant cultures.

Cultural ecology: The study of the cultural aspects of human interaction with the environment.

Cultural materialism: A practical, rather simplistic, functionalist approach to anthropology, with a focus on the specific hows and whys of culture.

Cultural relativism: The study of culture without any attempt to show scientifically that one is "better" than another; cultures interpreted nonjudgmentally.

Culture: Learned and shared behavior in humans.

A culture: A group of humans who share a common set of traits and (usually) identify themselves as a group separate from other groups.

Culture area: A geographic region where environment and cultures are similar.

Currency: A quantifiable unit of measure of cost and benefit used in optimization models.

Diet–breadth: An optimization model that predicts the order in which resources (generally foods) will be added to the diet.

Domesticated: The result of the process of domestication.

Domestication: A process by which organisms and/or landscapes are "controlled" (in agriculture, domestication means that the genetic makeup of an organism is purposefully altered by humans to their advantage).

Dry farming: The production of crops relying only on rainfall to water the crops.

Ecology: The study of the interactions between organism(s) and the environment.

Ecosystem: A bounded community (or communities) that includes both abiotic (basic elements) and biotic (producers and heterotrophs) components and that is tied together in a system.

Ecotone: The intersection of, and transition between, two ecozones, usually a more productive place than either of the ecozones.

Ecozone: An area defined by biotic communities and/or geographic criteria (short for environmental zone).

Empirical science: Knowledge system based on tangible data observable by other scientists ("hard" data, testable, reproducible).

Environment: The surroundings within which an organism interacts.

Environmental determinism: The view that the environment "dictates" what a culture must do and how it must adapt.

Environmental manipulation: Large-scale alteration of the environment by people to effect changes to the advantage of the culture.

Environmental zone: (See ecozone.) An area defined by biotic communities and/or geographic criteria.

Ethnobiology: The classification and knowledge of materials of biological origin (plants and animals) by a traditional culture.

Ethnobotany: The classification, knowledge, and use of plants by a traditional culture.

Ethnocentrism: The view that one's own group is superior to other groups.

Ethnoecology: The classification of the various components of the natural world, including both the abiotic (geography, astronomy, soils, etc.) and the biotic (botany, zoology, etc.), by a traditional culture.

Ethnography: The study of a particular group at a particular time.

Ethnology: The comparative study of culture.

Ethnomedicine: Knowledge and materials used by a traditional culture for medical purposes.

Ethnopharmacology: The classification, knowledge, and use of materials (plants, animals, and other substances) of a traditional culture for medical or other nondietary purposes.

Ethnoscience: Knowledge system (empirical and/or nonempirical) of a traditional culture.

Ethnozoology: The knowledge, use, and significance of animals in a traditional culture.

Evolution: Change over time (specific disciplines have more specific definitions).

Evolutionary ecology: Any evolutionary thinking about ecology, but commonly defined as an analytical approach that presumes that cultures behave as biological organisms do, that natural selection processes act on them, and that they adapt and evolve as organisms do.

Fat: Lipids that store energy within the body of an organism.

Feng shui: The Chinese science of proper arrangement of elements in a landscape to ensure harmony.

Fission-fusion: The splitting of a larger group into smaller ones that later rejoin to reform the original larger group (fission-fusion may occur for a variety of reasons, including as part of a seasonal round).

Food chain: The flow of nutrients through trophic levels; the sequence of which organism is consuming which.

Food web: Food chains linked together by some common thread.

Foragers: Hunter-gatherer groups in which the group is oriented toward moving people to resources (forager groups are typically viewed as being smaller and more mobile than collectors).

Garden: A small agricultural plot, tilled by hand.

Gathering: The collection of relatively small and nonmobile resources, such as wild plants, small land fauna, and shellfish.

Grazers: Animals whose food consists mainly of grasses and other low-growing plants.

Habitat: The place an organism lives; where its niche is located geographically in the environment.

Herbivore: Plant eater (some species may be specialized, e.g., fruit-eaters [frugivores]).

Herdsman husbandry: A pastoral technique used as a component of an agricultural system in which most members of the group are sedentary agriculturalists but some animals are raised at distant pastures tended by herdsmen.

Historical ecology: The study of ecology from a historical perspective to understand how ecological relationships and practices function or developed.

Horticulture: Low-intensity agriculture involving relatively small-scale fields, plots, and gardens; food raised primarily for personal consumption rather than for trade or a central authority.

Human biological ecology: The study of the biological adaptations of humans to their environment.

Human ecology: The broad field of study of human interaction with the environment.

Hunters and gatherers: Groups that make their *primary* living from the exploitation of "wild" foods.

Hunting: Actively looking for, killing, butchering, and consuming animals.

Intensive agriculture: Large-scale agriculture often involving the use of animal labor, equipment, and water diversion techniques; production of a surplus to feed specialists.

Irrigation: The purposeful diversion of water from its natural course onto agricultural fields.

Landscape: The face of the earth as modified, created, or perceived by people, including physical and cultural geography.

Liebig's Law of the Minimum: The law that states that some condition or resource will limit carrying capacity (the Convoy principle).

Life expectancy: A measure of the average age at death of all the people who died in a particular year.

Linear programming: A mathematical technique of calculating the optimal allocation of resources toward a defined goal in the face of multiple constraints.

Milch pastoralism: The production of milk as the major pastoral product.

Minerals: Inorganic substances needed by the body for combination with organic compounds.

Monoculture: Primary use of a single species of domesticated plant in a field.

Multilinear Cultural Evolution: The evolution of cultures along many lines.

Mutualism: A relationship between two species in which they both directly benefit.

Neoevolution: A revival of nineteenth-century unilineal cultural evolution.

New Ecology: The idea that human cultures formed only one of the population units interacting in the environment, placing humans within a unified science of ecology.

Niche: The role an organism plays in the environment; what it eats, where it lives, and how it reproduces.

Nomads: Another term for pastoralists.

Nomadic pastoralism: The type of pastoralism in which groups are small, mobile, and completely dependant on their animals.

Nutrition: Materials needed by the body for proper operation.

Omnivore: Species that have very broad diets (can eat a wide variety of animal and plant foods).

Optimization: The process of getting the best return for the investment made; using resources as efficiently as possible to accomplish goals.

Optimization models: Models used to explain some aspects of behavior related to the utilization of resources.

Passive environmental manipulation: Nonphysical (e.g., ritual) activities designed to maintain and control the environment.

Passive resource management: Nonphysical (e.g., ritual) management or control of resources (often abiotic) to maintain and/or increase productivity.

Pastoralism: The herding, breeding, consumption, and use of managed or domesticated animals, to the general exclusion of plants.

Patch-choice: An optimization model that predicts the order in which patches of resources (generally foods) will be utilized.

Permaculture: Permanent, sustainable agriculture involving complex systems with many species.

Physiological adaptations: Short-term biological changes in the body as a result of adaptation; for example, the development of a tan to protect the skin from ultraviolet radiation.

Political ecology: The study of the day-to-day conflicts, alliances, and negotiations that ultimately result in some sort of definitive behavior; how politics affects or structures resource use.

Polyculture: Planting multiple domesticated species in a field (thereby increasing diversity).

Possibilism: The idea that a variety of solutions are present in any environment and that a culture chooses a solution or solutions best suited to them.

Postmodermism: A paradigm that holds that human cultural behavior is essentially arbitrary and so can be interpreted in any number of arbitrary ways.

Protein: A complex combination of amino acids; required by the body for its constituent amino acids.

Rational choice theory: A theory that asserts that people decide how to achieve their goals on the basis of deliberate, individual consideration of all available information and that they are good calculators of their chances.

Refugia: A surviving remnant of a past biome.

Resource: Something used by an organism; may be either renewable or nonrenewable.

Resource management: The small-scale alteration of specific resources to effect changes to the advantage of the culture (see active and passive resource management).

Resource monitoring: The visitation of resource locations to determine present and future conditions.

Resource universe: In studies using optimization models, the resources that are available and utilized by the group being studied.

Scavenging: The act of obtaining animals that are already dead.

Seasonal round: A system of movement of people and/or resources about the landscape on the basis of the seasonal availability of resources and their geographic location.

Seasonal transhumance: The seasonal movement of herders and their animals from pasture to pasture while the rest of the population stays in one place; the term is sometimes applied to hunter-gatherers to describe a seasonal round.

Sedentary: Living in one place all the time.

Sedentary animal husbandry: A pastoral technique used as a component of an agricultural system in which full-time farmers will also raise some animals in and around their farms (examples include modern dairy farming and cattle feed lots).

Seminomadic pastoralism: The type of pastoralism in which animals are moved from pasture to pasture but some horticulture is practiced to supplement the animal products (this is the most common form of pastoralism).

Semisedentary pastoralism: The type of pastoralism in which some members of the group move around seasonally with their animals but in which horticulture forms an important aspect of the economy.

Silviculture: Agriculture emphasizing tree crops.

Slash-and-burn: A horticultural technique involving the cutting, drying, and burning of natural vegetation from a small plot and planting crops in the field (the field quickly loses its agricultural productivity, and another plot must be processed; sometimes called "shifting cultivation").

State: A society with a large population, complex social and political structures, complex record keeping, urban centers (cities), central authority, monumental architecture, specialization, and the legal control of the use of force.

Stewardship: Having responsibility for maintaining a resource; being a guardian.

Storage: Taking some resource and saving it for later use.

Strategy: An overall plan to achieve a long-term goal; in the context of this book, an overall subsistence system such as hunting and gathering.

Stress: A condition that forces change in a system.

Subsistence: A complex system that includes resources, technology, social and political organizations, settlement patterns, and all of the other aspects of making a living.

Succession: Patterned, developmental change in a plant community as it evolves to maturity.

Swidden: A sustainable horticultural system involving the rotation of fields where slash-and-burn is the primary technique used.

Symbiotic: A long-term, dependent relationship between two species.

Tactics: Methods or techniques used to accomplish a goal; the action component of a strategy.

Tethered nomadism: A pastoral group whose seasonal round is within a well-defined territory.

Tether: A restriction due to the distribution of a resource, e.g., having to stay within a certain walking distance of a spring.

Traditional culture: Generally a culture that is non-Western and nonindustralized.

Tragedy of the Commons: A situation in which an unregulated valuable resource is overexploited to the benefit of a few but to the detriment of the many.

Tribe: A society with a relatively large population, a number of villages, and leaders (chiefs) with some actual power.

Trophic pyramid: A description of the levels of relationship between producers, heterotrophs, and decomposers: what is eaten and how many conversions from solar energy have taken place.

Unilinear cultural evolution: The theory that cultures evolved upward along a single line.

Vitamins: Organic compounds used by the body to maintain certain functions but not manufactured by the body, so must be present in food.

World systems theory: The idea that the world's economy is interconnected and related, generally due to the linkages established by European expansion in the last several hundred years.

References

Abrams, Elliot M., and David J. Rue

1988 The Causes and Consequences of Deforestation among the Prehistoric Maya. *Human Ecology* 16(4):377–395.

Adams, Charles C.

1935 The Relation of General Ecology to Human Ecology. *Ecology* 16(3):316–335.

Adams, Robert M.

1966 *The Evolution of Urban Society*. Chicago: Aldine.

2001 Complexity in Archaic States. *Journal of Anthropological Archaeology* 20(3):345–360.

Alberta Society of Professional Biologists

1988 *Native People and Renewable Resource Management* (symposium volume). Edmonton: author.

Alcorn, Janis B.

1984 *Huastec Mayan Ethnobotany*. Austin: University of Texas Press.

Alcorn, Janis B., and Augusta Molnar

1996 Deforestation and Human-Forest Relationships: What Can We Learn from India? In: *Tropical Deforestation: The Human Dimension*, Leslie E. Sponsel, Thomas N. Headland, and Robert C. Bailey, eds., pp. 99–121. New York: Columbia University Press.

Anderson, Anthony B. (ed.)

1990 *Alternatives to Deforestation*. New York: Columbia University Press.

Anderson, E. N.

1972 The Life and Culture of Ecotopia. In: *Reinventing Anthropology*, Dell Hymes, ed., p. 264–283. New York: Random House.

1988 *The Food of China*. New Haven: Yale University Press.

1996 *Ecologies of the Heart*. New York: Oxford University Press.

2000 Is It All Politics? Political Ecology within Human Ecology. Paper presented at the annual meetings of the American Anthropological Association, San Francisco.

Anderson, E. N., and Marja Anderson

1973 *Mountains and Water: The Cultural Ecology of South Coastal China.* Taipei: Orient Cultural Service.

1978 *Fishing in Troubled Waters.* Taipei: Orient Cultural Service.

Anderson, E. N., and Sharon Burton

1992 Tragedies of the Commons: Malaysia. Paper presented at the Pacific Coast Division, American Association for the Advancement of Science, Santa Barbara.

Anderson, Leslie

1994 *The Political Ecology of the Modern Peasant: Calculation and Community.* Baltimore: Johns Hopkins.

Anderson, M. Kat

1999 The Fire, Pruning, and Coppice Management of Temperate Ecosystems for Basketry Material by California Indian Tribes. *Human Ecology* 27(1):79–113.

Arens, William

1979 *The Man-Eating Myth: Anthropology and Anthropophagy.* New York: Oxford University Press.

Arima, Eugene, and John DeWhirst

1990 Nootkans of Vancouver Island. In: *Handbook of North American Indians,* Vol. 7, Northwest Coast, Wayne Suttles, ed., pp. 391–411. Washington, DC: Smithsonian Institution.

Ascher, William

1999 *Why Governments Waste Natural Resources.* Baltimore: Johns Hopkins University Press.

Ashmore, Wendy

In Press Social Archaeologies of Landscape. In: *Blackwell Companion to Social Archaeology,* Lynn Meskell and Robert Preucel, eds. Oxford: Blackwell Publishers.

Askenasy, Hans

1994 *Cannibalism: from Sacrifice to Survival.* Amherst, NY: Prometheus Books.

Atran, Scott

1993 Itza Maya Tropical Agro-Forestry. *Current Anthropology* 34(5):633–700.

Avery, Dennis

1985 U.S. Farm Dilemma: The Global Bad News Is Wrong. *Science* 230:408–412.

Bailey, Robert C.

1991 *The Behavioral Ecology of Efe Pygmy Men in the Ituri Forest, Zaire.* University of Michigan, Museum of Anthropology, Anthropological Papers No. 86.

Bailey, Robert C., Genevieve Head, Mark Jenike, Bruce Owen, Robert Rechtman, and Elzbieta Zechenter

1989 Hunters and Gatherers in Tropical Rain Forests: Is It Possible? *American Anthropologist* 91(1):59–82.

Balée, William

1994 *Footprints of the Forest.* New York: Columbia University Press.

1998a (Ed.) *Advances in Historical Ecology.* New York: Columbia University Press.

1998b Historical Ecology: Premises and Postulates. In: *Advances in Historical Ecology,* William Balée, ed., pp. 13–29. New York: Columbia University Press.

1998c Introduction. In: *Advances in Historical Ecology,* William Balée, ed., pp. 1–10. New York: Columbia University Press.

Bar-Yosef, Ofer, and Richard H. Meadow

1995 The Origins of Agriculture in the Near East. In: *Last Hunters First Farmers: New Perspectives on the Prehistoric Transition to Agriculture,* T. Douglas Price and Anne Birgitte Gebauer, eds., pp. 39–94. Santa Fe: School of American Research Press.

Barkow, Jerome, Leda Cosmides, and John Tooby

1992 *The Adapted Mind.* New York: Oxford University Press.

Barth, Fredrik

1956 Ecological Relationships of Ethnic Groups in Swat, Northern Pakistan. *American Anthropologist* 58(6):1079–1099.

1964 *Nomads of South Persia: The Baseri Tribe of the Khamseh Confederacy.* New York: Humanities Press.

Bates, Daniel G.

1971 The Role of the State in Peasant–Nomad Mutualism. *Anthropological Quarterly* 44:109–131.

1998 *Human Adaptive Strategies: Ecology, Culture, and Politics.* Boston: Allyn and Bacon.

Bates, Daniel G., and Susan H. Lees

1996 Pastoralism. In: *Case Studies in Human Ecology,* Daniel G. Bates and Susan H. Lees, eds., pp. 153–157. New York: Plenum Press.

Baumhoff, M. A.

1981 The Carrying Capacity of Hunter-Gatherers. In: *Affluent Foragers: Pacific Coasts East and West,* Shuzo Koyama and David Hurst Thomas, eds., pp. 77–87. Osaka: Senri Ethnological Studies No. 9.

Beach, Hugh
1988 *The Saami of Lapland.* London: The Minority Rights Group, Report No. 55.

Begler, Elsie B.
1978 Sex, Status, and Authority in Egalitarian Society. *American Anthropologist* 80(3):571–588.

Belovsky, Gary E.
1987 Hunter-Gatherer Foraging: A Linear Programming Approach. *Journal of Anthropological Archaeology* 6(1):29–76.
1988 An Optimal Foraging-Based Model of Hunter-Gatherer Population Dynamics. *Journal of Anthropological Archaeology* 7(4):329–372.

Benedek, Emily
1992 *The Wind Won't Know Me: A History of the Navajo-Hopi Land Dispute.* New York: Alfred A. Knopf.

Benedict, Ruth
1934 *Patterns of Culture.* Boston: Houghton Mifflin.

Bennett, John W.
1976 *The Ecological Transition: Cultural Anthropology and Human Adaptation.* New York: Academic Press.
1992 *Human Ecology as Human Behavior.* New Brunswick, NJ: Transaction.

Berkes, Fikret
1999 *Sacred Ecology: Traditional Ecological Knowledge and Resource Management.* Philadelphia: Taylor and Francis.

Berlin, Brent, Dennis Breedlove, and Peter H. Raven
1974 *Principles of Tzeltal Plant Classification.* New York: Academic Press.

Bettinger, Robert L.
1980 Explanatory/Predictive Models of Hunter-Gatherer Adaptation. In: *Advances in Archaeological Method and Theory,* Vol. 3, Michael B. Schiffer, ed., pp. 189–255. New York: Academic Press.
1987 Archaeological Approaches to Hunter-Gatherers. *Annual Review of Anthropology* 16:121–142.
1991 *Hunter-Gatherers: Archaeological and Evolutionary Theory.* New York: Plenum Press.

Bettinger, Robert L., and Martin A. Baumhoff
1982 The Numic Spread: Great Basin Cultures in Competition. *American Antiquity* 47(3):485–503.

Binford, Lewis R.
1968 Post-Pleistocene Adaptations. In: *New Perspectives in Archaeology,* Sally Binford and Lewis R. Binford, eds., pp. 313–341. Chicago: Aldine.
1980 Willow Smoke and Dogs' Tails: Hunter-Gatherer Settlement Systems and Archaeological Site Formation. *American Antiquity* 45(1):4–20.

1990 Mobility, Housing, and Environment: A Comparative Study. *Journal of Anthropological Research* 46(2):119–152.

2001 *Constructing Frames of Reference: An Analytical Method for Archaeological Theory Building Using Ethnographic and Environmental Data Sets.* Berkeley: University of California Press.

Bird-David, Nurit

1990 The Giving Environment: Another Perspective on the Economic System of Gatherer-Hunters. *Current Anthropology* 31(2):189–196.

Birdsell, Joseph B.

1953 Some Environmental and Cultural Factors Influencing the Structuring of Australian Aboriginal Populations. *American Naturalist* 87(2):171–207.

Blackburn, Thomas C., and Kat Anderson (eds.)

1993 *Before the Wilderness: Environmental Management by Native Californians.* Ballena Press Anthropological Paper No. 40.

Blaustein, Andrew R., and David B. Wake

1995 The Puzzle of Declining Amphibian Populations. *Scientific American*, April, pp. 52–57.

Bliege Bird, Rebecca L., and Douglas W. Bird

1997 Delayed Reciprocity and Tolerated Theft: The Behavioral Ecology of Food-Sharing Strategies. *Current Anthropology* 38(1):49–78.

Blumenschine, Robert J., John A. Cavallo, and Salvatore D. Capaldo

1994 Competition for Carcasses and Early Hominid Behavioral Ecology: A Case Study and Conceptual Framework. *Journal of Human Evolution* 27(1,2,3):197–213.

Blumler, Mark A.

1992 Independent Inventionism and Recent Genetic Evidence on Plant Domestication. *Economic Botany* 46(1):98–111.

Boas, Franz

1927 *Primitive Art.* Cambridge: Harvard University Press.

1940 *Race, Language and Culture.* New York: MacMillan.

Bodenheimer, Friedrich S.

1951 *Insects as Human Food: A Chapter of the Ecology of Man.* The Hague: W. Junk.

Bodley, John H.

1999 *Victims of Progress* (4th ed.). Mountain View, CA: Mayfield Publishing.

2001 *Anthropology and Contemporary Human Problems.* Mountain View, CA: Mayfield.

Bol, Marsha C.

1998 *Stars Above, Earth Below: American Indians and Nature.* Niwot, CO: Roberts Reinhart Publishers.

Boserup, Ester

1965 *The Conditions of Agricultural Growth: The Economics of Agrarian Change Under Population Pressure.* Chicago: Aldine.

Boyd, Robert (ed.)

1999 *Indians, Fire, and the Land in the Pacific Northwest.* Corvallis: Oregon State University Press.

Bradfield, Richard Maitland

1995 *An Interpretation of Hopi Culture.* Derby, England: Privately printed.

Braidwood, Robert

1960 The Agricultural Revolution. *Scientific American* 203:130–141.

Bray, Francesca

1984 *Science and Civilization in China,* Vol. 6, Part 2: Agriculture. Cambridge: Cambridge University Press.

Bray, Warwick

2000 Ancient Food for Thought. *Nature* 408(6809):145–146.

Breedlove, Dennis, and Robert Laughlin

1992 *The Flowering of Man.* Washington, DC: Smithsonian Institution.

Brookfield, Harold C., and Paula Brown

1963 *Struggle for Land: Agriculture and Group Territories among the Chimbu of the New Guinea Highlands.* Melbourne: Oxford University Press.

Brown, Lester R.

1981 World Population Growth, Soil Erosion, and Food Security. *Science* 214:995–1002.

1994 Facing Food Insecurity. In: *State of the World 1994,* Lester R. Brown, ed., pp. 177–197. New York: Norton.

1996 *Tough Choices: Facing the Challenge of Food Scarcity.* New York: W. W. Norton & Company.

Brück, Joanna, and Melissa Goodman

1999 Introduction: Themes for a Critical Archaeology of Prehistoric Settlement. In: *Making Places in the Prehistoric World: Themes in Settlement Archaeology,* Joanna Brück and Melissa Goodman, eds., pp. 1–19. London: UCL Press.

Brugge, David M.

1994 *The Navajo-Hopi Land Dispute: An American Tragedy.* Albuquerque: University of New Mexico Press.

Brush, Stephen B.

1975 The Concept of Carrying Capacity for Systems of Shifting Cultivation. *American Anthropologist* 77(4):799–811.

Brush, Stephen B., and D. Stabinsky (eds.)

1996 *Intellectual Property Rights and Indigenous Knowledge.* Washington: Island Press.

Bryant, Carol, Anita Courtenay, Barbara Markesbery, and Kathleen DeWalt
1985 *The Cultural Feast: An Introduction to Food and Society.* St. Paul, MN: West Publishing Company.

Buck, J. L.
1937 *Land Utilization in China.* Chicago: University of Chicago Press.

Bullard, Robert
1990 *Dumping in Dixie.* Boulder, CO: Westview Press.

Burch, Ernest S., Jr., and Linda J. Ellanna (eds.)
1994a *Key Issues in Hunter-Gatherer Research.* Oxford: Berg Publishers.

Burch, Ernest S., Jr., and Linda J. Ellanna
1994b Introduction. In: *Key Issues in Hunter-Gatherer Research,* Ernest S. Burch, Jr., and Linda J. Ellanna, pp. 1–8. Oxford: Berg Publishers.

Burger, Kenneth J. B.
1993 *An Ecological Perspective of Feng Shui Landscape.* Taipei: Pacific Cultural Foundation.

Burton, Michael, and Douglas R. White
1984 Sexual Division of Labor in Agriculture. *American Anthropologist* 86(3):568–583.

Burton, Sharon
1992 Under What Circumstances Will a Commons Resource Management Group Arise? Paper presented at the annual meetings of the Southwestern Anthropological Association, Berkeley, CA.

Butzer, Karl W.
1976 *Early Hydraulic Civilization in Egypt: A Study in Cultural Ecology.* Chicago: University of Chicago Press.
1982 *Archaeology as Human Ecology.* Cambridge: Cambridge University Press.

Callicott, J. Baird, and Roger T. Ames (eds.)
1989 *Nature in Asian Traditions of Thought: Essays in Environmental Philosophy.* Albany: State University of New York Press.

Campbell, Bernard
1995 *Human Ecology* (2nd ed.). New York: Aldine.

Campbell, Kenneth L., and James W. Wood (eds.)
1994 *Human Reproductive Ecology: Interactions of Environment, Fertility, and Behavior.* Annals of the New York Academy of Sciences Vol. 709.

Candolle, Alphone de
1959 *Origin of Cultivated Plants* (orig. 1886). New York: Hafner.

Cantor, Norman F.
2001 *In the Wake of the Plague: The Black Death and the World It Made.* New York: The Free Press.

Caplan, Gerald L.
1970 *The Elites of Barotseland, 1878–1969: A Political History of Zambia's Western Province.* Berkeley: University of California Press.

Carkhuff, Robert R., and Bernard G. Berenson.
1977 *Beyond Counseling and Therapy.* New York: Holt, Rinehart and Winston.

Carlson, John B.
1993 Rise and Fall of the City of the Gods. *Discover* 46(6):58–69.

Carneiro, Robert L.
1960 Slash-and-Burn Agriculture: A Closer Look at Its Implications for Settlement Patterns. In: *Men and Cultures: Selected Papers of the Fifth International Congress of Anthropological and Ethnological Sciences,* Anthony F. C. Wallace, ed., pp. 229–234. Philadelphia: University of Pennsylvania Press.
1970 A Theory of the Origin of the State. *Science* 169:733–738.

Carter, W. E.
1969 *New Land and Old Traditions: Kekchi Cultivators in the Guatemalan Lowlands.* Gainesville: University of Florida, Latin American Monographs.

Cartmill, Matt
1972 Arboreal Adaptations and the Origin of the Order Primates. In: *The Functional and Evolutionary Biology of Primates,* Russell Tuttle, ed., pp. 97–122. Chicago: Aldine.

Cashdan, Elizabeth (ed.)
1990 *Risk and Uncertainty in Tribal and Peasant Economies.* Boulder, CO: Westview Press.

Chang, Kwang-chih
1986 *The Archaeology of Ancient China* (4th ed.). New Haven: Yale University Press.

Chase-Dunn, Christopher, and Kelly Mann
1998 *The Wintu and Their Neighbors: A Very Small World-System in Northern California.* Tucson: University of Arizona Press.

Chao, Kang
1986 *Man and Land in Chinese History.* Stanford: Stanford University Press.

Chayanov, A. V.
1966 *The Theory of Peasant Economy.* Homewood, IL: Richard D. Irwin, Inc.

Chen, Junshi, T. Colin Campbell, Li Junyao, and Richard Peto
1990 *Diet, Life-Style, and Mortality in China.* Ithaca, NY: Cornell University Press.

Cheu, S. P.
1952 Changes in Fat and Protein Content of the African Migratory Locust, *Locusta migratoria migratorioides* (R. and F.). *Bulletin of Entomological Research* 43:101–109.

Chew, Sing C.
2001 *World Ecological Degradation: Accumulation, Urbanization, and Deforestation 3000 B.C.–A.D. 2000.* Walnut Creek, CA: AltaMira Press.

Childe, V. Gordon
1936 *Man Makes Himself.* London: Watts.
1942 *What Happened in History.* Harmondsworth, Sussex: Pelican Books.

Chinese-American Joint Commission on Rural Reconstruction
1966 *Taiwan Agricultural Statistics 1901–1965.* Taipei.

Choe Chung-ho
1994 Korea's Landscape and Mindscape. *Koreana* 8(4):4–9.

Cipolla, Carlo
1967 *An Economic History of World Population.* Harmondsworth, Sussex: Penguin.

Clarke, Charlotte
1977 *Edible and Useful Plants of California.* Berkeley: University of California Press.

Cleland, Charles E.
1966 *The Prehistoric Animal Ecology and Ethnozoology of the Upper Great Lakes Region.* University of Michigan, Museum of Anthropology, Anthropological Papers No. 29.
1976 The Focal-Diffuse Model: An Evolutionary Perspective on the Prehistoric Cultural Adaptations of the Eastern United States. *Mid-Continental Journal of Archaeology* 1(1):59–76.

Clements, Frederic, and Victor E. Shelford
1939 *Bio-Ecology.* New York: Wiley and Sons.

Clemmer, Richard O.
1991 Crying for the Children of Sacred Ground: A Review Article on the Hopi-Navajo Land Dispute. *American Indian Quarterly* 15(2):225–230.

Clifford, James, and George Marcus (eds.)
1986 *Writing Culture.* Berkeley: University of California Press.

Cohen, Joel E.
1995 Population Growth and Earth's Human Carrying Capacity. *Science* 269:341–346.

Cohen, Mark N.
1977 *The Food Crisis in Prehistory: Overpopulation and the Origins of Agriculture.* New Haven: Yale University Press.

Cohen, Mark N., and George Armelagos (eds.)
1984 Paleopathology at the Origins of Agriculture, Editors Summary. In: *Paleopathology at the Origins of Agriculture,* Mark N. Cohen and George Armelagos, eds., pp. 585–602. Orlando: Academic Press.

Cohen, Ronald
1984 Warfare and State Formation: Wars Make States and States Make War. In: *Warfare, Culture, and Environment*, R. Brian Ferguson, ed., pp. 329–358. New York: Academic Press.

Cohen, Yehudi A.
1974 Culture as Adaptation. In: *Man in Adaptation: The Cultural Present*, Yehudi A. Cohen, ed., pp. 45–68. New York: Aldine.

Committee on Sustainable Agriculture and the Environment in the Humid Tropics, Board on Agriculture and Board on Science and Technology for International Development, National Research Council
1993 *Sustainable Agriculture and the Environment in the Humid Tropics.* Washington, DC: National Academy Press.

Conconi, Julieta R. E. de, Jose M. P. Moreno, Carlos M. Mayadon, Fernando R. Valdez, Manuel A. Perez, Esteban E. Prado, and Hector B. Rodriguez
1984 Protein Content of Some Edible Insects in Mexico. *Journal of Ethnobiology* 4(1):61–72.

Condon, Richard G., Julia Ogina, and Holman Elders
1996 *The Northern Copper Inuit.* Norman: University of Oklahoma Press.

Conference on Common Property Resource Management
1986 *Proceedings.* Prepared by Panel on Common Property Resource Management, Board on Science and Technology for International Development, Office of International Affairs, National Research Council, Washington, D.C.

Conklin, Harold C.
1957 *Hanunóo Agriculture: A Report on an Integrated System of Shifting Cultivation in the Philippines.* Rome: Food and Agricultural Organization of the United Nations.
1959 *Population-Land Balance Under Systems of Tropical Forest Agriculture.* Proceedings of the Ninth Pacific Science Congress of the Pacific Science Association, 1957, 7:63.

Connolly, Bob, and Robin Anderson
1987 *First Contact: New Guinea's Highlanders Encounter the Outside World.* New York: Penguin Books.

Cooke, Fadzilah Majid
1999 *The Challenge of Sustainable Forests: Forest Resource Policy in Malaysia, 1970–1995.* Honolulu: University of Hawaii Press.

Coon, Carleton
1951 *Caravan.* New York: Henry Holt.

Cornelius, C., G. Dandrifosse, and C. Jeuniaux
1976 Chitinolytic Enzymes of the Gastric Mucosa of *Perodictius potto* (Primate Prosimian): Purification and Enzyme Specificity. *International Journal of Biochemistry* 7:445–448.

Cowan, C. Wesley, and Patty Jo Watson (eds.)
1992 *The Origins of Agriculture: An International Perspective.* Washington, DC: Smithsonian Institution.

Cowgill, George
1975 On Causes and Consequences of Ancient and Modern Population Changes. *American Anthropologist* 77(3):505–524.

Cox, Paul, and Sandra Banack (eds.)
1991 *Islands, Plants, and Polynesians.* Portland: Dioscorides Press.

Cribb, Roger
1991 *Nomads in Archaeology.* Cambridge: Cambridge University Press.

Croll, Elisabeth, and David Parkin (eds.)
1992 *Bush Base, Forest Farm: Culture, Environment and Development.* London: Routledge.

Cronk, Lee
1991 Human Behavioral Ecology. *Annual Review of Anthropology* 20: 25–53.
1999 *That Complex Whole: Culture and the Evolution of Human Behavior.* Boulder, CO: Westview Press.

Cronk, Lee, Napoleon Chagnon, and William Irons (eds.)
2000 *Adaptation and Human Behavior: An Anthropological Perspective.* New York: Aldine.

Cronon, William
1983 *Changes in the Land.* New York: Hill and Wang.

Crosby, Alfred W.
1989 *America's Forgotten Pandemic: The Influenza of 1918.* Cambridge: Cambridge University Press.

Crown, Patricia L.
1990 The Hohokam of the American Southwest. *Journal of World Prehistory* 4(2):223–255.

Crumley, Carole L. (ed.)
1994 *Historical Ecology: Cultural Knowledge and Changing Landscapes.* Santa Fe: School of American Research Press.

Crumley, Carole L.
1998 Foreword. In: *Advances in Historical Ecology*, William Balée, ed. pp. ix–xiv. New York: Columbia University Press.

Crumley, Carole L., A. Elizabeth van Deventer, and Joseph J. Fletcher (eds.)
2001 *New Directions in Anthropology and Environment: Intersections.* Walnut Creek, CA: AltaMira Press.

Cushing, Frank
1920 *Zuni Breadstuffs.* New York: Museum of the American Indian (Heye Foundation), Indian Notes and Monographs, 7:7–642.

Dahlberg, Frances (ed.)
1981 *Woman the Gatherer.* New Haven: Yale University Press.

Daly, Herman, and Kenneth W. Townsend (eds.)
1993 *Valuing the Earth: Economics, Ecology, Ethics.* Cambridge: Massachusetts Institute of Technology.

Darwin, Charles
1859 *On the Origin of Species by Means of Natural Selection, or, The Preservation of Favoured Races in the Struggle for Life.* London: J. Murray.

Davis, Shelton H.
1977 *Victims of the Miracle: Development and the Indians of Brazil.* Cambridge: Cambridge University Press.

Dazhong, Wen, and David Pimentel
1984 Energy Inputs in Agricultural Systems of China. *Agriculture, Ecosystems, and Environment* 11:29–35.
1986a Seventeenth-Century Organic Agriculture in China: I: Cropping Systems in Jiaxing Region. *Human Ecology* 14(1):1–14.
1986b Seventeenth-Century Organic Agriculture in China: II: Energy Flows through and Agroecosystem in Jiaxing Region. *Human Ecology* 14(1):15–18.

DeFoliart, G. R., M. D. Finke, and M. L. Sunde
1982 Potential Value of the Mormon Cricket (Orthoptera:Tettigoniidae) Harvested as a High-Protein Feed for Poultry. *Journal of Economic Entomology* 75(5):848–852.

Denevan, William
1992 *The Pristine Myth: The Landscape of the Americas in 1492.* Annals of the Association of American Geographers 82:369–385.
2001 *Cultivated Landscapes of Native Amazonia and the Andes.* New York: Oxford University Press.

DeVore, Irven, and S. L. Washburn
1963 Baboon Ecology and Human Evolution. In: *African Ecology and Human Evolution,* F. Clark Howell and Francois Bourliére, eds., pp. 335–367. Chicago: Aldine.

DeWalt, Billie R.
1994 Using Indigenous Knowledge to Improve Agriculture and Natural Resource Management. *Human Organization* 53(2):123–131.

Dewar, Robert E.
1984 Environmental Productivity, Population Regulation, and Carrying Capacity. *American Anthropologist* 86(3):601–614.

Diamond, Jared
1995 Easter's End. *Discover* 16(8):62–69.

Diener, Paul, and Eugene Robkin

1978 Ecology, Evolution, and the Search for Cultural Origins: The Question of Islamic Pig Prohibition. *Current Anthropology* 19(3):493–540.

Dincauze, Dena F.

2000 *Environmental Archaeology: Principles and Practice.* Cambridge: Cambridge University Press.

Dissanayake, Ellen

1992 *Homo Aestheticus.* New York: Free Press.

Divale, William Tullio, and Marvin Harris

1976 Population, Warfare, and the Male Supremacy Complex. *American Anthropologist* 78(3):521–538.

Dobzhansky, Theodosius G.

1972 On the Evolutionary Uniqueness of Man. *Evolutionary Biology* 6:415–430.

Doolittle, William F.

2000 *Cultivated Landscapes of Native North America.* Oxford: Oxford University Press.

Douglas, Mary

1966 *Purity and Danger.* New York: Praeger.

1970 *Natural Symbols.* London: Barrie and Rockliff.

1975 *Implicit Meanings.* London: Routledge and Kegan Paul.

Dove, Michael R.

1981 Symbiotic Relationships between Human Populations and *Imperata cylindrica*: The Question of Ecosystematic Succession and Preservation in South Kalimantan. In: *Conservation Inputs from Life Sciences,* M. Nordin, ed., pp. 187–200. Kuala Lumpur: Universiti Kebangsaan Malaysia.

1993a A Revisionist View of Tropical Deforestation and Development. *Environmental Conservation* 20:17–25.

1993b Smallholder Rubber and Swidden Agriculture in Borneo: A Sustainable Adaptation to the Ecology and Economy of the Tropical Forest. *Economic Botany* 47(2):136–147.

Downing, Theodore E., Susanna B. Hecht, Henry A. Pearson, and Carmen Garcia-Downing

1992 *Development or Destruction: The Conversion of Tropical Forest to Pasture in Latin America.* Boulder, CO: Westview Press.

Downs, James F.

1972 *The Navajo.* New York: Holt, Rinehart and Winston (reprinted in 1984 by Waveland Press).

Drucker, Peter

1992 *The Ecological Vision.* New Brunswick, NJ: Transaction.

Drucker, Philip

1951 The Northern and Central Nootkan Tribes. *Bureau of American Ethnology Bulletin* 144.

Drucker, Philip, and Robert F. Heizer

1967 *To Make My Name Good: A Reexamination of the Southern Kawakuitl Potlatch.* Berkeley: University of California Press.

Dryzek, John

1988 *Rational Ecology.* Oxford: Basil Blackwell.

Duffy, Kevin

1996 *Children of the Forest: Africa's Mbuti Pygmies.* Prospect Heights, IL: Waveland Press.

Dufour, Darna L.

1987 Insects as Food: A Case Study from the Northwest Amazon. *American Anthropologist* 89(2):383–397.

Duhring, John L.

1984 Nutrition in Pregnancy. In: *Nutrition Reviews' Present Knowledge in Nutrition*, pp. 636–645. Washington, DC: Nutrition Foundation, Inc.

Duke, J., and E. Ayensu

1985 *Medicinal Plants of China.* New York: Reference Press.

Dunbar, Robin

1996 *Grooming, Gossip, and the Evolution of Language.* Cambridge: Harvard University Press.

Dunbar, Robin, Chris Knight, and Camilla Power (eds.)

1999 *The Evolution of Culture: An Interdisciplinary View.* New Brunswick, NJ: Rutgers University Press.

Durham, William H.

1981 Overview: Optimal Foraging Analysis in Human Ecology. In: *Hunter-Gatherer Foraging Strategies: Ethnographic and Archaeological Analyses*, Bruce Winterhalder and Eric Alden Smith, eds., pp. 218–231. Chicago: University of Chicago Press.

Durrenberger, Paul (ed.)

1984 *Chayanov, Peasants and Economic Anthropology.* New York: Academic Press.

Dwyer, Peter D.

1990 *The Pigs That Ate the Garden: A Human Ecology from Papua New Guinea.* Ann Arbor: University of Michigan Press.

Dyer, Christopher, and J. McGoodwin (eds.)

1994 *Folk Management in the World's Fisheries.* Boulder: University of Colorado Press.

Dyson-Hudson, Neville

1972 The Study of Nomads. In: *Perspectives on Nomadism*, William Irons and Neville Dyson-Hudson, eds., pp. 2–29. International Studies in Sociology and Social Anthropology, Vol. 13.

Dyson-Hudson, Rada, and Neville Dyson-Hudson

1980 Nomadic Pastoralism. *Annual Review of Anthropology* 9:15–61.

Earle, Timothy, and Andrew L. Christenson (eds.)

1980 *Modeling Change in Prehistoric Subsistence Economies.* New York: Academic Press.

Eaton, S. Boyd, and Stanley B. Eaton

1999 Hunter-Gatherers and Human Health. In: *The Cambridge Encyclopedia of Hunters and Gatherers*, Richard B. Lee and Richard Daly, eds., pp. 449–456. Cambridge: Cambridge University Press.

Ebeling, Walter

1985 *Handbook of Indian Foods and Fibers of Arid America.* Berkeley: University of California Press.

Edwards, Richard

1979 *Contested Terrain.* New York: Basic Books.

Ehrlich, Paul R.

2000 *Human Natures: Genes, Cultures and the Human Prospect.* Washington, DC: Island Press.

Ehrlich, Paul R., and Anne H. Ehrlich

1981 *Extinction: The Causes and Consequences of the Disappearance of Species.* New York: Random House.

1996 *Betrayal of Science and Reason: How Anti-Environmental Rhetoric Threatens Our Future.* Washington, DC: Island Press.

Ehrlich, Paul R., and Marcus Feldman

2003 Genes and Cultures: What Creates Our Behavioral Phenome? *Current Anthropology* 44(1):87–107.

Eisenberg, Nancy

1986 *Altruistic Emotion, Cognition, and Behavior.* Hillsdale, NJ: Lawrence Erlbaum Associates.

Ekvall, Robert B.

1968 *Fields on the Hoof: Nexus of Tibetan Nomadic Pastoralism.* New York: Holt, Reinhart and Winston.

Ellen, Roy R.

1982 *Environment, Subsistence and System.* Cambridge: Cambridge University Press.

1986 Ethnobiology, Cognition and the Structure of Prehension: Some General Theoretical Notes. *Journal of Ethnobiology* 6(1):83–98.

Elster, Jon

1987 *Rational Choice.* Cambridge: Cambridge University Press.

Elvin, Mark

1973 *The Pattern of the Chinese Past.* Stanford: Stanford University Press.

Ember, Carol

1978 Myths about Hunter-Gatherers. *Ethnology* 17(4):439–448.

Engels, Friedrich

1942 *The Origin of the Family, Private Property, and the State.* New York: International Publishers.

1966 *Anti-Duhring: Herr Eugen Duhring's Revolution in Science.* New York: International Publishers (original English edition 1894).

Erichsen-Brown, Charlotte

1979 *Uses of Plants for the Past 500 Years.* Aurora, Ontario, Canada: Breezy Creeks Press.

Erickson, Clark

1986 *Cómo Construir Waru Waru?* Puno, Peru: Proyecto Agrícola de los Campos Elevados.

Etkin, Nina

1988 Ethnopharmacology: Biobehavioral Approaches in the Anthropological Study of Indigenous Medicines. *Annual Review of Anthropology* 17:23–42.

Etkin, Nina L. (ed.)

1994 *Eating on the Wild Side: The Pharmacologic, Ecologic, and Social Implications of Using Noncultigens.* Tucson: University of Arizona Press.

Evangelou, Phylo

1984 *Livestock Development in Kenya's Maasailand: Pastoralists' Transition to a Market Economy.* Boulder, CO: Westview Press.

Evans-Pritchard, E. E.

1940 *The Nuer.* Oxford: Oxford University Press.

Evers, Larry, and Felipe Molina

1987 *Yaqui Deer Songs/Maso Bwikam.* Tucson: University of Arizona Press.

Fagan, Brian

1999 *Floods, Famines, and Emperors: El Nino and the Fate of Civilizations.* New York: Basic Books.

Fairhead, James, and Melissa Leach

1996 *Misreading the African Landscape.* Cambridge: Cambridge University Press.

Farberow, Norman (ed.)

1980 *The Many Faces of Suicide.* New York: McGraw-Hill.

1975 *Suicide in Different Cultures.* Baltimore: University Park Press.

Farnsworth, Norman R., and Djaja Doel Soejarto
1985 Potential Consequence of Plant Extinction in the United States on the Current and Future Availability of Prescription Drugs. *Economic Botany* 39(3):231–240.

Farris, Glenn J.
1982 Pine Nuts as an Aboriginal Food Source in California and Nevada: Some Contrasts. *Journal of Ethnobiology* 2(2):114–122.

Faust, Betty B.
1998 *Mexican Rural Development and the Plumed Serpent.* Westport, CT: Bergin and Garvey.

Fedick, Scott L.
1995 Indigenous Agriculture in the Americas. *Journal of Archaeological Research* 3(4):257–303.

Fedick, Scott (ed.)
1996 *The View from Yalahau.* University of California, UC Mexus.

Fedigan, Linda Marie
1986 The Changing Role of Women in Models of Human Evolution. *Annual Review of Anthropology* 15:25–66.

Feher-Elston, Catherine
1988 *Children of the Sacred Ground.* Flagstaff, AZ: Northland Press.

Feinman, Gary M., and Joyce Marcus (eds.)
1998 *Archaic States.* Santa Fe: School of American Research.

Feld, Steven, and Keith Basso (eds.)
1996 *Senses of Place.* Santa Fe: School of American Research.

Ferguson, R. Brian
1984 A Reexamination of the Causes of Northwest Coast Warfare. In: *Warfare, Culture, and Environment,* R. Brian Ferguson, ed., pp. 267–328. New York: Academic Press.

Feshbach, Murray, and Alfred Friendly, Jr.
1992 *Ecocide in the USSR.* New York: Basic Books.

Field, Barry C.
1994 *Environmental Economics: An Introduction.* New York: McGraw-Hill.

Finerman, Ruthbeth
1983 Experience and Expectation: Conflict and Change in Traditional Family Health Care among the Quichua of Saraguro. *Social Science and Medicine* 17(17):1291–1298.
1984 A Matter of Life and Death: Health Care Change in an Andean Community. *Social Science and Medicine* 18(4):329–334.

Firth, Raymond
1936 *We the Tikopia.* London: George Allen and Unwin.

1959 *Social Change in Tikopia.* London: George Allen and Unwin.

1965 *Economics of the New Zealand Maori.* London: Routledge, Kegan Paul.

1966 *Malay Fishermen: Their Peasant Economy.* New York: Norton.

Firth, Rosemary

1966 *Housekeeping among Malay Peasants.* London: Athlone Press.

Fischhoff, Baruch, Sara L. Lichtenstein, Paul Slovic, Stephen L. Derby, and Ralph L. Keeney

1981 *Acceptable Risk.* Cambridge: Cambridge University Press.

Fiske, Susan T., and Shelley E. Taylor

1991 *Social Cognition* (2nd ed.). New York: McGraw-Hill.

Flachsenberg, Henning, and Hugo A. Galletti

1998 Forest Management in Quintana Roo, Mexico. In: *Timber, Tourists, and Temples: Conservation and Development in the Maya Forest of Belize, Guatemala, and Mexico,* Richard Primack, David Bray, Hugo A. Galletti, and Ismael Ponciano, eds., pp. 47–60. Washington, DC: Island Press.

Flannery, Kent V.

1969 Origins and Ecological Effects of Domestication in Iran and the Near East. In: *The Domestication and Exploitation of Plants and Animals,* Peter Ucko and George Dimbleby, eds., pp. 73–100. London: Duckworth.

1972 The Cultural Evolution of Civilizations. *Annual Review of Ecology and Systematics* 3:399–426.

Flannery, Kent V. (ed.)

1982 *Maya Subsistence.* New York: Academic Press.

Flannery, Kent V., Joyce Marcus, and Robert G. Reynolds

1989 *The Flocks of the Wamani: A Study of Llama Herders on the Punas of Ayacucho, Peru.* San Diego: Academic Press.

Flannery, Timothy F.

1995 *The Future Eaters : An Ecological History of the Australasian Lands and People.* New York: George Braziller.

Fletcher, Alice, and Francis La Flesche

1911 The Omaha Tribe. In: *Annual Report of the Bureau of American Ethnology for the Years 1905–1906,* pp. 17–672. Washington, DC: Government Printing Office.

Flood, Josephine

1980 *The Moth Hunters: Aboriginal Prehistory of the Australian Alps.* Australian Institute of Aboriginal Studies New Series No. 14.

Fog, Agner

1999 *Cultural Selection.* Dordrecht: Kluwer Academic Publishers.

Foin, Theodore C., and William G. Davis

1987 Equilibrium and Nonequilibrium Models in Ecological Anthropology: An Evaluation of "Stability" in Maring Ecosystems in New Guinea. *American Anthropologist* 89(1):9–30.

Ford, Jesse, and Dennis Martinez (eds.)

2000 Traditional Ecological Knowledge, Ecosystem Science, and Resource Management. Special Section, *Ecological Applications* 10(5): 1249–1340.

Ford, Richard I. (ed.)

1994 *The Nature and Status of Ethnobotany* (2nd ed.). University of Michigan, Museum of Anthropology, Anthropological Papers No. 67.

2001 *Ethnobiology at the Millennium: Past Promise and Future Prospects.* University of Michigan, Museum of Anthropology, Anthropological Papers No. 91.

Forde, C. Daryll

1931 Hopi Agriculture and Land Ownership. *Journal of the Royal Anthropological Institute of Great Britain and Ireland* 61:357–405.

Forde, C. Daryll (ed.)

1934 *Habitat, Economy, and Society: A Geographical Introduction to Ethnology.* London: Methuen & Co.

Forman, Richard T. T., and Michel Godron

1986 *Landscape Ecology.* New York: John Wiley & Sons.

Foucault, Michel

1970 *The Order of Things.* New York: Random House.

Fox, James J.

1977 *Harvest of the Palm: Ecological Change in Eastern Indonesia*. Cambridge: Harvard University Press.

Frake, Charles

1962 Cultural Ecology and Ethnography. *American Anthropologist* 64(1):53–59.

Frank, Robert

1988 *Passions Within Reason.* Cambridge: Harvard University Press.

Fratkin, Elliot

1998 *Ariaal Pastoralists of Kenya: Surviving Drought and Development in Africa's Arid Lands.* Boston: Allyn and Bacon.

Froment, Alain

2001 Evolutionary Biology and Health of Hunter-Gatherer Populations. In: *Hunter-Gatherers: An Interdisciplinary Perspective*, Catherine Panter-Brick, Robert H. Layton, and Peter Rowley-Conwy, eds., pp. 239–266. Cambridge: Cambridge University Press.

Galletti, Hugo A.

1998 The Maya Forest of Quintana Roo: Thirteen Years of Conservation and Community Development. In: *Timber, Tourists, and Temples: Conservation and Development in the Maya Forest of Belize, Guatemala, and Mexico*, Richard Primack, David Bray, Hugo A. Galletti, and Ismael Ponciano, eds., pp. 33–46. Washington, DC: Island Press.

Galvin, Kathleen A., D. Layne Coppock, and Paul W. Leslie

1994 Diet, Nutrition, and the Pastoral Strategy. In: *African Pastoralist Systems: An Integrated Approach*, Elliot Fratkin, Kathleen A. Galvin, and Eric Abella Roth, eds., pp. 113–131. Boulder, CO: Lynne Rienner Publishers.

Garber, P. A.

1987 Foraging Strategies among Living Primates. *Annual Review of Anthropology* 16:339–364.

Gardner, Paul S.

1992 Diet Optimization Models and Prehistoric Subsistence Change in the Eastern Woodlands. Ph.D. dissertation, University of North Carolina, Chapel Hill.

Gardner, Robert, and Karl Heider

1969 *Gardens of War*. New York: Random House.

Geertz, Clifford

1963 *Agricultural Involution*. Berkeley: University of California Press.

1984 Anti-Anti-Relativism. *American Anthropologist* 86(2):263–278.

Ghai, Dharam, Azizur Rahman Khan, Eddy Lee, and Simir Radwan (eds.)

1979 *Agrarian Systems and Rural Development*. New York: Holmes and Meier.

Giampietro, Mario, Sandra G. F. Bukkens, and David Pimentel

1993 Labor Productivity: A Biophysical Definition and Assessment. *Human Ecology* 21(3):229–260.

Giddens, Anthony

1984 *The Constitution of Society*. Berkeley: University of California Press.

Gidwitz, Tom

2002 Pioneers of the Bajo. *Archaeology* 55(1):28–35.

Gillis, Anna Maria

1991 Should Cows Chew Cheatgrass on Commonlands? *BioScience* 41(10):668–675.

Gladwin, Christina, and Kathleen Truman (eds.)

1988 *Food and Farm*. Washington: Society for Economic Anthropology.

Glassow, Michael A.

1978 The Concept of Carrying Capacity in the Study of Culture Process. In: *Advances in Archaeological Method and Theory*, Vol. 1, Michael B. Schiffer, ed., pp. 31–48. New York: Academic Press.

Gleick, Peter
1998 *The World's Water.* Washington, DC: Island Press.
Gluckman, Max
1941 *Economy of the Central Barotse Plain.* Livingstone: The Rhodes-Livingstone Papers No. 7.
1951 The Lozi of Barotseland in North-Western Rhodesia. In: *Seven Tribes of Central Africa,* Elizabeth Colson and Max Gluckman, eds., pp. 1–93. Manchester: University of Manchester Press.
Goldschmidt, Walter
1969 *Kambuya's Cattle.* Berkeley: University of California Press.
1979 A General Model for Pastoral Social Systems. In: *Pastoral Production and Society: Proceedings of the International Meeting on Nomadic Pastoralism,* C. Lefebure, ed., pp. 15–27. Cambridge: Cambridge University Press.
Goldstein, Melvyn C., and Cynthia M. Beall
1990 *Nomads of Western Tibet: The Survival of a Way of Life.* Berkeley: University of California Press.
Golley, Frank B.
1993 *A History of the Ecosystems Concept in Ecology.* New Haven: Yale University Press.
Goodall, Jane
1986 *The Chimpanzees of Gombe.* Cambridge: Harvard University Press.
Goodenough, Ward
1953 *Native Astronomy in the Central Carolines.* Philadelphia: University Museum, University of Pennsylvania.
Goodenough, Ward (ed.)
1964 *Explorations in Cultural Anthropology.* New York: McGraw-Hill.
Goodland, R.
1975 History of "Ecology." *Science* 188:313.
Goodman, Alan H., Darna L. Dufour, and Gretel H. Pelto (eds.)
2000 *Nutritional Anthropology: Biocultural Perspectives on Food and Nutrition.* Mountain View, CA: Mayfield Publishing.
Goodstein, Eban S.
1999 *The Trade-Off Myth: Fact and Fiction about Jobs and the Environment.* Washington, DC: Island Press.
Goody, Jack
1982 *Cooking, Cuisine and Class.* Cambridge: Cambridge University Press.
Goudie, Andrew
1994 *The Human Impact on the Natural Environment* (4th ed.). Cambridge: Massachusetts Institute of Technology Press.

Gould, Richard A.
1969 *Yiwara: Foragers of the Australian Desert.* New York: Scribners.
1982 To Have and Not to Have: The Ecology of Sharing among Hunter-Gatherers. In: *Resource Managers: North American and Australian Hunter-Gatherers,* Nancy M. Williams and Eugene S. Hunn, eds., pp. 69–91. Boulder, CO: Westview Press.
Grady, Denise
1993 Death at the Corners. *Discover* 14(12):82–91.
Gragson, Ted L., and Ben G. Blount (eds.)
1999 *Ethnoecology: Knowledge, Resources, and Rights.* Athens: University of Georgia Press.
Grant, Campbell, James W. Baird, and J. Kenneth Pringle
1968 *Rock Drawings of the Coso Range, Inyo County, California.* Maturango Museum Publication No. 4.
Grayson, Donald K.
2001 The Archaeological Record of Human Impacts on Animal Populations. *Journal of World Prehistory* 15(1):1–68.
Green, Donald, and Ian Shapiro
1994 *Pathologies of Rational Choice Theory.* New Haven: Yale University Press.
Greenberg, James B., and Thomas K. Park
1994 Political Ecology. *Journal of Political Ecology* 1:1–12.
Gregg, Susan Alling
1988 *Forager and Farmers: Population Interaction and Agricultural Expansion in Prehistoric Europe.* Chicago: University of Chicago Press.
Griaule, Marcel
1965 *Conversations with Ogotemmeli.* Oxford: Oxford University Press.
Griffin, Keith
1974 *The Political Economy of Agrarian Change.* London: Macmillan.
Griffin, Keith (ed.)
1984 *Institutional Reform and Economic Development in the Chinese Countryside.* Boulder: M. E. Sharpe.
Grove, A. T., and Oliver Rackham
2001 *The Nature of Mediterranean Europe.* New Haven: Yale University Press.
Guha, R. (ed.)
1994 *Social Ecology.* Delhi: Oxford University Press.
Guillet, David
1987 Terracing and Irrigation in the Peruvian Highlands. *Current Anthropology* 28(4):409–430.
Gunda, Bela
1984 *The Fishing Culture of the World* (2 vols.). Budapest: Akademiai Kiado.

Hack, John T.
1942 *The Changing Physical Environment of the Hopi Indians of Arizona.* Papers of the Peabody Museum of American Archaeology and Ethnology 35(1).

Halperin, Rhoda and James Dow (eds.)
1977 *Peasant Livelihood: Studies in Economic Anthropology and Cultural Ecology.* New York: St. Martin's Press.

Hamilton, William J. III
1987 Omnivorous Primate Diets and Human Overconsumption of Meat. In: *Food and Evolution: Toward a Theory of Human Food Habits,* Marvin Harris and Eric B. Ross, eds., pp. 117–132. Philadelphia: Temple University Press.

Hammett, Julia E.
1997 Interregional Patterns of Land Use and Plant Management in Native North America. In: *People, Plants, and Landscapes: Studies in Paleoethnobotany,* Kristen J. Gremillion, ed., pp. 195–216. Tuscaloosa: University of Alabama Press.

Hammond, Kenneth R., John Rohrbaugh, Jeryl Mumpower, and Leonard Adelman
1977 Social Judgment Theory: Applications in Policy Formation. In: *Human Judgment and Decision Processes in Applied Settings,* Martin Kaplan and Steven Schwartz, eds., pp. 1–27. New York: Academic Press.

Handlin Smith, Joanna
1987 Benevolent Societies: The Reshaping of Charity During the Late Ming and Early Ch'ing. *Journal of Asian Studies* 46(2):309–337.

Hardesty, Donald L.
1975 The Niche Concept: Suggestions for Its Use in Studies of Human Ecology. *Human Ecology* 3(2):71–85.
1977 *Ecological Anthropology.* New York: John Wiley & Sons.
1997 *The Archaeology of the Donner Party.* Reno: University of Nevada Press.

Hardin, Garrett
1968 *The Tragedy of the Commons.* Science 162:1243–1248.
1991 The Tragedy of the *Unmanaged* Commons. In: *Commons without Tragedy,* Robert V. Andelson, ed., pp. 162–185. Savage, MD: Barnes and Noble.

Harding, Robert S. O.
1981 An Order of Omnivores: Nonhuman Primate Diets in the Wild. In: *Omnivorous Primates: Gathering and Hunting in Human Evolution,* Robert S. O. Harding and Geza Teleki, eds., pp. 191–214. New York: Columbia University Press.

Harding, Thomas G., David Kaplan, Marshall D. Sahlins, and Elman R. Service
1960 *Evolution and Culture.* Ann Arbor: University of Michigan Press.

Harlan, Jack R.
1992 *Crops and Man.* Madison, WI: American Society of Agronomy and Crop Science Society of America.
Harner, Michael
1977 The Ecological Basis for Aztec Sacrifice. *American Ethnologist* 4(1):117–135.
Harris, David, and Gordon Hillman (eds.)
1989 *Foraging and Farming: The Evolution of Plant Exploitation.* Cambridge: Cambridge University Press.
Harris, Marvin
1966 The Cultural Ecology of India's Sacred Cattle. *Current Anthropology* 7(1):51–66.
1968 *The Rise of Anthropological Theory.* New York: Crowell.
1974 *Cows, Pigs, Wars and Witches.* New York: Random House.
1977 *Cannibals and Kings.* New York: Random House.
1979 *Cultural Materialism.* New York: Random House.
1981 *America Now.* New York: Simon and Schuster.
1984 A Cultural Materialist Theory of Band and Village Warfare: The Yanomamo Test. In: *Warfare, Culture, and Environment,* R. Brian Ferguson, ed., pp. 111–140. New York: Academic Press.
1985 *Good to Eat: Riddles of Food and Culture.* New York: Simon and Schuster.
Harrison, P. D., and B. L. Turner
1978 *Pre-Hispanic Maya Agriculture.* Albuquerque: University of New Mexico Press.
Hart, Terese B., and John A. Hart
1986 The Ecological Basis of Hunter-Gatherer Subsistence in African Rain Forests: The Mbuti of Eastern Zaire. *Human Ecology* 14(1):29–55.
Hassan, Fekri A.
1980 The Growth and Regulation of Human Population in Prehistoric Times. In: *Biosocial Mechanisms of Population Regulation,* Mark N. Cohen, Roy S. Malpass, and Harold G. Klein, eds., pp. 305–319. New Haven: Yale University Press.
Hatley, Tom, and John Kappelman
1980 Bears, Pigs, and Plio-Pleistocene Hominids: A Case for the Exploitation of Belowground Resources. *Human Ecology* 8(4):371–387.
Hawkes, Kristen
1993 Why Hunter-Gatherers Work. *Current Anthropology* 34(4):341–361.
Hawkes, Kristen, and James F. O'Connell
1985 Optimal Foraging Models and the Case of the !Kung. *American Anthropologist* 87(2):401–405.

Hawkes, Kristen, Kim Hill, and James F. O'Connell
1982 Why Hunters Gather: Optimal Foraging and the Aché of Eastern Paraguay. *American Ethnologist* 9(2):379–398.

Hayami, Yujiro, and Masao Kikuchi
1981 *Asian Village Economy at the Crossroads.* Tokyo: University of Tokyo and Johns Hopkins University.

Hayami, Yujiro, and Vernon Ruttan
1985 *Agricultural Development* (2nd ed.). Baltimore: Johns Hopkins University.

Hayden, Brian
1981 Research and Development in the Stone Age: Technological Traditions among Hunter-Gatherers. *Current Anthropology* 22(5):519–548.
1995 A New Overview of Domestication. In: *Last Hunters-First Farmers: New Perspectives on the Prehistoric Transition to Agriculture,* T. Douglas Price and Anne Birgitte Gebauer, eds., pp. 273–299. Santa Fe: School of American Research Press.

He Bochuan
1991 *China on the Edge: The Crisis of Ecology and Development.* San Francisco: China Books and Periodicals.

Headland, Thomas, and Lawrence A. Reid
1989 Hunter-Gatherers and Their Neighbors from Prehistory to the Present. *Current Anthropology* 30(1):43–66.

Headley, Lee
1982 *Suicide in Asia and the Near East.* Berkeley: University of California Press.

Heider, Karl
1970 *The Dugum Dani.* Viking Fund Publications in Anthropology No. 49.
1979 *Grand Valley Dani: Peaceful Warriors.* New York: Holt, Rinehart and Winston.

Heiser, Charles, Jr.
1985 *Of Plants and People.* Norman: University of Oklahoma.

Helm, June
1965 Bilaterality in the Socio-territorial Organization of the Arctic Drainage Dene. *Ethnology* 4(4):361–385.

Helvarg, David
1994 *The War against the Greens.* Washington, DC: Sierra Club.

Henderson, Junius, and John P. Harrington
1914 *Ethnozoology of the Tewa Indians.* Bureau of American Ethnology Bulletin 56.

Henry, Donald O.
1995 *Prehistoric Cultural Ecology and Evolution: Insights from Southern Jordan.* New York: Plenum Press.

Herskovits, M. J.
1926 The Cattle Complex in East Africa. *American Anthropologist* 28:230–272, 361–388, 494–528, 633–664.

Higham, Charles
1995 The Transition to Rice Cultivation in Southeast Asia. In: *Last Hunters–First Farmers: New Perspectives on the Prehistoric Transitio n to Agriculture*, T. Douglas Price and Anne Birgitte Gebauer, eds., pp. 127–155. Santa Fe: School of American Research Press.

Hill, Kim, and Kristen Hawkes
1983 Neotropical Hunting among the Aché of Eastern Paraguay. In: *Adaptive Responses of Native Amazonians*, Raymond B. Hames and William T. Vickers, eds., pp. 139–188. New York: Academic Press.

Hill, Kim, and A. Magdalena Hurtado
1996 *Aché Life History: The Ecology and Demography of a Foraging People.* New York: Aldine.

Ho, Ping-ti
1975 *The Cradle of the East.* Hong Kong and Chicago: Chinese University of Hong Kong and University of Chicago Press.
1988 The Origins of Chinese Agriculture. Paper presented at the Fifth International Conference on the History of Science in China, San Diego.

Hogg, Gary
1966 *Cannibalism and Human Sacrifice.* New York: Citadel Press.

Hooi, Alexis John
2002 Congo's Civil War Imperils Ituri Livelihood. *Cultural Survival Voices* 1(4):1, 8.

Hopkins, Raymond F., Donald J. Puchala, and Ross B. Talbot
1979 *Food, Politics, and Agricultural Development: Case Studies in the Public Policy of Rural Modernization.* Boulder, CO: Westview Press.

Hrdy, Sarah
1999 *Mother Nature.* New York: Pantheon.

Huang, H. T.
2002 Hypolactasia and the Chinese Diet. *Current Anthropology* 43(5):809–820.

Huang, Philip
1990 *The Peasant Family and Rural Development in the Yangzi Delta, 1350–1988.* Stanford: Stanford University Press.

Hunn, Eugene S.
1977 *Tzeltal Folk Zoology.* New York: Academic Press.
1979 The Abominations of Leviticus Revisited: A Commentary on Anomaly in Symbolic Anthropology. In: *Classifications in the Social Context,*

Roy Ellen and David Reason, eds., pp. 103–116. New York: Academic Press.

1981 On the Relative Contribution of Men and Women to Subsistence among Hunter-Gatherers of the Columbia Plateau: A Comparison with Ethnographic Atlas Summaries. *Journal of Ethnobiology* 1(1):124–134.

1991 *Nch'i Wana, the Big River.* Seattle: University of Washington Press.

Hunt, Robert C.

1988 Size and Structure of Authority in Canal Irrigation Systems. *Journal of Anthropological Research* 44(4):335–355.

Huntington, Ellsworth

1945 *Mainsprings of Civilization.* New York: Wiley.

Hurtado, Ana Magdalena, Kristen Hawkes, Kim Hill, and Hillard Kaplan

1985 Female Subsistence Strategies among Aché Hunter-Gatherers of Eastern Paraguay. *Human Ecology* 13(1):1–28.

Ingold, Tim

1980 *Hunters, Pastoralists, and Ranchers.* Cambridge Studies in Social Anthropology No. 28.

1987 *The Appropriation of Nature: Essays on Human Ecology and Social Relations.* Iowa City: University of Iowa Press.

Irons, William, and Lee Cronk

2000 Two Decades of a New Paradigm. In: *Adaptation and Human Behavior: An Anthropological Perspective,* Lee Cronk, Napoleon Chagnon, and William Irons, eds., pp. 3–26. New York: Aldine.

Isaac, Glynn

1978a Food Sharing and Human Evolution: Archaeological Evidence from the Plio-Pleistocene of East Africa. *Journal of Anthropological Research* 34(3):311–325.

1978b The Food-Sharing Behavior of Protohuman Hominids. *Scientific American* 238(4):90–108.

Iverson, Peter

1990 *The Navajos.* New York: Chelsea House Publishers.

Jacobs, Lynn

1991 *Waste of the West: Public Lands Ranching.* Tucson: author.

Jenike, Mark R.

2001 Nutritional Ecology: Diet, Physical Activity, and Body Size. In: *Hunter-Gatherers: An Interdisciplinary Perspective,* Catherine Panter-Brick, Robert H. Layton, and Peter Rowley-Conwy, eds., pp. 205–238. Cambridge: Cambridge University Press.

Jilek, Wolfgang

1982 *Indian Healing.* Surrey, BC: Hancock House.

Jochim, Michael A.
1976 *Hunter-Gatherer Subsistence and Settlement: A Predictive Model.* New York: Academic Press.
1981 *Strategies for Survival: Cultural Behavior in an Ecological Context.* New York: Academic Press.
1983 Optimization Models in Context. In: *Archaeological Hammers and Theories,* James A. Moore and Authur S. Keene, eds., pp. 157–172. New York: Academic Press.
1998 *A Hunter-Gatherer Landscape: Southwest Germany in the Late Paleolithic and Mesolithic.* New York: Plenum Press.

Johannes, R. E.
1981 *Words of the Lagoon.* Berkeley: University of California Press.

Johns, Timothy
1990 *The Origins of Human Diet and Medicine: Chemical Ecology.* Tucson: University of Arizona Press.

Johnson, Allen W., and Timothy Earle
2000 *The Evolution of Human Societies* (2nd ed.). Stanford: Stanford University Press.

Johnston, Francis E. (ed.)
1987 *Nutritional Anthropology.* New York: Alan R. Liss, Inc.

Jonaitis, Aldona
1986 *Art of the Northern Tlingit.* Seattle: University of Washington.

Jones, Kevin T., and David B. Madsen
1989 Calculating the Cost of Resource Transportation: A Great Basin Example. *Current Anthropology* 30(4):529–534.

Kammer, Jerry
1980 *The Second Long Walk: The Navajo–Hopi Land Dispute.* Albuquerque: University of New Mexico Press.

Kant, Immanuel
1978 *Anthropology from a Pragmatic Point of View.* Tr. by Victor Lyle Dowdell. Carbondale: Southern Illinois University Press. (German original, 1798.)

Kaus, Andrea
1992 Common Ground: Ranchers and Researchers in the Mapimi Biosphere Reserve. Ph.D. dissertation, University of California, Riverside.

Kay, Charles E., and Randy T. Simmons (eds.)
2002 *Wilderness and Political Ecology: Aboriginal Influences and the Original State of Nature.* Salt Lake City: University of Utah Press.

Kay, Richard F.
1981 The Nut-Crackers–A New Theory of the Adaptations of the Ramapithecinae. *American Journal of Physical Anthropology* 55:141–151.

Kay, Richard F., and Wendy Sue Sheine

1979 On the Relationship between Chitin Particle Size and Digestibility in the Primate *Galago senegalensis. American Journal of Physical Anthropology* 50:301–308.

Kearney, Michael

1984 *World View.* Novato, CA: Chandler and Sharp.

1986 From the Invisible Hand to Visible Feet: Anthropological Studies of Migration and Development. *Annual Reviews in Anthropology* 15:331–361.

Keegan, William F.

1986 The Optimal Foraging Analysis of Horticultural Production. *American Anthropologist* 88(1):92–107.

Keene, Authur S.

1979 Economic Optimization Models and the Study of Hunter-Gatherer Subsistence Settlement Systems. In: *Transformations: Mathematical Approaches to Culture Change,* Colin Renfrew and Kenneth L. Cooke, eds., pp. 369–404. New York: Academic Press.

1981 *Prehistoric Foraging in a Temperate Forest: A Linear Programming Model.* New York: Academic Press.

1983 Biology, Behavior, and Borrowing: A Critical Examination of Optimal Foraging Theory in Archaeology. In: *Archaeological Hammers and Theories,* James A. Moore and Authur S. Keene, eds., pp. 137–155. New York: Academic Press.

1985a Constraints on Linear Programming Applications in Archaeology. In: *For Concordance in Archaeological Analysis: Bridging Data Structure, Quantitative Technique, and Theory,* Christopher Carr, ed., pp. 239–273. Prospect Heights, IL: Waveland Press.

1985b Nutrition and Economy: Models for the Study of Prehistoric Diet. In: *The Analysis of Prehistoric Diets,* Robert I. Gilbert, Jr., and James H. Mielke, eds., pp. 155–190. New York: Academic Press.

Keirsey, D., and Marilyn Bates

1978 *Please Understand Me.* Del Mar, CA: Prometheus Nemesis.

Kelly, Robert L.

1995 *The Foraging Spectrum: Diversity in Hunter-Gatherer Lifeways.* Washington, DC: Smithsonian Institution.

Kempton, Willett

1981 *The Folk Classification of Ceramics.* New York: Academic Press.

Kennard, Edward A.

1979 Hopi Economy and Subsistence. In: *Handbook of North American Indians,* Vol. 9, Southwest, Alfonso Ortiz, ed., pp. 554–563. Washington, DC: Smithsonian Institution.

Keyes, Charles (ed.)
1983 Peasant Strategies in Asian Societies: Moral and Rational Economic Approaches–A Symposium. *Journal of Asian Studies* 42(4):753–867.

Khazanov, A. M.
1984 *Nomads and the Outside World.* Cambridge: Cambridge University Press.

Kiernan, Michael J., and Freese, Curtis H.
1997 Mexico's Plan Piloto Forestal: The Search for Balance between Socioeconomic and Ecological Sustainability. In: *Harvesting Wild Species: Implications for Biodiversity Conservation*, Curtis H. Freese, ed., pp. 93–131. Baltimore: Johns Hopkins University Press.

King, F. H.
1911 *Farmers of Forty Centuries.* New York: Mrs. F. H. King.

Kinsley, David
1995 *Ecology and Religion: Ecological Spirituality in Cross-Cultural Perspective.* Englewood Cliffs, NJ: Prentice-Hall.

Kintz, Ellen
1990 *Life under the Tropical Canopy: Tradition and Change among the Yucatec Maya.* New York: Holt, Rinehart and Winston.

Kirch, Patrick V.
1980 The Archaeological Study of Adaptation: Theoretical and Methodological Issues. In: *Advances in Archaeological Method and Theory*, Vol. 3, Michael B. Schiffer, ed., pp. 101–156. New York: Academic Press.
1994 *The Wet and the Dry: Irrigation and Agricultural Intensification in Polynesia.* Chicago: University of Chicago Press.
1997 Microcosmic Histories: Island Perspectives on "Global" Change. *American Anthropologist* 99(1):30–42.

Kirk, Ruth
1986 *Wisdom of the Elders.* Victoria, BC: British Columbia Provincial Museum.

Kitcher, Philip
1993 *The Advancement of Science.* New York: Oxford University Press.

Klee, Gary (ed.)
1980 *World Systems of Traditional Resource Management.* New York: V. H. Winston and Sons.

Köhler-Rollefson, Ilse
1988 The Aftermath of the Levantine Neolithic Revolution in the Light of Ecological and Ethnographic Evidence. *Paleoriént* 14:87–93.

Kohn, Alfie
1990 *The Brighter Side of Human Nature.* New York: Basic Books.

Kormondy, Edward J.
1996 *Concepts of Ecology* (4th ed.). Upper Saddle River, NJ: Prentice Hall.

Kormondy, Edward J., and Daniel E. Brown
1998 *Fundamentals of Human Ecology*. Upper Saddle River, NJ: Prentice Hall.
Kortlandt, Adriaan
1978 The Ecosystems in Which the Incipient Hominids Could Have Evolved. *Recent Advances in Primatology* 3:503–506.
Kottak, Conrad P.
1999 The New Ecological Anthropology. *American Anthropologist* 101(1):23–35.
Koyama, Shuzo, and David Hurst Thomas (eds.)
1981 *Affluent Foragers: Pacific Coasts East and West*. Osaka: Senri Ethnological Studies No. 9.
Krader, Lawrence
1959 The Ecology of Nomadic Pastoralism. *International Social Science Journal* 11(4):499–510.
Krech, Shepard, III
1999 *The Ecological Indian: Myth and History*. New York: W. W. Norton & Company.
Kroeber, Alfred L.
1939 *Cultural and Natural Areas of Native North America*. University of California Publications in American Archaeology and Ethnology 38.
1944 *Configurations of Culture Growth*. Berkeley: University of California Press.
1953 *Cultural and Natural Areas of Native North America*. Berkeley: University of California Press.
Kroeber, Alfred L., and Clyde Kluckhohn
1952 *Culture: A Critical Review of Concepts and Definitions*. Cambridge: Peabody Museum Papers, 47:1.
Kronenfeld, David B.
1979 Innate Language? *Language Science* 1(2):209–239.
Kronenfeld, David B., James Armstrong, and Stan Wilmoth
1985 Exploring the Internal Structure of Linguistic Categories: An Extensionist Semantic View. In: *Directions in Cognitive Anthropology*, Janet Dougherty, ed., pp. 91–109. Urbana: University of Illinois Press.
Kropotkin, Peter
1904 *Mutual Aid, a Factor in Evolution*. London: Wm. Heinemann.
Kuhn, Thomas
1962 *The Structure of Scientific Revolutions*. Chicago: University of Chicago Press.
Kurland, Jeffrey, and Stephen Beckerman
1985 Optimal Foraging and Hominid Evolution: Labor and Reciprocity. *American Anthropologist* 87(1):73–93.

Kurz, Richard B., Jr.
1987 Contributions of Women to Subsistence in Tribal Societies. In: *Research in Economic Anthropology*, Vol. 8, Barry L. Isaac, ed., pp. 31–59. Greenwich, CN: JAI Press.

Lacerenza, Deborah
1988 An Historical Overview of the Navajo Relocation. *Cultural Survival Quarterly* 12(3):3–6.

Laird, Carobeth
1976 *The Chemehuevis.* Banning, CA: Malki Museum Press.
1984 *Mirror and Pattern.* Banning, CA: Malki Museum Press.

Laird, Sarah A. (ed.)
2002 *Biodiversity and Traditional Knowledge: Equitable Partnerships in Practice.* London: Earthscan.

Lakoff, George
1987 *Women, Fire and Dangerous Things.* Chicago: University of Chicago.

Lamb, H. H
1995 *Climate, History and the Modern World* (2nd ed.). London: Routledge.

Lamberg-Karlovsky, Martha (ed.)
2000 *The Breakout: The Origins of Civilization.* Peabody Museum Monographs No. 9.

Lando, Richard
1983 The Spirits Aren't So Powerful Any More: Spirit Belief and Irrigation Organization in Northern Thailand. *Journal of the Siam Society* 71(1/2):121–148.

Lang, George
1982 *The Cuisine of Hungary.* New York: Atheneum.

Lang, Susan
1989 The World's Healthiest Diet. *American Health*, Sept. 1989:105–112.

Langer, Ellen
1983 *The Psychology of Control.* Beverly Hills, CA: Sage.

Langness, L. L.
1987 *The Study of Culture* (2nd. ed.). Novato: Chandler and Sharp.

Lanner, E.
1981 *The Pinyon Pine.* Reno: University of Nevada Press.
1996 *Made for Each Other.* New York: Oxford University Press.

Lansing, J. Stephen
1987 Balinese "Water Temples" and the Management of Irrigation. *American Anthropologist* 89(2):326–341.
1991 *Priests and Programmers: Technologies of Power in the Engineered Landscape of Bali.* Princeton, NJ: Princeton University Press.

Lansing, J. Stephen, and James N. Kremer

1993 Emergent Properties of Balinese Water Temple Networks: Coadaptation on a Rugged Fitness Landscape. *American Anthropologist* 95(1):97–114.

Laughlin, William S.

1968 Hunting: An Integrated Biobehavior System and Its Evolutionary Importance. In: *Man the Hunter*, Richard B. Lee and Irven DeVore, eds., pp. 304–320. Chicago: Aldine.

Lawton, Harry W., Philip J. Wilke, Mary DeDecker, and William M. Mason

1976 Agriculture among the Paiute of Owens Valley. *The Journal of California Anthropology* 3(1):13–51.

Layton, Robert H.

2001 Hunter-Gatherers, Their Neighbors and the Nation State. In: *Hunter-Gatherers: An Interdisciplinary Perspective*, Catherine Panter-Brick, Robert H. Layton, and Peter Rowley-Conwy, eds., pp. 292–321. Cambridge: Cambridge University Press.

Leach, E. R.

1964 Animal Categories of Verbal Abuse. In: *New Directions in the Study of Language*, Eric Lenneberg, ed., pp. 23–64. Cambridge: Massachusetts Institute of Technology Press.

Leacock, Eleanor

1978 Women's Status in Egalitarian Society: Implications for Social Evolution. *Current Anthropology* 19(2):247–275.

Lee, Richard B.

1965 Subsistence Ecology of !Kung Bushmen. Ph.D. dissertation, University of California, Berkeley.

1968 What Hunters Do for a Living, or, How to Make Out on Scarce Resources. In: *Man the Hunter*, Richard B. Lee and Irven DeVore, eds., pp. 30–48. Chicago: Aldine.

1972 The !Kung Bushmen of Botswana. In: *Hunters and Gatherers Today*, M. G. Bicchieri, ed., pp. 327–368. New York: Holt, Rinehart and Winston.

1982 Lactation, Ovulation, Infanticide, and Women's Work: A Study of Hunter-Gatherer Population Regulation. In: *Biosocial Mechanisms of Population Regulation*, Mark N. Cohen, Roy S. Malpass, and Harold G. Klein, eds., pp. 321–348. New Haven: Yale University Press.

1984 *The Dobe !Kung*. New York: Holt, Rinehart and Winston.

Lee, Richard B., and Richard Daly (eds.)

1999 *The Cambridge Encyclopedia of Hunters and Gatherers*. Cambridge: Cambridge University Press.

Lee, Richard B., and Irven DeVore (eds.)

1968 *Man the Hunter*. Chicago: Aldine.

Lee, Robert G.
1987 Community Fragmentation: Implications for Future Wild Fire Management. In: *Proceedings of the Symposium on Wildland Fire 2000*, pp. 5–14. Berkeley: U.S. Forest Service, Pacific Southwest Forest and Range Experiment Station.

Lees, Susan H., and Daniel G. Bates
1974 The Origins of Specialized Nomadic Pastoralism: A Systematic Model. *American Antiquity* 39(2):187–193.

Lehr, Jay H. (ed.)
1992 *Rational Readings on Environmental Concerns.* New York: Van Nostrand Reinhold.

Lentz, David (ed.)
2000 *Imperfect Balance.* New York: Columbia University Press.

Lerner, M.
1980 *The Belief in a Just World: A Fundamental Delusion.* New York: Plenum Press.

Le Roy Ladurie, Emmanuel
1971 *Times of Feast, Times of Famine.* Barbara Bray, tr. New York: Doubleday.

Leslie, Paul W., James R. Bindon, and Paul T. Baker
1984 Caloric Requirements of Human Populations: A Model. *Human Ecology* 12(2):137–162.

Levi-Strauss, Claude
1962 *La Pensee Sauvage.* Paris: Plon.
1963 *Structural Anthropology* (Fr. orig. 1958). New York: Basic Books.
1964–1971 *Mythologiques.* Paris: Plon.

Lewis, Henry T.
1973 *Patterns of Indian Burning in California: Ecology and Ethnohistory.* Ballena Press Anthropological Papers No. 1.
1982 Fire Technology and Resource Management in Aboriginal North America and Australia. In: *Resource Managers: North American and Australian Hunter-Gatherers*, Nancy M. Williams and Eugene S. Hunn, eds., pp. 45–67. Boulder, CO: Westview Press.

Lewis, Martin
1992 *Green Delusions.* Raleigh, NC: Duke University Press.

Lewis, Walter, and Elvin-Lewis, Memory
1977 *Medical Botany.* New York: Wiley.

Libecap, Gary
1989 *Contracting for Property Rights.* Cambridge: Cambridge University Press.

Linden, Eugene
1991 Lost Tribes, Lost Knowledge. *Time*, Sept. 23, pp. 46–55.

Linton, Sally

1971 Woman the Gatherer: Male Bias in Anthropology. In: *Women in Cross-Cultural Perspective: A Source Book*, Sue-Ellen Jacobs, ed., pp. 9–21. Urbana: University of Illinois, Department of Urban & Regional Planning.

Lizot, Jacques

1977 Population, Resources, and Warfare among the Yanomami. *Man* 12(3/4):497–517.

Locke, John

1975 *An Essay Concerning Human Understanding* (orig. 1685). Oxford: Oxford University Press.

Loewe, Michael, and Edward L. Shaughnessy

1999 *The Cambridge History of Ancient China*. Cambridge: Cambridge University Press.

Lopez, Kevin Lee

1992 Returning to Fields. *American Indian Culture and Research Journal* 16:165–174.

Lovejoy, C. Owen

1981 The Origin of Man. *Science* 211:341–350.

1984 The Natural Detective. *Natural History* 10:24–28.

Low, Bobbi S.

1993 Behavioral Ecology of Conservation in Traditional Societies. Paper presented at the annual meetings of the American Anthropological Association, Washington, D.C.

Lowie, Robert

1937 *History of Ethnological Theory*. New York: Rinehart & Co.

MacArthur, Robert

1972 *Geographical Ecology*. New York: Harper and Row.

Mace, Ruth

1993 Transitions between Cultivation and Pastoralism in Sub-Saharan Africa. *Current Anthropology* 34(4):363–382.

MacLean, William C.

1984 Nutrition in Infancy. In: *Nutrition Reviews' Present Knowledge in Nutrition*, pp. 619–635. Washington, DC: Nutrition Foundation, Inc.

MacNeill, J.

1992 *The Mountains of the Mediterranean World*. Oxford: Oxford University Press.

MacNeish, Richard

1992 *The Origins of Agriculture and Settled Life*. Norman: University of Oklahoma Press.

Madsen, David B.
1982 Get It Where the Gettin's Good: A Variable Model of Great Basin Subsistence and Settlement Based on Data from the Eastern Great Basin. In: *Man and Environment in the Great Basin,* David B. Madsen and James F. O'Connell, eds., pp. 207–226. Society for American Archaeology Papers No. 2.
1986 Leap and Grab: Energetic Efficiency Tests of Cricket Use. Paper presented at the Great Basin Anthropological Conference, Las Vegas.
Madsen, David B., and James E. Kirkman
1988 Hunting Hoppers. *American Antiquity* 53(3):593–604.
Madsen, David B., and Dave N. Schmitt
1998 Mass Collecting and the Diet Breadth Model: A Great Basin Example. *Journal of Archaeological Science* 25(5):445–455.
Malinowski, Bronislaw
1922 *Argonauts of the Western Pacific.* New York: Dutton.
1935 *Coral Gardens and Their Magic.* Chicago: American Book Co.
1944 *A Scientific Theory of Culture.* Chapel Hill: University of North Carolina Press.
Mallory, J. P.
1989 *In Search of the Indo-Europeans.* London: Thames and Hudson.
Malthus, Thomas
1960 *On Population* (orig. 1798). New York: Random House.
Mamdani, Mahmood
1972 *The Myth of Population Control.* New York: Monthly Review Press.
Mann, Alan E.
1981 Diet and Human Evolution. In: *Omnivorous Primates,* Robert S. Harding and Geza Teleki, eds., pp. 10–36. New York: Columbia University Press.
Marcus, Joyce, and Gary M. Feinman
1998 Introduction. In: *Archaic States,* Gary M. Feinman and Joyce Marcus, eds., pp. 3–13. Santa Fe: School of American Research.
Marshall, Fiona
1990 Origins of Specialized Pastoral Production in East Africa. *American Anthropologist* 92(4):873–894.
Marten, Gerald G. (ed.)
1986 *Traditional Agriculture in Southeast Asia.* Honolulu: East-West Center.
Martin, Calvin
1978 *Keepers of the Game.* Berkeley: University of California Press.
Martin, Gary
1995 *Ethnobotany.* London: Chapman and Hall.
Martin, Laura
1986 "Eskimo Words for Snow": A Case Study in the Genesis and Decay of an Anthropological Example. *American Anthropologist* 88(2):418–442.

Martin, Paul S., and Christine R. Szuter
1999 War Zones and Game Sinks in Lewis and Clark's West. *Conservation Biology* 13(1):36–45.
Mason, Otis
1894 Technogeography, or the Relation of the Earth to the Industries of Mankind. *American Anthropologist* 7(2):137–161.
Matson, P. A., W. J. Parton, A. G. Power, and M. J. Swift
1997 Agricultural Intensification and Ecosystem Properties. *Science* 227:504–509.
McCabe, J. Terrence, Scott Perkin, and Claire Schofield
1992 Can Conservation and Development Be Coupled among Pastoral People? An Examination of the Maasai of the Ngorongoro Conservation Area, Tanzania. *Human Organization* 51(4):353–366.
McCay, Bonnie, and James Acheson (eds.)
1987 *The Question of the Commons.* Tucson: University of Arizona.
McCorriston, Joy
1992 The Early Development of Agriculture in the Ancient Near East: An Ecological and Evolutionary Study. Ph.D. dissertation, Yale University.
McCorriston, Joy, and Frank Hole
1991 The Ecology of Seasonal Stress and the Origins of Agriculture in the Near East. *American Anthropologist* 93(1):46–69.
McCreedy, Marion
1994 The Arms of the *Dibouka.* In: *Key Issues in Hunter-Gatherer Research,* Ernest S. Burch, Jr., and Linda J. Ellanna, pp. 15–34. Oxford: Berg Publishers.
McFeat, Tom (ed.)
1966 *Indians of the North Pacific Coast.* Seattle: University of Washington Press.
McGovern, Thomas H., Gerald F. Bigelow, Thomas Amorosi, and Daniel Russell
1988 Northern Islands, Human Error, and Environmental Degradation. *Human Ecology* 16(3):225–270.
McGrew, W. C.
1979 Evolutionary Implications of Sex Differences in Chimpanzee Predation and Tool Use. In: *The Great Apes,* D. A. Hamburg and E. R. McCown, eds., pp. 440–463. Menlo Park, CA: Benjamin/Cummings.
1981 The Female Chimpanzee as a Human Evolutionary Prototype. In: *Woman the Gatherer,* Frances Dahlberg, ed., pp. 35–73. New Haven: Columbia University Press.
McIntosh, Roderick J., Joseph A. Tainter, and Susan Keech McIntosh (eds.)
2000 *The Way the Wind Blows: Climate, History, and Human Action.* New York: Columbia University Press.
McNeish, Richard
1992 *Origins of Agriculture and Domestication.* Norman: University of Oklahoma Press.

Medin, Douglas L., and Scott Atran (eds.)
1999 *Folkbiology*. Cambridge: Massachusetts Institute of Technology.
Meggers, Betty J.
1954 Environmental Limitation on the Development of Culture. *American Anthropologist* 56(5):801–824.
Meggitt, Mervyn
1977 *Blood Is Their Argument*. Palo Alto, CA: Mayfield.
Meillassoux, Claude
1981 *Maidens, Meal, and Money*. Cambridge: Cambridge University Press.
Meilleur, Brien A.
1996 Forests and Polynesian Adaptations. In: *Tropical Deforestation: The Human Dimension*, Leslie E. Sponsel, Thomas N. Headland, and Robert C. Bailey, eds., pp.76–94. New York: Columbia University Press.
Mencius
1979 *Mencius*. D. Lau, tr. Harmondsworth, Sussex: Penguin.
Merleau-Ponty, Maurice
1962 *The Phenomenology of Perception*. London: Routledge, Kegan Paul.
1963 *The Structure of Behavior*. Boston, IL: Beacon Press.
1964 *Signs*. Evanston: Northwestern University Press.
1968 *The Visible and the Invisible*. Chicago: Northwestern University Press.
Merry, Sally
2003 Human Rights Law and the Demonization of Culture. *Anthropology News* 44(3):4–5.
Metcalfe, Duncan, and K. Renee Barlow
1992 A Model for Exploring the Optimal Trade-off between Field Processing and Transport. *American Anthropologist* 94(2):340–356.
Meyer, William B.
1996 *Human Impact on the Earth*. Cambridge: Cambridge University Press.
Mills, C. Wright
1959 *The Sociological Imagination*. Oxford: Oxford University Press.
Minnich, Richard
1983 Fire Mosaics in Southern California and Northern Baja California. *Science* 219:1287–1294.
Minnis, Paul E. (ed.)
2000 *Ethnobotany: A Reader*. Norman: University of Oklahoma Press.
Mintz, Sidney
1985 *Sweetness and Power*. New York: Viking.
Mischel, Walter
1984 On the Predictability of Behavior and the Structure of Personality. In: *Personality and the Prediction of Behavior*, Robert A. Zucker, Joel Aronoff, and A. I. Rabin, eds., pp. 269–305. New York: Academic Press.

Mithen, Steven J.
1990 *Thoughtful Foragers: A Study of Prehistoric Decision Making.* Cambridge: Cambridge University Press.
Moerman, Daniel E.
1986 *Medicinal Plants of Native America* (2 vols.). University of Michigan Museum of Anthropology Technical Reports No. 19.
Moldenke, Harold N., and Alma L. Moldenke
1952 *Plants of the Bible.* Waltham, MA: Chronica Botanica.
Molles, Manuel C., Jr.
1999 *Ecology: Concepts and Applications.* Boston: McGraw-Hill.
Molnar, Stephen, and Iva M. Molnar
2000 *Environmental Change and Human Survival: Some Dimensions of Human Ecology.* Upper Saddle River, NJ: Prentice Hall.
Montaigne, Michel de.
1943 *The Complete Essays of Montaigne.* Donald M. Frame, tr. and ed. Stanford: Stanford University Press.
Montesquieu, Charles, Baron
1949 *The Spirit of the Laws* (orig. 1748). New York: Hafner.
Moore, James A.
1981 The Effects of Information Networks in Hunter-Gatherer Societies. In: *Hunter-Gatherer Foraging Strategies: Ethnographic and Archaeological Analyses,* Bruce Winterhalder and Eric Alden Smith, eds., pp. 194–217. Chicago: University of Chicago Press.
1983 The Trouble with Know-It-Alls: Information as a Social and Ecological Resource. In: *Archaeological Hammers and Theories,* James A. Moore and Authur S. Keene, eds., pp. 173–191. New York: Academic Press.
Moore, Omar Khayyam
1957 Divination–A New Perspective. *American Anthropologist* 59(1):69–74.
Moran, Emilio F.
1993a Deforestation and Land Use in the Brazilian Amazon. *Human Ecology* 21(1):1–21.
1993b *Through Amazonian Eyes: The Human Ecology of Amazonian Populations.* Iowa City: University of Iowa Press.
1996 Deforestation in the Brazilian Amazon. In: *Tropical Deforestation: The Human Dimension,* Leslie E. Sponsel, Thomas N. Headland, and Robert C. Bailey, eds., pp.149–164. New York: Columbia University Press.
Moran, Emilio F. (ed.)
1990 *The Ecosystem Approach in Anthropology.* Ann Arbor: University of Michigan.
Morgan, Lewis Henry
1851 *League of the Ho-dé-no-sau-nee, Iroquois.* Rochester, NY: Sage & Brother, Publishers.

1871 *Systems of Consanguinity and Affinity of the Human Family.* Smithsonian Institution, Contributions to Knowledge 17(2).

1877 *Ancient Society.* New York: Henry Holt.

1882 *Houses and House-Life of the American Aborigines.* Washington, DC: Government Printing Office.

Morrell, Mike

1985 *The Gitksan and Wet'suwet'en Fishery in the Skeena River System.* Hazelton, B.C., Canada: Gitksan-Wet'suwet'en Tribal Council.

Morton, A.

1981 *History of Botanical Science.* New York: Academic Press.

Moseley, Michael

1975 *The Maritime Origins of Andean Civilization.* Menlo Park, CA: Cummings Publishing Company.

Murdock, George P.

1968 The Current Status of the World's Hunting and Gathering Peoples. In: *Man the Hunter,* Richard B. Lee and Irven DeVore, eds., pp. 13–20. Chicago: Aldine.

1969 Ethnographic Atlas: A Summary. *Ethnology* 6(2):109–236.

Murphy, Earl

1967 *Governing Nature.* Chicago: Quadrangle Books.

Myers, Isabel Briggs, and Peter B. Myers

1980 *Gifts Differing.* Palo Alto, CA: Consulting Psychologists Press.

Nabhan, Gary

1985 *Gathering the Desert.* Tucson: University of Arizona Press.

1989 *Enduring Seeds.* San Francisco: North Point Press.

Nabhan, Gary, Amadeo Rea, Karen Reichhardt, Eric Mellink, and Charles F. Hutchinson

1982 Papago Influences on Habitat and Biotic Diversity: Quitovac Oasis Ethnoecology. *Journal of Ethnobiology* 2(2):124–143.

Nagengast, Carole, and Terence Turner

1997 Introduction: Universal Human Rights Versus Cultural Relativity. *Journal of Anthropological Research* 53(3):269–272.

Nash, Roderick

1973 *Wilderness and the American Mind.* New Haven: Yale University Press.

National Research Council.

1989 *Lost Crops of the Incas.* Washington, DC: National Academy Press.

Nazarea, Virginia D. (ed.)

1999 *Ethnoecology: Situated Knowledge/Located Lives.* Tucson: University of Arizona Press.

Nazarea, Virginia D.

1998 *Cultural Memory and Biodiversity.* Tucson: University of Arizona Press.

Neihardt, John
1932 *Black Elk Speaks*. New York: William Morrow.

Nelson, Richard K.
1983 *Make Prayers to the Raven*. Chicago: University of Chicago Press.

Nerlove, Sara B.
1974 Woman's Workload and Infant Feeding Practices: A Relationship with Demographic Implications. *Ethnology* 13(2):207–214.

Netting, Robert
1974 Agrarian Ecology. *Annual Review of Anthropology* 3:21–56.
1981 *Balancing on an Alp*. Cambridge: Cambridge University Press.
1986 *Cultural Ecology* (2nd ed.). Prospect Heights, IL: Waveland Press.

Nichols, Deborah
1987 Risk and Agricultural Intensification during the Formative Period in the Northern Basin of Mexico. *American Anthropologist* 89(3): 596–616.

Nietschmann, Bernard
1973 *Between Land and Water*. New York: Seminar Press.

Nisbett, Richard, Eugene Borgida, Harvey Reed, and Rick Crandall
1976 Popular Induction: Information Is Not Necessarily Informative. In: *Cognition and Social Behavior*, John S. Carroll and John W. Payne, eds., pp. 113–133. New York: Academic Press.

Nisbett, Richard, and Lee Ross
1980 *Human Inference: Strategies and Shortcomings of Social Judgment.* Englewood Cliffs, NJ: Prentice Hall.

Nishida, Toshisada
1973 The Ant-Gathering Behavior by the Use of Tools among Wild Chimpanzees of the Mahale Mountains. *Journal of Human Evolution* 2:357–370.
1987 Local Traditions and Cultural Transmission. In: *Primate Societies*, Barbara B. Smuts, Dorothy L. Cheney, Robert M. Seyfarth, Richard W. Wrangham, and Thomas T. Stuhsaker, eds., pp. 462–474. Chicago: University of Chicago Press.

Nishida, Toshisada, and Mariko Hiraiwa-Hasegawa
1987 Chimpanzees and Bonobos: Cooperative Relationships among Males. In: *Primate Societies*, Barbara B. Smuts, Dorothy L. Cheney, Robert M. Seyfarth, Richard W. Wrangham, and Thomas T. Stuhsaker, eds., pp. 165–177. Chicago: University of Chicago Press.

Noss, Andrew J.
1997 The Economic Importance of Communal Net Hunting among the BaAka of the Central African Republic. *Human Ecology* 25(1):71–89.

O'Connell, James F., and Kristen Hawkes

1984 Food Choice and Foraging Strategies among the Alyawara. *Journal of Anthropological Research* 40(4):504–535.

Odum, Eugene P.

1953 *Fundamentals of Ecology*. Philadelphia: W. B. Saunders.

1975 *Ecology*. New York: Holt, Rinehart and Winston.

1993 *Ecology and Our Endangered Life-Support Systems* (2nd ed.). Sunderland, MA: Sinauer Associates, Inc.

Oldfield, Margery, and Janis Alcorn

1987 Conservation of Traditional Agroecoystems. *BioScience* 37:199–208.

Olol-Dapash, Meitamei

2002 Mau Forest Destruction: Human and Ecological Disaster in the Making. *Cultural Survival Voices* 1(3):1, 9.

Orans, Martin

1975 Domesticating the Functional Dragon: An Analysis of Piddocke's Potlatch. *American Anthropologist* 77(2):312–328.

Orians, Gordon H., and Nolan E. Pearson

1979 On the Theory of Central Place Foraging. In: *Analysis of Ecological Systems*, David J. Horn, Gordon R. Stairs, and Roger D. Mitchell, eds., pp. 155–177. Columbus: Ohio State University.

Orlove, Benjamin S., and Stephen B. Brush

1996 Anthropology and the Conservation of Biodiversity. *Annual Review of Anthropology* 25:329–352.

Ortiz, Alfonso (ed.)

1983 *Handbook of North American Indians*, Vol. 10, Southwest. Washington, DC: Smithsonian Institution.

Ortiz de Montellano, Bernard R.

1978 Aztec Cannibalism: An Ecological Necessity? *Science* 200:611–617.

Ostrom, Elinor

1990 *Governing the Commons*. Cambridge: Cambridge University Press.

Ott, Sandra

1993 *The Circle of Mountains: A Basque Sheepherding Community*. Reno: University of Nevada Press.

Overal, William Leslie

1990 Introduction to Ethnozoology: What It Is or Could Be. In: *Proceedings of the First International Congress of Ethnobiology*, Darrell A. Posey, William Leslie Overal, Charles R. Clement, Mark J. Plotkin, Elaine Elisabetsky, Clarice Novaes de Mota, and José Flàvio Pessôa de Barros, eds., pp. 127–129. Belém, Brazil: Museu Paraense Emílio Goeldi.

Piatelli-Palmerini, Massimo
1994 *Inevitable Illusions: How Mistakes of Reason Rule Our Minds.* New York: John Wiley and Sons.

Panter-Brick, Catherine, Robert H. Layton, and Peter Rowley-Conwy
2001a Lines of Inquiry. In: *Hunter-Gatherers: An Interdisciplinary Perspective,* Catherine Panter-Brick, Robert H. Layton, and Peter Rowley-Conwy, eds., pp. 1–11. Cambridge: Cambridge University Press.

Panter-Brick, Catherine, Robert H. Layton, and Peter Rowley-Conwy (eds.)
2001b *Hunter-Gatherers: An Interdisciplinary Perspective.* Cambridge: Cambridge University Press.

Parezo, Nancy J.
1996 The Diné (Navajos): Sheep Is Life. In: *Paths of Life: American Indians of the Southwest and Northern Mexico,* Thomas E. Sheridan and Nancy J. Parezo, eds., pp. 3–33. Tucson: University of Arizona Press.

Park, Robert Ezra
1936 Human Ecology. *American Journal of Sociology* 42:1–15.

Parker, R., and H. Toots
1980 Trace Elements in Bones as Paleobiological Indicators. In: *Fossils in the Making,* A. K. Behrensmeyer and A. P. Hill, eds., pp. 197–207. Chicago: University of Chicago Press.

Parlowe, Anita
1988 *Cry, Sacred Ground.* Washington D.C.: Christic Institute.

Pearson, Richard, and Anne Underhill
1987 The Chinese Neolithic: Recent Trends in Research. *American Anthropologist* 89(4):807–822.

Pemberton, Robert W.
1988 The Use of the Thai Giant Waterbug, *Lethocereus indicus* (Hemiptera:Belostomatidae), as Human Food in China. *Pan-Pacific Entomologist* 64:81–82.

Perrings, Charles
1987 *Economy and Environment.* Cambridge: Cambridge University Press.

Peters, David Urlin
1960 *Land Use in Barotseland.* Rhodes-Livingstone Institute Communication No. 19.

Peterson, Christopher, Steven Maier, and Martin E. P. Seligman
1993 *Learned Helplessness.* New York: Oxford University Press.

Piddocke, Stuart
1965 The Potlatch System of the Southern Kwakiutl: A New Perspective. *Southwestern Journal of Anthropology* 21(3):244–264.

Pimentel, David, S. Christ, L. Spritz, L. Fitton, R. Safouer, R. Blair, C. Harvey, P. Resosudarmo, K. Sinclair, D. Kurtz, and M. McNair

1995 Environmental and Economic Costs of Soil Erosion and Conservation Benefits. *Science* 267:1117–1122.

Pimentel, David, M. Herdendorf, S. Eisenfeld, L. Olander, M. Carroquino, C. Corson, J. McDade, Y. Chung, W. Cannon, J. Roberts, L. Bluman, and J. Gregg

1994 Achieving a Secure Energy Future: Environmental and Economic Issues. *Ecological Economics* 9(3):201–219.

Pimentel, David, Michael McNair, Louise Buck, Marcia Pimentel, and Jeremy Kamil

1997 The Value of Forests to World Food Security. *Human Ecology* 25(1):91–120.

Pinkerton, Evelyn (ed.)

1989 *Cooperative Management of Local Fisheries: New Directions for Improved Management and Community Development.* Vancouver: University of British Columbia.

Pinkerton, Evelyn, and Martin Weinstein

1995 *Fisheries That Work: Sustainability through Community-Based Management.* Vancouver: David Suzuki Foundation.

Piperno, Dolores, and Deborah Pearsall

1998 *The Origins of Agriculture in the Neotropics.* New York: Academic Press.

Plotkin, Mark J.

2000 *Medicine Quest: In Search of Nature's Healing Secrets.* New York: Viking.

Plotkin, Mark J., and Lisa Famolare (eds.)

1992 *Sustainable Harvest and Marketing of Rain Forest Products.* Washington, DC: Island Press.

Ponting, Clive

1991 *A Green History of the World.* New York: Penguin.

Popkin, Samuel

1979 *The Rational Peasant.* Berkeley: University of California Press.

Posey, Darrell Addison

2001 Intellectual Property Rights and the Sacred Balance: Some Spiritual Consequences from the Commercialization of Traditional Resources. In: *Indigenous Traditions and Ecology: The Interbeing of Cosmology and Community,* John A. Grom, ed., pp. 3–23. Cambridge: Harvard University Press.

Postel, Sandra

1994 Carrying Capacity: Earth's Bottom Line. In: *State of the World 1994,* Lester R. Brown, ed., pp. 3–21. New York: Norton.

Pottier, Johan

1999 *The Anthropology of Food.* Cambridge: Polity Press.

Potts, Richard

1988 On an Early Hominid Scavenging Niche. *Current Anthropology* 29(1):153–155.

Preston, William L.

1997 Large Game in Colonial California: Precolumbian Mirror or Mirage? Paper presented at the annual meeting of the Association of American Geographers, Ft. Worth, TX.

Price, David H.

1994 Wittfogel's Neglected Hydraulic/Hydroagricultural Distinction. *Journal of Anthropological Research* 50(2):187–204.

Price, David

1995 Energy and Human Evolution. *Population and Environment* 16(4):301–319.

Price, T. Douglas, and Anne Birgitte Gebauer (eds.)

1995a *Last Hunters–First Farmers: New Perspectives on the Prehistoric Transition to Agriculture.* Santa Fe: School of American Research Press.

Price, T. Douglas, and Anne Birgitte Gebauer

1995b New Perspectives on the Transition to Agriculture. In: *Last Hunters–First Farmers: New Perspectives on the Prehistoric Transition to Agriculture,* T. Douglas Price and Anne Birgitte Gebauer, eds., pp. 3–19. Santa Fe: School of American Research Press.

Price, T. Douglas, and James A. Brown (eds.)

1985 *Prehistoric Hunter-Gatherers: The Emergence of Cultural Complexity.* Orlando: Academic Press.

Prucha, Francis Paul

1984 *The Great Father: The United States Government and the American Indians* (2 vols.). Lincoln: University of Nebraska Press.

Pulliam, H. Ronald

1981 On Predicting Human Diets. *Journal of Ethnobiology* 1(1):61–68.

Purseglove, J.

1972 *Tropical Crops* (2 vols.). London: Longmans.

Pyke, G. H., H. R. Pulliam, and E. L. Charnov

1977 Optimal Foraging: A Selective Review of Theory and Tests. *Quarterly Review of Biology* 52(2):137–154.

Pyne, Stephen J.

1995 *World Fire: The Culture of Fire on Earth.* New York: Henry Holt.

1998 Forged in Fire: History, Land, and Anthropogenic Fire. In: *Advances in Historical Ecology,* William Balée, ed., pp. 64–103. New York: Columbia University Press.

Quinn, Naomi

1975 Decision Models of Social Structure. *American Ethnologist* 2(1): 19–45.

1978 Do Mfantse Fish Sellers Estimate Probabilities in Their Heads? *American Ethnologist* 5(2):206–226.

Radcliffe-Brown, A. R.

1957 *A Natural Science of Society.* New York: Free Press.

Randall, Robert

1977 Change and Variation in Samal Fishing: Making Plans to Make a Living in the Southern Philippines. Ph.D. dissertation, University of California, Berkeley.

Rappaport, Roy A.

1967 *Pigs for the Ancestors: Ritual in the Ecology of a New Guinea People.* New Haven: Yale University Press.

1971 The Sacred in Human Evolution. *Annual Review of Ecology and Systematics* 2:23–44.

1984 *Pigs for the Ancestors: Ritual in the Ecology of a New Guinea People* (2nd ed.). New Haven: Yale University Press.

1999 *Ritual and Religion in the Making of Humanity.* New Haven: Yale University Press.

Read, Piers Paul

1974 *Alive: The Story of the Andes Survivors.* Philadelphia: J.B. Lippincott Co.

Redding, Richard W.

1988 A General Explanation of Subsistence Change: From Hunting and Gathering to Food Production. *Journal of Anthropological Archaeology* 7(1):56–97.

Redfield, Robert, and Alfonso Villa Rojas

1934 *Chan Kom, A Maya Village.* Washington, DC: Carnegie Institute of Washington.

Redford, Kent H.

1990 The Ecologically Noble Savage. *Orion Nature Quarterly* 9:25–29.

Redford, Kent H., and Jane A. Mansour (eds.)

1996 *Traditional Peoples and Biodiversity Conservation in Large Tropical Landscapes.* Arlington, VA: Nature Conservency.

Redman, Charles L.

1999 *Human Impact on Ancient Environments.* Tucson: University of Arizona Press.

Reed, Charles (ed.)

1977 *Origins of Agriculture.* Hague: Mouton.

Reichel-Dolmatoff, G.

1971 *Amazonian Cosmos.* Chicago: University of Chicago Press.

1976 Cosmology and Ecological Analysis: A View from the Rain Forest. *Man* 11:307–316.

Reidhead, Van A.

1979 Linear Programming Models in Archaeology. *Annual Review of Anthropology* 8:543–578.

Repetto, Robert

1992 Accounting for Environmental Assets. *Scientific American* 266(6):94–100.

Richards, Audrey

1939 *Land, Labour and Diet in Northern Rhodesia: An Economic Study of the Bemba Tribe.* Oxford: Oxford University Press.

1948 *Hunger and Work in a Savage Tribe* (orig. 1932). London: Routledge, Kegan Paul.

Richerson, Peter J.

1977 Ecology and Human Ecology: A Comparison of Theories in the Biological and Social Sciences. *American Ethnologist* 4(1):1–26.

Richerson, Peter J., and Robert Boyd

1992 Cultural Inheritance and Evolutionary Ecology. In: *Evolutionary Ecology and Human Behavior,* Eric Alden Smith and Bruce Winterhalder, eds., pp. 61–92. New York: Aldine.

Richerson, Peter J., Robert Boyd, and Robert L. Bettinger

2001 Was Agriculture Impossible During the Pleistocene but Mandatory During the Holocene?: A Climate Change Hypothesis. *American Antiquity* 66(3):387–411.

Ridington, Robin

1981 Technology, World View, and Adaptive Strategy in a Northern Hunting Society. *Canadian Review of Sociology and Anthropology* 19(4):469–481.

Rifkin, Jeremy

1992 *Beyond Beef.* New York: Dutton.

Rindos, David

1984 *The Origins of Agriculture: An Evolutionary Pers*pective. Orlando: Academic Press.

Robarchek, Clayton

1987 Primitive Warfare and the Ratomorphic Image of Mankind. *American Anthropologist* 91(4):903–920.

Robarchek, Clayton, and Carole Robarchek

1998 *Waorani: The Contexts of Violence and War.* Fort Worth, TX: Harcourt Brace.

Robbins, Wilfred William, John Peabody Harrington, and Barbara Freire-Marreco

1916 *Ethnobotany of the Tewa Indians.* Bureau of American Ethnology Bulletin 55.

Robinson, John G., and Elizabeth L. Bennett (eds.)

2000 *Hunting for Sustainability in Tropical Forests.* New York: Columbia University Press.

Robson, John R. K., and D. E. Yen

1976 Some Nutritional Aspects of the Philippine Tasaday Diet. *Ecology of Food and Nutrition* 5:83–89.

Roemer, John

1982 *A General Theory of Exploitation and Class.* Cambridge: Harvard University Press.

1988 *Free to Lose: An Introduction to Marxist Economic Philosophy.* Cambridge: Harvard University Press.

Roemer, John (ed.)

1986 *Analytical Marxism.* Cambridge: Cambridge University Press.

Rogers, Carl

1961 *On Becoming a Person.* Boston: Houghton Mifflin.

Roosevelt, A. C.

1992 Secrets of the Forest. *The Sciences* 32(6):22–28.

Rose, Lisa, and Fiona Marshall

1996 Meat Eating, Hominid Sociality, and Home Bases Revisited. *Current Anthropology* 37(2):307–338.

Ruddle, Kenneth, and T. Akimichi (eds.)

1984 *Maritime Institutions in the Western Pacific.* Osaka: National Museum of Ethnology, Seri Ethnological Studies, l7.

Ruddle, Kenneth, and Ray Chesterfield

1977 *Education for Traditional Food Procurement in the Orinoco Delta.* Berkeley: University of California Press.

Ruddle, Kenneth, and Gongfu Zhong

1988 *Integrated Agriculture-Aquaculture in South China: The Dike-Pond System of the Zhujiang Delta.* Cambridge: Cambridge University Press.

Rudel, T. K., with Bruce Horowitz

1993 *Tropical Deforestation: Small Farmers and Land Clearing in the Ecuadorian Amazon.* New York: Columbia University Press.

Sadr, Karim

1991 *The Development of Nomadism in Ancient Northeast Africa.* Philadelphia: University of Pennsylvania Press.

Sahlins, Marshall

1972 *Stone Age Economics.* Chicago: Aldine.

1976 *Culture and Practical Reason.* Chicago: University of Chicago.

Sahlins, Marshall, and Elman Service

1960 *Evolution and Culture.* Ann Arbor: University of Michigan Press.

Saitoti, Tepilit Ole, and Carol Beckwith

1980 *Maasai.* New York: Harry N. Abrams Inc.

Saladin d'Anglure, Bernard

1984 Inuit of Quebec. In: *Handbook of North American Indians,* Vol. 5, Arctic, David Damas, ed., pp. 476–507. Washington, DC: Smithsonian Institution.

Salaman, Redcliffe

1949 *The History and Social Influence of the Potato.* Cambridge: Cambridge University Press.

Salisbury, Richard

1986 *A Homeland for the Cree.* Kingston (Ontario) and Montreal: McGill-Queens University Press.

Salzman, Philip Carl (ed.)

1980 *When Nomads Settle: Processes of Sedentarization as Adaptation and Response.* New York: Praeger.

Sanday, Peggy R.

1986 *Devine Hunger: Cannibalism as a Cultural System.* Cambridge: Cambridge University Press.

Sanderson, Stephen

1999 *Social Transformations: A General Theory of Historical Development.* Lanham, MD: Rowman and Littlefield.

Sargent, Frederick, II (ed.)

1974 *Human Ecology.* Amsterdam: North-Holland Publishing Company.

Sarma, Akkaraju

1977 *Approaches to Paleoecology.* Dubuque, IA: Wm. C. Brown Company.

Satir, Virginia

1983 *Conjoint Family Therapy.* Palo Alto, CA: Science and Behavior Books.

Sauer, Carl

1925 *The Morphology of Landscape.* University of California Publications in Geography 2:19–54.

1952 *Agricultural Origins and Dispersals.* New York: American Geographic Society.

Schalk, Randall F.

1981 Land Use and Organizational Complexity among Foragers on Northwestern North America In: *Affluent Foragers: Pacific Coasts East and West,* Shuzo Koyama and David Hurst Thomas, eds., pp. 53–75. Osaka: Senri Ethnological Studies No. 9.

Schmitt, Jean-Claude

1983 *The Holy Greyhound* (Fr. orig. 1979). Cambridge: Cambridge University Press.

Schueller, Gretel H.

2001 Eat Locally: Think Globally. *Discover* 22(5):70–77.

Schultz, Theodore

1964 *Transforming Traditional Agriculture.* New Haven: Yale University Press.

Schulz, Richard

1976 Some Life and Death Consequences of Perceived Control. In: *Cognition and Social Behavior,* John S. Carroll and John W. Payne, eds., pp. 135–153. New York: Academic Press.

Schwarz, Maureen Trudelle
1995 The Explanatory and Predictive Power of History: Coping with the "Mystery Illness," 1993. *Ethnohistory* 42(3):375–401.
1997 Unraveling the Anchoring Cord: Navajo Relocation, 1974 to 1996. *American Anthropologist* 99(1):43–55.
Scott, James
1976 *The Moral Economy of the Peasant.* New Haven: Yale University Press.
1985 *Weapons of the Weak.* New Haven: Yale University Press.
Sen, Amartya
1984 *Resources, Values and Development.* Oxford: Oxford University Press.
Serpell, James (ed.)
1995 *The Domestic Dog.* Cambridge: Cambridge University Press.
Service, Elman R.
1962 *Primitive Social Organization: An Evolutionary Perspective.* New York: Random House.
Shapiro, Roy D.
1984 *Optimization Models for Planning and Allocation: Text and Cases in Mathematical Programming.* New York: John Wiley & Sons.
Shapiro, Warren
1991 Claude Levi-Strauss Meets Alexander Goldenweiser: Boasian Anthropology and the Study of Totemism. *American Anthropologist* 93(3):599–610.
Sharer, Robert
1995 *The Ancient Maya* (5th ed.). Stanford: Stanford University Press.
Shepardson, Mary
1982 The Status of Navajo Women. *American Indian Quarterly* 6(1–2):149–169.
Shipman, Pat
1986 Scavenging or Hunting in Early Hominids: Theoretical Framework and Tests. *American Anthropologist* 88(1):27–43.
Shnirelman, Victor A.
1994 *Cherchez le Chien*: Perspectives on the Economy of the Traditional Fishing-Oriented People of Kamchatka. In: *Key Issues in Hunter-Gatherer Research*, Ernest S. Burch, Jr., and Linda J. Ellanna, pp. 169–188. Oxford: Berg Publishers.
Shrader-Frechette, K.
1991 *Risk and Rationality.* Berkeley: University of California Press.
Silberbauer, George B.
1972 The G/wi Bushmen. In: *Hunters and Gatherers Today*, M. G. Bicchieri, ed., pp. 271–326. New York: Holt, Rinehart and Winston.
Simmonds, N. W. (ed.)
1976 *Evolution of Crop Plants.* London: Longmans.

Simmons, I. G.
1969 Evidence of Vegetation Changes Associated with Mesolithic Man in Britain. In: *The Domestication and Exploitation of Plants and Animals*, Peter J. Ucko and G. W. Dimbleby, eds., pp. 111–119. Chicago: Aldine.

Simms, Steven R.
1984 Aboriginal Great Basin Foraging Strategies: An Evolutionary Analysis. Ph.D. dissertation, University of Utah.
1992 Wilderness as a Human Landscape. In: *Wilderness Tapestry: An Eclectic Approach to Preservation*, Samuel I. Zeveloff, L. Mikel Vause, and William H. McVaugh, eds., pp. 183–201. Reno: University of Nevada Press.

Simon, Herbert A.
1960 *The New Science of Management Decision*. New York: Harper.

Simoons, Frederick J.
1967 *Eat Not This Flesh*. Madison: University of Wisconsin Press.
1979 Questions in the Sacred Cow Controversy. *Current Anthropology* 20(3):467–493.

Smil, Vaclav
2000 *Feeding the World*. Cambridge: Massachusetts Institute of Technology Press.

Smith, Adam
1920 *The Wealth of Nations* (orig. 1776). New York: Dutton.

Smith, Andrew B.
1992 Origins and Spread of Pastoralism in Africa. *Annual Review of Anthropology* 21:125–141.

Smith, Bruce D.
1995 Seed Plant Domestication in Eastern North America. In: *Last Hunters–First Farmers: New Perspectives on the Prehistoric Transition to Agriculture*, T. Douglas Price and Anne Birgitte Gebauer, eds., pp. 193–213. Santa Fe: School of American Research Press.

Smith, Carol A. (ed.)
1976 *Regional Analysis*. New York: Academic Press.

Smith, Eric Alden
1983 Anthropological Applications of Optimal Foraging Theory: A Critical Review. *Current Anthropology* 24(5):625–651.
1991 *Inujjuamiut Foraging Strategies*. Seattle: University of Washington Press.
2000 Three Styles in the Evolutionary Analysis of Human Behavior. In: *Adaptation and Human Behavior: An Anthropological Perspective*. Lee Cronk, Napoleon Chagnon, and William Irons, eds., pp. 27–46. New York: Aldine.

Smith, Eric Alden, and Bruce Winterhalder
1992 Natural Selection and Decision-Making: Some Fundamental Principles. In: *Evolutionary Ecology and Human Behavior*, Eric Alden Smith and Bruce Winterhalder, eds., pp. 25–60. New York: Aldine.

Smith, J. Russell
1950 *Tree Crops: A Permanent Agriculture.* New York: Devin Adair.
Smith, Michael E., and T. Jeffrey Price
1994 Aztec-Period Agricultural Terraces in Morelos, Mexico: Evidence for Household-level Agricultural Intensification. *Journal of Field Archaeology* 21(2):169–179.
Smith, Thomas
1977 *Nakahara: Family Farming and Population in a Japanese Village, 1717–1830.* Stanford: Stanford University Press.
Snyder, Gary
1969 *Earth House Hold.* New York: New Directions.
Spear, Thomas, and Richard Waller (eds.)
1993 *Being Maasai: Ethnicity and Identity in East Africa.* London: James Curry.
Spencer, Arthur
1978 *The Lapps.* New York: Crane, Russak & Company, Inc.
Spencer, J. E.
1966 *Shifting Cultivation in Southeast Asia.* Berkeley: University of California Press.
Spencer, Paul
1988 *The Maasai of Matapato: A Study of Rituals or Rebellion.* Blomington: Indiana University Press.
Speth, John D.
2002 Were Our Ancestors Hunters or Scavengers? In: *Archaeology: Original Readings in Method and Practice,* Peter N. Peregrine, Carol R. Ember, and Melvin Ember, eds., pp. 151–166. Upper Saddle River, NJ: Prentice Hall.
Speth, John D., and Katherine A. Spielman
1983 Energy Source, Protein Metabolism, and Hunter-Gatherer Subsistence Strategies. *Journal of Anthropological Archaeology* 2(1):1–31.
Spielmann, Katherine A., and James F. Eder
1994 Hunters and Farmers: Then and Now. *Annual Review of Anthropology* 23:303–323.
Sponsel, Leslie E.
2001 Do Anthropologists Need Religion, and Vice Versa? Adventures and Dangers in Spiritual Ecology. In: *New Directions in Anthropology and Environment*, Carole Crumley, ed., pp. 177–202. Walnut Creek, CA: AltaMira Press.
Sponsel, Leslie E., Robert C. Bailey, and Thomas N. Headland
1996a Anthropological Perspectives on the Causes, Consequences, and Solutions of Deforestation. In: *Tropical Deforestation: The Human Dimension,* Leslie E. Sponsel, Thomas N. Headland, and Robert C. Bailey, eds., pp. 3–52. New York: Columbia University Press.

Sponsel, Leslie E., Thomas N. Headland, and Robert C. Bailey (eds.)
1996b *Tropical Deforestation: The Human Dimension.* New York: Columbia University Press.

Spooner, Brian
1973 The Cultural Ecology of Pastoral Nomads. In: *Current Topics in Anthropology, Vol. 8, Theory, Methods, and Content,* Addison-Wesley Module in Anthropology No. 45. Reading, MA: Addison-Wesley Publishing Company.

Sproat, Gilbert M.
1868 *Scenes and Studies of Savage Life.* London: Smith, Elder.

Squire, Larry R.
1987 *Memory and Brain.* Oxford: Oxford University Press.

Stagner, Ross
1989 *A History of Psychological Theories.* New York: Macmillan.

Stahl, Ann B.
1984 Hominid Dietary Selection before Fire. *Current Anthropology* 25(2):151–168.

Stanford, Craig B., and Henry T. Bunn (eds.)
2001 *Meat-eating and Human Evolution.* Oxford: Oxford University Press.

Sternberg, Robert J.
1985 Human Intelligence: The Model Is the Message. *Science* 230:1111–1118.

Steward, Julian H.
1936 The Economic and Social Basis of Primitive Bands. In: *Essays in Honor of Alfred Lewis Kroeber,* Robert H. Lowie, ed., pp. 331–350. Berkeley: University of California Press.
1938 *Basin-Plateau Aboriginal Sociopolitical Groups.* Bureau of American Ethnology Bulletin 120.
1955 *Theory of Culture Change.* Urbana: University of Illinois Press.

Stinson, Sara
1992 Nutritional Adaptation. *Annual Review of Anthropology* 21:143–170.

Stonich, Susan
1993 *I Am Destroying the Land!: The Political Ecology of Poverty and Environmental Destruction in Honduras.* Boulder, CO: Westview Press.

Stonich, Susan, and Billie R. DeWalt
1996 The Political Ecology of Deforestation in Honduras. In: *Tropical Deforestation: The Human Dimension,* Leslie E. Sponsel, Thomas N. Headland, and Robert C. Bailey, eds., pp.187–215. New York: Columbia University Press.

Stott, Philip, and Sian Sullivan (eds.)
2001 *Political Ecology: Science, Myth and Power.* London: Edward Arnold.

Stroup, Richard, and Jane Shaw

1992 The Free Market and the Environment. In: *Rational Readings on Environmental Concerns*, Jay H. Lehr, ed., pp. 267–277. New York: Van Nostrand Reinhold.

Struever, Stuart (ed.)

1972 *Prehistoric Agriculture.* New York: Doubleday/American Museum of Natural History.

Stuart, James W.

1978 Subsistence Ecology of the Isthmus Nahuat Indians of Southern Veracruz, Mexico. Ph.D. dissertation, University of California, Riverside.

Suttles, Wayne

1987 *Coast Salish Essays.* Seattle: University of Washington Press.

Sutton, Mark Q.

1984 The Productivity of *Pinus monophylla* and the Modeling of Great Basin Subsistence Strategies. *Journal of California and Great Basin Anthropology* 6(2):240–246.

1988 *Insects as Food: Aboriginal Entomophagy in the Great Basin.* Ballena Press Anthropological Paper No. 33.

2000 Strategy and Tactic in the Analysis of Archaeological Hunter-Gatherer Systems. *North American Archaeologist* 21(3):217–231.

Suzuki, Akira

1966 On the Insect-eating Habits among Wild Chimpanzees Living in the Savanna Woodland of Western Tanzania. *Primates* 7(4):481–487.

Swezey, Sean L., and Robert F. Heizer

1977 Ritual Management of Salmonid Fish Resources in California. *Journal of California Anthropology* 4(1):6–29.

Tanaka, Jiro

1976 Subsistence Ecology of the Central Kalahari San. In: *Kalahari Hunter-Gatherers*, Richard B. Lee and Irven DeVore, eds., pp. 98–119. Cambridge: Harvard University Press.

Tannenbaum, Nicola

1984 The Misuse of Chayanov: "Chayanov's Rule" and Empiricist Bias in Anthropology. *American Anthropologist* 86(4):927–942.

Tanner, Adrian

1979 *Bringing Home Animals.* New York: St. Martin's Press.

Tanner, Nancy

1981 *On Becoming Human.* Cambridge: Cambridge University Press.

Tanner, Nancy, and Adrienne Zihlman

1976 Women in Evolution, Part I: Innovation and Selection in Human Origins. *Signs* 1(3, part 1):585–608.

Taylor, Ronald L.
1975 *Butterflies in My Stomach.* Santa Barbara, CA: Woodbridge Press.

Teleki, Geza
1973 The Omnivorous Chimpanzee. *Scientific American* 228:32–42.
1974 Chimpanzee Subsistence Technology: Materials and Skills. *Journal of Human Evolution* 3:575–594.
1981 The Omnivorous Diet and Eclectic Feeding Habits of Chimpanzees in Gombe National Park, Tanzania. In: *Omnivorous Primates: Gathering and Hunting in Human Evolution*, Robert S. O. Harding and Geza Teleki, eds., pp. 303–343. New York: Columbia University Press.

Terborgh, John
1989 *Where Have All the Birds Gone?* Princeton, NJ: Princeton University Press.

Testart, Alain
1982 The Significance of Food Storage among Hunter-Gatherers: Residence Patterns, Population Densities, and Social Inequalities. *Current Anthropology* 23(5):523–537.
1988 Some Major Problems in the Social Anthropology of Hunter-Gatherers. *Current Anthropology* 29(1):1–31.

Thomas, David Hurst
1981 Complexity among Great Basin Shoshoneans: The World's Least Affluent Hunter-Gatherers? In: *Affluent Foragers: Pacific Coasts East and West*, Shuzo Koyama and David Hurst Thomas, eds., pp. 19–52. Osaka: Senri Ethnological Studies No. 9.
1983 On Steward's Models of Shoshonean Sociopolitical Organization: A Great Bias in the Basin? In: *The Development of Political Organization in Native North America*, Elizabeth Tooker, ed., pp. 59–68. Washington, DC: The American Ethnological Society.
1986 Contemporary Hunter-Gatherer Archaeology in America. In: *American Archaeology: Past and Future*, David J. Meltzer, Don D. Fowler, and Jeremy A. Sabloff, eds., pp. 237–276. Wasington, DC: Smithsonian Institution Press.

Thomson, A. K., R. E. Hytten, and W. Z. Billewicz
1970 The Energy Cost of Human Lactation. *British Journal of Nutrition* 24:565–572.

Timbrook, Jan, John R. Johnson, and David D. Earle
1982 Vegetation Burning by the Chumash. *Journal of California and Great Basin Anthropology* 4(2):163–186.

Tindale, Norman B.
1972 The Pitjandjara. In: *Hunters and Gatherers Today*, M. G. Bicchieri, ed., pp. 217–268. New York: Holt, Rinehart and Winston.

Tobey, Ronald C.
1981 *Saving the Prairies.* Berkeley: University of California Press.

Toledo, Victor M.
1992 What Is Ethnobiology? Origins, Scope, and Implications of a Rising Discipline. *Ethnoecológica* 1(1):5–21.

Townsend, Patricia K.
2000 *Environmental Anthropology: From Pigs to Policies.* Prospect Heights, IL: Waveland Press.

Tucker, Richard P.
2000 *Insatiable Appetite: The United States and the Ecological Degradation of the Tropical World.* Berkeley: University of California Press.

Turnbull, Colin M.
1962 *The Forest People: A Study of the Pygmies of the Congo.* New York: Simon and Schuster.
1983 *The Mbuti Pygmies: Change and Adaptation.* New York: Holt, Reinhart and Winston.

Turner, B. L., and Peter D. Harrison
1983 *Pulltrouser Swamp: Ancient Maya Habitat, Agriculture, and Settlement in Northern Belize.* Austin: University of Texas Press.

Turner, Christy G., II, and Jacqueline A. Turner
1999 *Man Corn: Cannibalism and Violence in the Prehistoric American Southwest.* Salt Lake City: University of Utah Press.

Turner, Jonathan, and Alexandra Maryanski
1979 *Functionalism.* Menlo Park, CA: Benjamin/Cummings.

Turner, Nancy J.
1985 "The Importance of a Rose": Evaluating the Cultural Significance of Plants in Thompson and Lillooet Interior Salish. *American Anthropologist* 90(2):272–290.

Turner, Nancy J., and Barbara S. Efrat
1982 *Ethnobotany of the Hesquiat People of the West Coast of British Columbia.* Victoria: British Columbia Provincial Museum, Cultural Recovery Papers 2.

Turner, Nancy J., Laurence C. Thompson, M. Terry Thompson, and Annie Z. York
1990 *Thompson Ethnobotany.* Victoria, BC: Royal British Columbia Museum.

Turner, Victor W.
1952 *The Lozi Peoples of North-Western Rhodesia.* London: International African Institute.
1967 *The Forest of Symbols.* Ithaca, NY: Cornell University Press.

Tversky, Amos, and Kahnemaqn, Daniel
1981 The Framing of Decisions and the Psychology of Choice. *Science* 211:453–458.

Tyler, Stephen (ed.)
1968 *Cognitive Anthropology.* New York: Holt, Rinehart and Winston.

Ucko, Peter, and George Dimbleby (eds.)
1969 *The Domestication and Exploitation of Plants and Animals.* Chicago: Aldine.

Ugan, Andrew, Jason Bright, and Alan Rogers
2003 When Is Technology Worth the Trouble? *Journal of Archaeological Science* 30(10):1315–1329.

Underhill, Anne P.
1997 Current Issues in Chinese Neolithic Archaeology. *Journal of World Prehistory* 11(2):103–160.

United States Dept. of Agriculture
1963 *Composition of Foods.* Agriculture Handbook No. 8. Washington, DC: USDA.

Vale, Thomas R.
1982 *Plants and People: Vegetation Change in North America.* Washington DC: Association of American Geographers.

Vayda, Andrew P.
1986 Holism and Individualism in Ecological Anthropology. *Reviews in Anthropology* 13:295–313.
1993 Ecosystems and Human Actions. In: *Humans as Components of Ecosystems,* Mark J. McDonnell and Steward T. A. Pickett, eds., pp. 61–71. New York: Springer-Verlag.
1995 Failures of Explanation in Darwinian Ecological Anthropology. *Philosophy of the Social Sciences* 25:219–249, 360–375.
1996 *Methods and Explanations in the Study of Human Actions and Their Environmental Effects.* Jakarta, Indonesia: CIFOR-WWF.

Vayda, Andrew P. (ed.)
1969 *Environment and Cultural Behavior.* Garden City, NY: Doubleday.

Vayda, Andrew P., and Bonnie J. McCay
1975 New Directions in Ecology and Ecological Anthropology. *Annual Review of Anthropology* 4:293–306.

Vayda, Andrew P., and Roy A. Rappaport
1968 Ecology, Cultural and Noncultural. In: *Introduction to Cultural Anthropology: Essays in the Scope and Methods of the Science of Man,* James A. Clifton, ed., pp. 477–497. Boston: Houghton Mifflin.

Vayda, Andrew, and Bradley Walters
1999 Against Political Ecology. *Human Ecology* 27(1):167–179.

Vecsey, Christopher, and Robert Venables (eds.)
1980 *American Indian Environments.* Syracuse: Syracuse University Press.

Vietmeyer, Noel D.
1986 Lesser-known Plants of Potential Use in Agriculture and Forestry. *Science* 232:1379–1384.

Vitousek, Peter M., Harold A. Mooney, Jane Lubchenco, and Jerry M. Melillo

1997 Human Domination of Earth's Ecosystem. *Science* 277:494–499.

Volhard, Ewald

1939 *Kannibalismus.* Stuttgart: Strecker and Strecker Verlag.

Walens, Stanley

1981 *Feasting with Cannibals.* Princeton, NJ: Princeton University Press.

Wallace, A. F. C.

1970 *Culture and Personality.* New York: Random House.

Waller, Richard, and Neal W. Sobania

1994 Pastoralism in Historical Perspective. In: *African Pastoralist Systems: An Integrated Approach*, Elliot Fratkin, Kathleen A. Galvin, and Eric Abella Roth, eds., pp. 45–68. Boulder, CO: Lynne Rienner Publishers.

Wallerstein, Immanuel

1976 *The Modern World-System.* New York: Academic Press.

Wardwell, Allen

1996 *Tangible Visions: Northwest Coast Indian Shamanism and Its Art.* New York: The Monacelli Press with the Corvus Press.

Watson, Patty Jo

1995 Explaining the Transition to Agriculture. In: *Last Hunters–First Farmers: New Perspectives on the Prehistoric Transition to Agriculture*, T. Douglas Price and Anne Birgitte Gebauer, eds., pp. 21–37. Santa Fe: School of American Research Press.

Weatherford, Jack

1988 *Indian Givers: How the Indians of the Americas Transformed the World.* New York: Crown Publishers.

1991 *Native Roots: How the Indians Enriched America.* New York: Fawcett Columbine.

Weeks, Priscilla

1992 The State vs. The People: Strategies of Two Texas Communities. Paper presented at the annual meetings of the American Anthropological Association, San Francisco.

Weins, John A.

1976 Population Responses to Patchy Environments. *Annual Review of Ecology and Systematics* 7:81–120.

Weiss, Mark L., and Alan E. Mann

1990 *Human Biology and Behavior.* Glenview, IL: Scott, Foresman/Little Brown.

Weissleder, W. (ed.)

1978 *The Nomadic Alternative: Modes and Models of Interaction in the African-Asian Steppes.* The Hague: Mouton.

Wen Duzhong, and David Piementel

1986a Seventeenth Century Organic Agriculture in China, Part I: Cropping Systems in Jiaxing Region. *Human Ecology* 14(1):1–14.

1986b Seventeenth Century Organic Agriculture in China, Part II: Energy Flows through an Agrosystem in Jiaxing Region. *Human Ecology* 14(1):15–28.

Werner, Emmy, and Ruth Smith

1982 *Vulnerable but Invincible: A Longitudinal Study of Resilient Children and Youth.* New York: McGraw-Hill.

Werner, Louis

1992 Cultivating the Secrets of Aztec Gardens. *Americas* 44(6):6–15.

Western, David, and R. Michael Wright (eds.)

1994 *Natural Connections: Perspectives in Community-Based Conservation.* Washington D.C.: Island Press.

White, Benjamin

1983 "Agricultural Involution" and Its Critics: Twenty Years After. *Bulletin of Concerned Asian Scholars* 15(3):18–31.

White, Leslie

1949 *The Science of Culture.* New York: Grove Press.

White, Richard

1997 Indian People and the Natural World: Asking the Right Questions. In: *Rethinking American Indian History*, Donald L. Fixico, ed., pp. 87–100. Albuquerque: University of New Mexico Press.

White, Tim D.

1992 *Prehistoric Cannibalism at Mancos 5MTUMR-2346.* Princeton, NJ: Princeton University Press.

Whitmore, Thomas, and B. L. Turner

2001 *Cultivated Landscapes of Middle America on the Eve of Conquest.* New York: Oxford University Press.

Whitney, Eleanor Noss, and Sharon Rady Rolfes

1996 *Understanding Nutrition* (7th ed.). Minneapolis/St. Paul: West Publishing.

Whittle, Alasdair

1996 *Europe in the Neolithic: The Creation of New Worlds* (2nd ed.). Cambridge: Cambridge University Press.

Wilk, Richard R.

1991 *Household Ecology: Economic Change and Domestic Life among the Kekchi Maya of Belize.* Tucson: University of Arizona Press.

Wilke, Philip J.

1988 Bow Staves Harvested from Juniper Trees by Indians of Nevada. *Journal of California and Great Basin Anthropology* 10(1):3–31.

Wilke, Philip J., Robert L. Bettinger, Thomas F. King, and James F. O'Connell
1972 Harvest Selection and Domestication in Seed Plants. *Antiquity* 46(183):203–208.

Wilken, Gene
1987 *Good Farmers*. Berkeley: University of California Press.

Will, Pierre-Etienne
1990 *Bureaucracy and Famine in Eighteenth-Century China*. Stanford: Stanford University Press.

Will, Pierre-Etienne, and Bin Wong
1991 *Nourish the People: The State Civilian Granary System in China, 1650–1850*. Ann Arbor: University of Michigan Press.

Williams, Dee Mack
1996 Grassland Enclosures: Catalyst of Land Degradation in Inner Mongolia. *Human Organization* 55(3):307–313.

Williams, Gerald W.
2002 Aboriginal Use of Fire. In: *Wilderness and Political Ecology: Aboriginal Influences and the Original State of Nature*, Charles E. Kay and Randy T. Simmons, eds., pp. 179–214. Salt Lake City: University of Utah Press.

Wilson, Lynn, Jr.
1967 The Historical Roots of Our Ecologic Crisis. *Science* 155:1203–1207.

Winkelman, Michael
1998 Aztec Human Sacrifice: Cross-Cultural Assessments of the Ecological Hypothesis. *Ethnology* 37(3):285–298.

Winterhalder, Bruce
1981 Optimal Foraging Strategies and Hunter-Gatherer Research in Anthropology: Theory and Models. In: *Hunter-Gatherer Foraging Strategies: Ethnographic and Archaeological Analyses*, Bruce Winterhalder and Eric Alden Smith, eds., pp. 13–35. Chicago: University of Chicago Press.
1986 Optimal Foraging: Simulation Studies of Diet Choice in a Stochastic Environment. *Journal of Ethnobiology* 6(1):205–223.
2001 The Behavioural Ecology of Hunter-Gatherers. In: *Hunter-Gatherers: An Interdisciplinary Perspective*, Catherine Panter-Brick, Robert H. Layton, and Peter Rowley-Conwy, eds., pp. 12–38. Cambridge: Cambridge University Press.

Winterhalder, Bruce, and Carol Goland
1997 An Evolutionary Ecology Perspective on Diet Choice, Risk, and Plant Domestication. In: *People, Plants, and Landscapes: Studies in Paleoethnobotany*, Kristen J. Gremillion, ed., pp. 123–160. Tuscaloosa: University of Alabama Press.

Winterhalder, Bruce, Flora Lu, and Bram Tucker
1999 Risk-Sensitive Adaptive Tactics: Models and Evidence from Subsistence Studies in Biology and Anthropology. *Journal of Archaeological Research* 7(4):301–348.

Winterhalder, Bruce, and Eric Alden Smith
1992 Evolutionary Ecology and the Social Sciences. In: *Evolutionary Ecology and Human Behavior*, Eric Alden Smith and Bruce Winterhalder, eds., pp. 3–23. New York: Aldine.

Winterhalder, Bruce, and Eric Alden Smith (eds.)
1981 *Hunter-Gatherer Foraging Strategies: Ethnographic and Archaeological Analyses*. Chicago: University of Chicago Press.

Winthrop, Kathryn
2001 Historical Ecology: Landscapes of Change in the Pacific Northwest. In: *New Directions in Anthropology and Environment*, Carole Crumley, ed., pp. 203–222. Walnut Creek, CA: AltaMira Press.

Wishart, David J.
1979 The Dispossession of the Pawnee. *Annals of the Association of American Geographers* 69(3):382–401.

Wissler, Clark
1926 *The Relation of Nature to Man in Aboriginal America*. Oxford: Oxford University Press.

Witherspoon, Gary
1983 Navajo Social Organization. In: *Handbook of North American Indians, Vol. 10, Southwest*, Alfonso Ortiz, ed., pp. 524–535. Washington, DC: Smithsonian Institution.

Wittfogel, Karl
1957 *Oriental Despotism*. New Haven: Yale University Press.

Wittwer, Sylvan, Yu Youtai, Sun Han, and Wang Lianzheng
1987 *Feeding a Billion: Frontiers of Chinese Agriculture*. East Lansing: Michigan State University Press.

Wolf, Eric
1972 Ownership and Political Ecology. *Anthropological Quarterly* 45(3):201–205.
1982 *Europe and the People Without History*. Berkeley: University of California Press.

Wolpoff, Milford H.
1999 *Paleoanthropology* (2nd ed.). Boston: McGraw-Hill.

Wood, John J.
1985 Navajo Livestock Reduction. *Nomadic Peoples* 9:21–31.

Wyman, Leland C., and Flora L. Bailey
1964 *Navaho Indian Ethnoentomology*. University of New Mexico Publications in Anthropology No. 12.

Yamamoto, Norio

1985 The Ecological Complementarity of Agro-Pastoralism: Some Comments. In: *Andean Ecology and Civilization: An Interdisciplinary Perspective on Andean Ecological Complementarity*, Shozo Masuda, Izumi Shimada, and Craig Morris, eds., pp. 85–99. Tokyo: University of Tokyo Press.

Zigas, Vincent

1990 *Laughing Death: The Untold Story of Kuru.* Clifton, NJ: Humana Press.

Zihlman, Adrienne L.

1981 Women as Shapers of the Human Adaptation. In: *Woman the Gatherer*, Francis Dahlberg, ed., pp. 75–120. New Haven: Yale University Press.

Zohary, Daniel

1982 *Plants of the Bible.* Cambridge: Cambridge University Press.

Zohary, Daniel, and Maria Hopf

1988 *Domestication of Plants in the Old World.* Oxford: Oxford University Press.

Index

About the Authors

Mark Q. Sutton began his career in anthropology in 1968. While still in high school, he took advantage of the opportunity to participate in archaeological excavations conducted by the local community college. He went on to earn a B.A. (1972), M.A. (1977), and a Ph.D. (1987) in anthropology. He has worked as an archaeologist for the U.S. Air Force, the U.S. Bureau of Land Management, and various private consulting firms and taught at a number of community colleges and universities. Since 1987 he has been at California State University, Bakersfield, where he is now a professor of anthropology. Dr. Sutton works on understanding hunter-gatherer adaptations to arid environments but has also investigated entomophagy, prehistoric diet and technology, and optimal foraging theory. Dr. Sutton has worked at more than one hundred sites in western North America and has published over 130 books, monographs, and papers on archaeology and anthropology.

E. N. Anderson is professor of anthropology at the University of California, Riverside. He holds a Ph.D. in anthropology from the University of California, Berkeley. His research has been focused on uses and management of plants and animals by local communities. In particular, he has studied fishing and fisheries in Hong Kong, Malaysia, Singapore, Tahiti, and British Columbia, and agriculture and forest management in Mexico and the United States. His current research is largely with the contemporary Maya of Quintana Roo, Mexico. He is married and has five children and three grandchildren, as well as two muttish sheepdogs.